Undergraduate Texts in Mathematics

Undergraduate Texts in Mathematics

Undergraduate Texts in Mathematics are generally aimed at third- and fourth-year undergraduate mathematics students at North American universities. These texts strive to provide students and teachers with new perspectives and novel approaches. The books include motivation that guides the reader to an appreciation of interrelations among different aspects of the subject. They feature examples that illustrate key concepts as well as exercises that strengthen understanding.

More information about this series at http://www.springer.com/series/666

Peter R. Mercer

More Calculus of a Single Variable

Peter R. Mercer
Mathematics Department
SUNY Buffalo State
Buffalo, NY, USA

ISSN 0172-6056 ISSN 2197-5604 (electronic)
ISBN 978-1-4939-4681-5 ISBN 978-1-4939-1926-0 (eBook)
DOI 10.1007/978-1-4939-1926-0
Springer New York Heidelberg Dordrecht London

Printed on acid-free paper

Springer is part of Springer Science+Business Media (www.springer.com)

For Mari and Rex,
Pina,
Sarah, Hannah, and John

Preface

Many college students take a couple of courses in calculus. Afterwards, they either (i) take no more calculus, (ii) take calculus of several variables, or (iii) take real analysis. This book offers another option, not exclusive from options (ii) or (iii).

In the first two college calculus courses, much attention is given (naturally) to preparing students for things to come. But typically there is little time devoted to appreciating the bigger picture or for generally admiring the scenery. Many of the most appealing aspects of the subject are often left for students to pick up on their own. Unfortunately, however, students seldom do so.

I have taught undergraduate real analysis and graduate real analysis for teachers for over 15 years. These courses have evolved in a direction which attempts to address these concerns, and this book is a product of the evolution. The main goal is to see how beautifully things fit together, while admiring the scenery along the way. There are a lot of things in this book that experts in real and classical analysis already know; part of the idea here is that there is no reason why a good calculus student should not know them as well.

The book could be used as a text for a third course in calculus of a single real variable, as a supplementary text for a first course in real analysis, or as a reference for anyone who teaches calculus. The book is almost entirely self-contained, but readers would be best to have already taken the equivalent of two one-semester courses in single variable calculus. Some familiarity with sequences and some experience with proofs would be beneficial, though not entirely necessary.

I have presented ideas in a manner which emphasizes *breadth* as much as *depth*. Throughout the text and in the exercises, alternative approaches to many topics are taken. Such explorations are frequently more meaningful than simply aiming for generalization. Indeed, different arguments offer different insights. But whenever I have diverged from what is customary, I have given the usual treatments their due attention.

Many of the methods, examples, and exercises in the text are adapted from papers in the recent mathematics literature, chiefly: *The American Mathematical Monthly*,

The College Mathematics Journal, *Mathematics Magazine*, *The Mathematical Gazette*, and the *International Journal of Mathematics and Mathematical Sciences in Education and Technology*. I hope that readers will be encouraged to read these and other mathematics journals. At the very least, they will find therein solutions to many of the exercises.

I have also tried to emphasize *pattern*: pattern of development, pattern of proof, pattern of argument, and pattern of generalization. In calculus many threads are related in many ways, but in the end it is a coherent subject. Theorems are often named to emphasize pattern, for example the Cauchy–Schwarz Inequality and the Cauchy–Schwarz Integral Inequality, Jensen's Inequality and Jensen's Integral Inequality, the Mean Value Theorem for Sums and the Mean Value Theorem for Integrals, etc.

- The real numbers are introduced carefully, but with an eye on economy. One can be something of an expert in calculus without necessarily knowing all there is to know about the real numbers. As presented here, the "completeness axiom" for the real numbers is the *Increasing Bounded Sequence Property*, rather than the Least Upper Bound Property or Cauchy-completeness, though the latter two notions are explored in some exercises. The Archimedean Property of the real numbers is used freely without explicit mention, but it too is addressed in a few exercises. The word "compact" is never used. The *Nested Interval Property*, a close cousin of the Increasing Bounded Sequence Property, also plays an important part, with bisection algorithms getting their fair attention.
- Important throughout the entire book is the pair of inequalities

$$\left(1+\frac{1}{n}\right)^n < \mathrm{e} < \left(1+\frac{1}{n}\right)^{n+1} \qquad \text{for } n = 1, 2, 3, \ldots$$

  where e is Euler's number $\mathrm{e} \cong 2.71828$. These estimates are frequently revisited, refined, and extended. Inequalities in general play a prominent role as well. The most important of these are Bernoulli's Inequality, the Arithmetic Mean – Geometric Mean Inequality (the AGM Inequality for short), $1 + x \le \mathrm{e}^x$, and Jensen's Inequality.
- Considerable emphasis is given to the symbiotic relationship between the exponential function and calculus itself. The former, for example, gives meaning to functions involving real exponents, it enables us to extend Bernoulli's Inequality and the AGM Inequality, and many of their consequences.
- Considerable attention is devoted to three consequences of the Intermediate Value Theorem which are often overlooked: the Universal Chord Theorem, the Average Value Theorem for Sums and its weighted version, the Mean Value Theorem for Sums. The latter two are so named because of their integral analogues, the Average Value Theorem and the Mean Value Theorem for Integrals. In obtaining these, the Extreme Value Theorem is indispensable. The relationship between sums and integrals is emphasized throughout.

- The definite integral is developed as an extension of the notion of the average value of a continuous function evaluated at N points. Proving that the average value of a continuous function exists is deferred to Appendix A; the proof uses some rather sophisticated ideas which are not used elsewhere. The definite integral's relationship with area is then discussed. Readers who go on to study mathematical analysis will see that the integral as an average is a more enduring theme than the integral as area.
- Chapter 7 (Other Mean Value Theorems) contains some results which have a flavor similar to that of the Mean Value Theorem. Subsequent chapters are independent of this one.
- Chapter 12 (Classic Examples) is also independent of the rest of the book, except that Wallis's product in Sect. 12.1 is used in Chap. 13 to obtain the constant $\sqrt{2\pi}$ which appears in Stirling's formula.
- Some important series are studied, for example, Geometric series, p-series, the Alternating Harmonic series, the Gregory-Leibniz series, and some Taylor series. But series in general are not covered systematically. For example, there is no treatment of power series, tests for convergence, radius of convergence, etc.
- Quadrature rules are studied as means for doing calculus and studying inequalities, rather than being used for conventional numerical methods. Indeed, the quadrature rules are usually applied to a function whose definite integral is known. Particular attention is given to the Trapezoid and Midpoint Rules applied to convex/concave functions.
- Motivated largely by the Mean Value Theorem for the Second Derivative, error terms are studied in Chap. 14. An inequality can often be recast as an equality which contains an error term. Jensen's Inequality, the AGM Inequality, Young's Integral Inequality (among others), and quadrature rules are considered in this way.

Many of the topics in the book, even the better-known ones, have not been collected elsewhere in any single volume. And a good number of these have been published heretofore only in journals. As a result, this book is not in direct competition with any others. Still, there is naturally some overlap between this and other books. Most notably:

A Primer of Real Functions, by R.P. Boas Jr. (Math. Assoc. of America, 1981).

Excursions in Classical Analysis: Pathways to Advanced Problem Solving and Undergraduate Research, by H. Chen (Math. Assoc. of America, 2010).

The Cauchy-Schwarz Master Class, by J.M. Steele (Math. Assoc. of America and Cambridge University Press, 2004).

Inequalities, by G.H. Hardy, J.E. Littlewood & G. Polya (Cambridge Mathematical Library, 2nd edition, 1988).

Analytic Inequalities, by D.S. Mitrinović (Springer-Verlag, New York, 1970).

Mean Value Theorems and Functional Equations, by P.K. Sahoo & T. Riedel (World Scientific, Singapore, 1998).

I have tried to write the sort of book that I would use, that I would like to own as a reference, and that I would fairly recommend to others. To borrow a phrase from G. H. Hardy: I can hardly have failed completely, the subject matter being so attractive that only extravagant incompetence could make it dull.

Acknowledgments

I am fortunate to have learned mathematics from some first-rate mathematicians. I am particularly grateful to my best calculus and analysis professors at the University of Guelph: Pal Fischer, John A. Holbrook, George Leibbrandt, and Alexander McD. Mercer. At the University of Toronto: Ian Graham (who was also my thesis advisor) and Joe Repka. And at UNC Chapel Hill: Joseph Cima and Norberto Kerzman.

Alexander Mercer, who is also my father, has been my greatest influence in all matters mathematical. His encouragement for this project in particular cannot be overstated. Indeed, he and my mother, Mari Mercer, have forever been supportive of all my endeavors. I am so very thankful.

My father also read large portions of the manuscript and made countless wise suggestions. My wonderful colleague Tina Carter and my excellent student Thomas Morse III read large portions of the manuscript as well, likewise making many fine suggestions.

My friends (also colleagues at Buffalo State College) Daniel W. Cunningham and Thomas Giambrone have been a great help to me over the years. These are two guys who can be relied upon for virtually anything. I am particularly indebted to Dan, for general advice about this book, for additional proof reading, and for teaching me most of what I know about *LaTeX*.

I am grateful to all those at Springer Mathematics who have helped me along the way. My editors, Kaitlin Leach and Eugene Ha, have provided exceptional guidance. The superb yet anonymous reviewers have devoted considerable time and thought. They have improved the book immeasurably.

Such fine support notwithstanding, there surely remain some typographical errors, oversights, and even blunders – for which I take sole responsibility. I will be happy to be informed of any errata, via the email address below.

Finally, I thank my wife Pina and children Sarah, Hannah, and John for their unwavering patience throughout this project. They could always guess the reason, whenever their husband/father disappeared to the basement for hours upon hours.

Buffalo, USA
October 13, 2014

Peter R. Mercer
mercerpr@buffalostate.edu

Contents

Chapter 1
The Real Numbers

Everything is vague to a degree you do not realize till you have tried to make it precise.

—Bertrand Russell

We assume that the reader has some familiarity with the set of real numbers, which we denote by **R**. We review interval notation, absolute value, rational and irrational numbers, and we say a few things about sequences. The main point of this chapter is to acquaint the reader with two very important properties of **R**: the *Increasing Bounded Sequence Property*, and the *Nested Interval Property*.

1.1 Intervals and Absolute Value

For two real numbers $a < b$, we write

$$\begin{aligned}[a,b] &= \{x \in \mathbf{R} : a \le x \le b\},\\ (a,b) &= \{x \in \mathbf{R} : a < x < b\},\\ (-\infty,b] &= \{x \in \mathbf{R} : x \le b\}, \text{ and}\\ (a,+\infty) &= \{x \in \mathbf{R} : a < x\},\end{aligned}$$

along with the obvious definitions for $[a,b)$ etc.

For $a < b$, the distance from a to b is $b-a$. One half of this distance is $\frac{b-a}{2}$. The midpoint of the interval $[a,b]$ is $c = \frac{a+b}{2}$. It satisfies

$$a + \frac{b-a}{2} = c = b - \frac{b-a}{2}.$$

These are simple but useful observations. See **Fig. 1.1.**

The distance between any two real numbers is measured via the absolute value function. For $x \in \mathbf{R}$, the **absolute value** of x is given by

$$|\,x\,| = \sqrt{x^2}.$$

P.R. Mercer, *More Calculus of a Single Variable*, Undergraduate Texts in Mathematics, DOI 10.1007/978-1-4939-1926-0_1

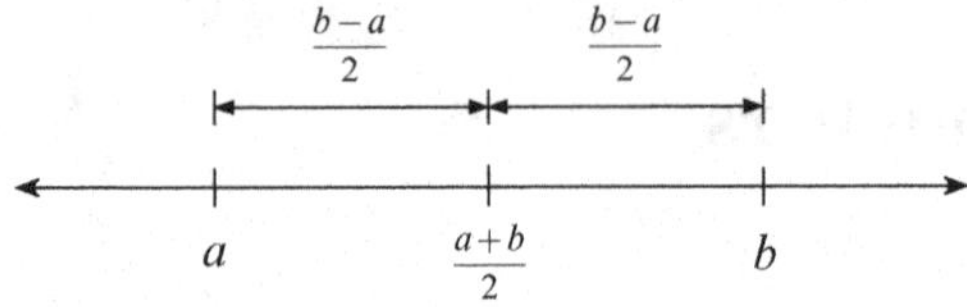

Fig. 1.1 $a + \frac{b-a}{2} = \frac{a+b}{2} = b - \frac{b-a}{2}$

Here we agree to take the *nonnegative* square root of x^2. Therefore,

$$-|x| \leq x \leq |x|.$$

Since $|x-a| = \sqrt{(x-a)^2}$, we see that $|x-a|$ is the *distance that x is from a*. So for $r > 0$, we have

$$|x-a| < r \quad \Leftrightarrow \quad x \in (a-r, a+r).$$

We might say that "x is within r of a." See **Fig. 1.2.**

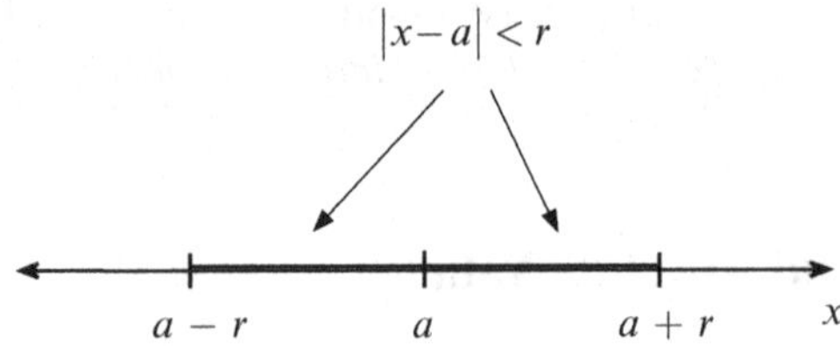

Fig. 1.2 $|x-a| < r \Leftrightarrow x \in (a-r, a+r)$

A few basic but important facts about absolute value are as follows.

Lemma 1.1. *Let $x, y \in \mathbf{R}$. Then*

(i) $|xy| = |x||y|$,
(ii) $|x-y| = |y-x|$,
(iii) $|x+y| \leq |x| + |y|$,
(iv) $|x-y| \geq \big||x| - |y|\big|$.

Proof. For (i), $|xy| = \sqrt{(xy)^2} = \sqrt{x^2y^2} = \left(\sqrt{x^2}\right)\left(\sqrt{y^2}\right) = |x||y|$.
For (ii), we need only observe that $(x-y)^2 = (y-x)^2$.
For (iii),

$$\begin{aligned} |x+y|^2 &= (x+y)^2 = x^2 + 2xy + y^2 \\ &= |x|^2 + 2xy + |y|^2 \leq |x|^2 + 2|xy| + |y|^2 \\ &= |x|^2 + 2|x||y| + |y|^2 \qquad (\text{by (i)}) \\ &= \left(|x| + |y|\right)^2. \end{aligned}$$

Then taking (nonnegative) square roots, we get $|x+y| \leq |x| + |y|$.

For (iv), we write $|x| = |(x-y)+y|$ and apply (iii) to obtain

$$|x| \le |x-y| + |y|, \qquad \text{which gives} \qquad |x| - |y| \le |x-y|.$$

Now we reverse the roles of x and y, and use (ii), to get

$$|y| - |x| \le |x-y|.$$

Taken together, these last two inequalities read

$$\pm(\,|x| - |y|\,) \le |x-y|.$$

That is,

$$\big|\,|x| - |y|\,\big| \le |x-y|.$$

□

In Lemma 1.1, item (iii) is known as the **triangle inequality**, which is very useful. We'll see in Sect. 2.3 why it gets this name. Item (iv) is also useful; it is called the **reverse triangle inequality**.

Remark 1.2. The trick used in the proof of item (iv) in Lemma 1.1, of subtracting y and adding y, then using the triangle inequality, is very common in calculus and real analysis. ○

1.2 Rational and Irrational Numbers

We denote by **N** the set of natural numbers:

$$\mathbf{N} = \{1, 2, 3, 4, \ldots\}.$$

The set **N** is closed under addition and multiplication, but it is not closed under subtraction—that is, the difference of two natural numbers need not be a natural number.

Appending to **N** all differences of all pairs of elements from **N**, we get the set of integers:

$$\mathbf{Z} = \{\ldots, -2, -1, 0, 1, 2, 3, \ldots\}.$$

The set **Z** is closed under addition, multiplication, and subtraction. But **Z** it is not closed under division—that is, the quotient of two integers need not be an integer.

Appending to **Z** all quotients of all pairs of elements from **Z** (with nonzero denominators) we get the set of rational numbers:

$$\mathbf{Q} = \left\{ \frac{p}{q} : p, q \in \mathbf{Z}, \text{ and } q \neq 0 \right\}.$$

The set $\mathbf{Q}$ is closed under addition, multiplication, subtraction, and division. (The reader should agree that it is indeed closed under division.) But as the following lemma shows, $\mathbf{Q}$ is not closed under the operation of taking square roots—that is, the square root of a rational number need not be a rational number.

Lemma 1.3. *There is no rational number x such that $x^2 = 2$. That is, $\sqrt{2}$ is an irrational number.*

Proof. (e.g., [10, 21]) We prove by contradiction. Suppose that $x = p/q$, where $p, q \in \mathbf{Z}$, with $q \neq 0$, is such that $x^2 = 2$. Then $2q^2 = p^2$. If we factor p into a product of prime numbers, then there is a certain number of $2's$ in the product. Whatever this certain number is, p^2 then has an *even* number of $2's$ in its product of primes. Likewise, q^2 has an *even* number of $2's$ in its product of primes. But then $2q^2$ must have an *odd* number of $2's$ in its product of primes. Therefore $2q^2 = p^2$ cannot hold and we have a contradiction, as desired. □

For other proofs of Lemma 1.3, see Exercises 1.13 and 1.14. Essentially the same argument as given above shows that the square root of any prime number (e.g., $\sqrt{3}$, or $\sqrt{5}, \ldots$ etc.) is irrational. Therefore the square root of any natural number that is not itself a perfect square, is irrational.

Extending this argument a little further, we can see that the nth root of any prime number (like $\sqrt{2}$ or $\sqrt{3}$, but also $\sqrt[3]{2}$, or $\sqrt[3]{3}$, ..., or $\sqrt[4]{2}$, or $\sqrt[4]{3}$, ... etc.) is irrational. Therefore the nth root of any natural number that is not itself a perfect nth power, is irrational. (See Exercise 1.15.)

In older textbooks, numbers like $\sqrt[3]{2}, \sqrt[3]{3}, \sqrt[4]{2}, \sqrt[4]{3}$ etc. are called *surds*. For other proofs that many surds are irrational, see Exercises 1.16 and 1.17.

Remark 1.4. The reader will be aware of the usual English meaning of the word *irrational*. The English meaning of the word *surd* is something like *lacking sense*.∘

To expand $\mathbf{Q}$ to include numbers like $\sqrt{2}$, and indeed all the surds, a reasonable idea now might be to append to $\mathbf{Q}$ all nth roots of all rational numbers. But still, important numbers like e and π would remain excluded. (We'll say a little more about this later.) So instead we take a different approach.

Observe that any number whose decimal expansion (i.e., base 10) either terminates or eventually repeats, is a rational number. For example,

$$0.825 = \frac{33}{40}, \quad 0.\overline{2} = \frac{2}{9}, \quad 3.2\overline{142857} = \frac{45}{14}, \quad \text{and} \quad 51.821\overline{571428} = \frac{362{,}751}{7{,}000}.$$

The reader is probably familiar with this observation, although a proper proof is somewhat cumbersome. We explore this in Exercise 1.19. (And we'll see it again in Sect. 2.1.)

Remark 1.5. The converse of this observation is also true: the decimal expansion of any rational number either terminates or repeats. See Exercise 1.20. ∘

So since $\sqrt{2}$ is irrational (by Lemma 1.3), its decimal expansion neither terminates nor repeats. It begins 1.414213562373095... and continues endlessly with no repeating pattern. This gives us an ida of how to proceed.

We append to **Q** all nonterminating nonrepeating decimals. This gives the set of real numbers **R**:

$$\mathbf{R} = \{\, \textit{all} \text{ decimals: terminating, repeating, or otherwise} \,\}.$$

As we have already indicated with our pictures, a *model* for **R** is the familiar number line: $x \in \mathbf{R}$ corresponds to some specific point on the number line. But a real number can have *two different* decimal expansions. For example, if $x = 0.\overline{9}$ then $10x = 9.\overline{9}$, and subtracting the first equation from the second, we get $9x = 9$. Therefore, $x = 1.0$ (as well). Likewise $3.5\overline{9} = 3.6$, $-6.0237\overline{9} = -6.0238$, etc. Fortunately, ambiguities of this sort are the only ones that exist.

1.3 Sequences

Our description of the real numbers thus far has been very qualitative. To describe the real numbers precisely (and to remove our dependency on base 10, or on any other base), one must use *sequences*.

A **sequence** is a function $a : \mathbf{N} \to \mathbf{R}$. That is, the domain of the function is **N**. As such its set of values, or *terms* can be written as

$$\{\, a(1),\ a(2),\ a(3),\ \ldots\, \}.$$

But it is more customary to write $\{a_1, a_2, a_3, \ldots\}$, or $\{a_n\}_{n=1}^{\infty}$, or simply $\{a_n\}$.

Remark 1.6. For a sequence $\{a_n\}$, the domain doesn't really *need* to be **N**. Sometimes it is convenient to start at index zero, or somewhere else. For example $\{a(0), a(1), a(2), a(3), \ldots\}$, or $\{a(5), a(6), a(7), \ldots\}$, or $\{a(2), a(4), a(6), \ldots\}$ etc. Or a sequence may be *finite*, for example, $\{\,1, 3, 5, 7, 9\,\}$. Sequences that we consider will generally *not* be finite, unless otherwise stated; the domain of the sequence is usually clear from the context. ∘

Saying that the sequence $\{a_n\}$ *converges to* A means intuitively that a_n gets closer and closer to A, as n gets larger and larger. When $\{a_n\}$ converges to A, we often write simply: $a_n \to A$.

Example 1.7. The reader is surely comfortable accepting that the sequence

$$\{a_n\} = \left\{\frac{1}{n}\right\}$$

converges to 0. That is,

$$\frac{1}{n} \to 0\,.$$

Some other examples are

$$\frac{1}{\sqrt{n}} \to 0\,, \quad \frac{1}{n^2} \to 0\,, \quad \frac{1}{n^{1/10}} \to 0\,,$$

$$\text{indeed} \quad \frac{1}{n^p} \to 0 \quad \text{for any } p > 0\,,$$

$$\frac{n}{n+3} = \frac{n/n}{n/n+3/n} \to \frac{1}{1+0} = 1\,,$$

$$\frac{\sqrt{n}}{n+3} = \frac{\sqrt{n}/n}{n/n+3/n} \to \frac{0}{1+0} = 0\,, \quad \text{and}$$

$$\frac{\sqrt{n}+1}{2\sqrt{n}} = \frac{\sqrt{n}/\sqrt{n}+1/\sqrt{n}}{2\sqrt{n}/\sqrt{n}} \to \frac{1+0}{2} = \frac{1}{2}.$$ ◇

As we have done in Example 1.7, one can often think along the lines of the intuitive notion of convergence of a sequence. But sometimes more care is required.

Precisely, $\{a_n\}$ **converges to** A means that for *any* $\varepsilon > 0$ (no matter how small) we can find $N > 0$ (which is typically large) such that

$$\text{for } n > N, \text{ it is the case that } |a_n - A| < \varepsilon.$$

That is: for this N, *each of the terms* $a_{N+1}, a_{N+2}, a_{N+3}, \ldots$ is within ε of A. When no explicit mention of A is necessary, we might simply say that $\{a_n\}$ **converges**, or $\{a_n\}$ is **convergent**.

Example 1.8. We show that $\frac{1}{\sqrt{n}} \to 0$, as claimed in Example 1.7. Suppose that $\varepsilon > 0$ is given. First of all,

$$\left|\frac{1}{\sqrt{n}} - 0\right| = \left|\frac{1}{\sqrt{n}}\right| = \frac{1}{\sqrt{n}}.$$

And we can easily make $\frac{1}{\sqrt{n}} < \varepsilon$:

$$\frac{1}{\sqrt{n}} < \varepsilon \quad \Leftrightarrow \quad n > \frac{1}{\varepsilon^2}.$$

So we choose $N = 1/\varepsilon^2$. Then for $n >$ this N, we have $\left|\frac{1}{\sqrt{n}} - 0\right| < \varepsilon$ as desired. ◇

Example 1.9. Consider the sequence

$$\{a_n\} = \left\{\frac{n+100}{3n^2}\right\} = \left\{\frac{1}{3n} + \frac{100}{3}\frac{1}{n^2}\right\}.$$

It appears that the terms of this sequence are getting closer and closer to 0 as n gets larger and larger. We prove that indeed $\frac{n+100}{3n^2} \to 0$. Suppose that $\varepsilon > 0$ is given. First,

$$\left|\frac{n+100}{3n^2} - 0\right| = \frac{n+100}{3n^2}.$$

The idea now is to show that $\frac{n+100}{3n^2}$ is $<$ *something*, wherein the *something* can easily be made $< \varepsilon$. There are any number of ways to proceed from here. (This is part of the reason why showing that a sequence converges can be tricky.) We observe that for n large enough, we shall have $\frac{1}{3n} + \frac{100}{3}\frac{1}{n^2} < \frac{1}{n}$. More precisely,

$$\frac{n+100}{3n^2} < \frac{1}{n} \quad \text{for } n > 50. \tag{1.1}$$

And

$$\frac{1}{n} < \varepsilon \quad \Leftrightarrow \quad n > \frac{1}{\varepsilon}.$$

The inclination now might be to choose $N = 1/\varepsilon$. But to make *every* step of the above analysis valid, we must actually choose $N =$ the *larger* of $1/\varepsilon$ and 50. That is, $N = \max\{50, 1/\varepsilon\}$. (If $\varepsilon = 1/10$ for example, then simply taking $N = 10$ does not guarantee that (1.1) holds.) Then for $n >$ this N, we have $\left|\frac{n+100}{3n^2} - 0\right| < \varepsilon$ as desired. ◇

The reader should agree that a proof that $\left\{\frac{1}{n}\right\}$ converges to 0 is more or less contained in Example 1.9.

Example 1.10. Consider the sequence

$$\{a_n\} = \left\{\frac{3n-1}{4n+2}\right\}.$$

Since $\left\{\frac{3n-1}{4n+2}\right\} = \left\{\frac{3-1/n}{4+2/n}\right\}$, it appears that the terms are getting closer and closer to $3/4$ as n gets larger and larger. We prove that indeed $\frac{3n-1}{4n+2} \to \frac{3}{4}$. Let $\varepsilon > 0$ be given. Here,

$$\left|\frac{3n-1}{4n+2} - \frac{3}{4}\right| = \left|\frac{12n-4-12n-6}{4(4n+2)}\right| = \left|\frac{-10}{4(4n+2)}\right| = \frac{10}{4(4n+2)} = \frac{5}{4(2n+1)},$$

and this is about as much simplifying as can be done. The idea again is to show that $\frac{5}{4(2n+1)}$ is $<$ *something*, wherein the *something* can easily be made $< \varepsilon$. And again, there are any number of ways to proceed. Observe that

$$\frac{5}{4(2n+1)} < \frac{5}{2n+1} < \frac{5}{2n},$$

and

$$\frac{5}{2n} < \varepsilon \quad \Leftrightarrow \quad n > \frac{5}{2\varepsilon}.$$

So we choose $N = 5/(2\varepsilon)$. Then for $n >$ this N, we have $\left|\frac{3n-1}{4n+2} - \frac{3}{4}\right| < \varepsilon$ as desired.

Note: This example can be regarded as finished, but let us show another way that we might have proceeded. We could have observed that

$$\frac{5}{4(2n+1)} = \frac{5}{8n+4} < \frac{5}{8n} < \frac{1}{n}.$$

And here,

$$\frac{1}{n} < \varepsilon \quad \Leftrightarrow \quad n > \frac{1}{\varepsilon}.$$

In this case we would choose $N = 1/\varepsilon$. Then for $n >$ this N, we have $\left|\frac{3n-1}{4n+2} - \frac{3}{4}\right| < \varepsilon$ as desired.

The latter approach yielded $N = 1/\varepsilon$. So the $N = 5/(2\varepsilon)$ chosen in the former approach is larger than it really needs to be. But no matter, we just wanted to find any N that works. (If something is true from Monday onwards and it is true from Thursday onwards, then it is (obviously) true from Thursday onwards.) ◇

Example 1.11. Consider the sequence

$$\{a_n\} = \left\{\frac{1}{2n^2 - 1{,}001}\right\}.$$

It appears that the terms are getting closer and closer to 0 as n gets larger and larger. We prove that indeed $\frac{1}{2n^2-1{,}001} \to 0$. Suppose that $\varepsilon > 0$ is given. First off,

$$\left|\frac{1}{2n^2 - 1{,}001} - 0\right| = \left|\frac{1}{2n^2 - 1{,}001}\right|.$$

Now for $n > 22$ we have $2n^2 - 1{,}001 > 0$ and so for such n,

$$\left|\frac{1}{2n^2 - 1{,}001}\right| = \frac{1}{2n^2 - 1{,}001}.$$

The idea again is to show that $\frac{1}{2n^2-1{,}001}$ is $<$ *something*, wherein the *something* can easily be made $< \varepsilon$. And again, there are any number of ways to proceed. Notice that

$$\frac{1}{2n^2 - 1{,}001} < \frac{1}{n^2} \quad \text{not for all } n, \text{ but for } n > 31\,.$$

And

$$\frac{1}{n^2} < \varepsilon \quad \Leftrightarrow \quad n > \frac{1}{\sqrt{\varepsilon}}.$$

Now to make *every* step of our analysis valid, we choose

$$N = \max\left\{22,\, 31,\, \frac{1}{\sqrt{\varepsilon}}\right\} = \max\left\{31,\, \frac{1}{\sqrt{\varepsilon}}\right\}.$$

Then for $n >$ this N, we have $\left|\frac{1}{2n^2-1{,}001} - 0\right| < \varepsilon$ as desired.

Note: Again we could regard this example as finished, but let us show another way that we might have proceeded. An application of the quadratic formula shows that for $n > 22$ we have

$$\frac{1}{2n^2 - 1{,}001} < \frac{1}{n}.$$

In this case we would choose $N = \max\{22, 1/\varepsilon\}$. Then for $n >$ this N, we have $\left|\frac{1}{2n^2-1{,}001} - 0\right| < \varepsilon$ as desired. Whether the N in this latter approach is larger than the N of the former approach, depends on ε. ◇

The reader should look again at Example 1.9 and find a way to proceed different from the way given there, thus obtaining (probably) a different N.

The following is an important fundamental fact about sequences, which is almost obvious. One uses it routinely without explicit mention.

Lemma 1.12. *If $a_n \to A_1$ and $a_n \to A_2$, then $A_1 = A_2$.*

Proof. We show that for any given $\varepsilon > 0$, *no matter how small*, it is the case that $|A_1 - A_2| < \varepsilon$. For then we must have $A_1 = A_2$. So let $\varepsilon > 0$ be arbitrary. By the triangle inequality (i.e., item (iii) of Lemma 1.1) and item (ii) of Lemma 1.1,

$$|A_1 - A_2| = |A_1 - a_n + a_n - A_2| \le |A_1 - a_n| + |a_n - A_2| = |a_n - A_1| + |a_n - A_2|.$$

These last two terms are getting small as n gets large, since $a_n \to A_1$ and $a_n \to A_2$. So things look good. To make their sum $< \varepsilon$, we proceed as follows. Since $a_n \to A_1$, there is N_1 such that

$$|a_n - A_1| < \frac{\varepsilon}{2} \quad \text{for} \quad n > N_1.$$

Since $a_n \to A_2$, there is N_2 such that

$$\left|a_n - A_2\right| < \frac{\varepsilon}{2} \quad \text{for} \quad n > N_2.$$

So for $n > N = \max\{N_1, N_2\}$,

$$\left|A_1 - A_2\right| < \frac{\varepsilon}{2} + \frac{\varepsilon}{2} = \varepsilon,$$

as desired. □

Remark 1.13. In the proof of Lemma 1.12 we used that trick again: subtracting a_n and adding a_n, then applying the triangle inequality. ○

Lemma 1.12 (see Exercise 1.22 for another proof) says that if $a_n \to A$, then the **limit** A is unique. The limit is typically denoted by

$$A = \lim_{n\to\infty} a_n\,.$$

Of course, not all sequences are convergent. We say that the sequence $\{a_n\}$ **diverges**, or is **divergent**, if there exists no $A \in \mathbf{R}$ for which $a_n \to A$.

Example 1.14. Consider the sequence

$$\{a_n\} = \{(-1)^{n+1}\} = \{\,1,\ -1,\ 1,\ -1,\ 1,\ -1,\ \ldots\}.$$

For *any* real number A we have, by the triangle inequality:

$$|\,a_{n+1} - a_n\,| = |\,a_{n+1} - A + A - a_n\,| \le |a_{n+1} - A| + |a_n - A|\,.$$

If $a_n \to A$ then for any $\varepsilon > 0$, we can make the right-hand side above $< \varepsilon$, by taking n large enough. But this is impossible since $|\,a_{n+1} - a_n\,| = 2$ for every n. Therefore $\{a_n\}$ diverges. ◇

Remark 1.15. In Example 1.14 above we used that trick again: subtracting A and adding A, then applying the triangle inequality. ○

Now back to the real numbers, for a moment. We have seen that the decimal expansion of $\sqrt{2}$ begins 1.414213562… . So let us associate with $\sqrt{2}$ the sequence

$$\{a_n\} = \{\,1.4,\ 1.41,\ 1.414,\ 1.4142,\ 1.41421,\ 1.414213,\ 1.4142135,\ \ldots\}.$$

This sequence is *increasing*: each term (after the first) is larger than the previous term. This sequence is also *bounded above*: each term is less than 1.5 say, or less than 2, or less than 1.42, etc. Now each term a_n of this sequence is a rational number, and

$$\left|a_n - \sqrt{2}\right| < \frac{1}{10^n} \to 0.$$

So, by its very construction, the sequence converges to $\sqrt{2}$, which is irrational. So **Q** is *not* closed under the operation of taking the limit of a sequence which is increasing and bounded above.

This suggests a way to extend the rational numbers to include the irrational numbers: we append to **Q** all limits of all sequences of rational numbers which are increasing and bounded above. But we must do some more groundwork in order to make these ideas precise.

1.4 Increasing Sequences

A sequence $\{a_n\}$ is **increasing** if $a_n \le a_{n+1}$ for $n = 1, 2, \ldots$. And $\{a_n\}$ is **strictly increasing** if $a_n < a_{n+1}$ for $n = 1, 2, \ldots$.

A sequence $\{a_n\}$ is **decreasing** if $a_n \ge a_{n+1}$ for $n = 1, 2, \ldots$. (That is, $\{-a_n\}$ is increasing.) And $\{a_n\}$ is **strictly decreasing** if $a_n > a_{n+1}$ for $n = 1, 2, \ldots$.

A sequence $\{a_n\}$ is **bounded above** if there exists a number U such that $a_n \le U$ for $n = 1, 2, \ldots$. The number U is called an **upper bound** for $\{a_n\}$.

A sequence $\{a_n\}$ is **bounded below** if there exists a number L such that $L \le a_n$ for $n = 1, 2, \ldots$. (In which case, $\{-a_n\}$ is bounded above.) The number L is called a **lower bound** for $\{a_n\}$.

Remark 1.16. If a sequence $\{a_n\}$ has an upper bound U, then U is not unique: the number $U + 1$, for example, also serves as an upper bound for $\{a_n\}$. So does $U + 1/10$, as does $U + 1{,}000$, etc. Likewise, if $\{a_n\}$ has a lower bound L, then L is not unique. ○

Example 1.17. The sequence $\{a_n\} = \{\sqrt{n}\}$ is not bounded above: there is no U for which $a_n \le U$ for $n = 1, 2, \ldots$. It is bounded below, by $L = 1$ (and by $L = 1/2$, and by $L = 0$, and by $L = -10$ etc.). This sequence is strictly increasing. ◇

Example 1.18. The sequence $\{a_n\} = \{\frac{n}{n+1}\}$ is bounded above, by $U = 1$ for example. It is also bounded below, by $L = 1/2$ for example. This sequence is also strictly increasing (as the reader may verify). ◇

A sequence $\{a_n\}$ is **bounded** if $\{a_n\}$ is bounded above *and* bounded below. That is, there are numbers L, U such that $a_n \in [L, U]$ for every $n \in \mathbf{N}$. Setting $M = \max\{|L|, |U|\}$, we see that this is equivalent to saying that there exists a number M such that $|a_n| \le M$ for every $n \in \mathbf{N}$.

Example 1.19. Since $-4/3 \le (-1)^n - 1/3 \le 2/3$ for every $n \in \mathbf{N}$, the sequence $\{a_n\} = \{(-1)^n - 1/3\}$ is bounded above and bounded below. As such $\{a_n\}$ is bounded: $|a_n| \le 4/3$ for every $n \in \mathbf{N}$. This sequence is neither increasing nor decreasing. ◇

Example 1.20. The terms of the sequence $\{a_n\} = \{\frac{n^2}{n+1}\} = \{\frac{n}{1+1/n}\}$ appear to be getting arbitrarily large as n increases. We show that indeed this sequence is not bounded above. Let M be any given number (to be thought of as large). Then

$$\frac{n^2}{n+1} > M \quad \Leftrightarrow \quad n^2 - Mn - M > 0.$$

Using the quadratic formula, we see that

$$n > \tfrac{M+\sqrt{M^2+4M}}{2} \quad \Rightarrow \quad n^2 - Mn - M > 0.$$

So by taking n larger than $\frac{M+\sqrt{M^2+4M}}{2}$, we can make $\{a_n\} > M$. That is, we can make $\{a_n\}$ as large as we like. Therefore $\{a_n\} = \{\frac{n^2}{n+1}\}$ is not bounded above. ◇

If a sequence is increasing without any upper bound, or decreasing without any lower bound, then its terms cannot be getting arbitrarily close to any particular number. This, in its contrapositive form, is the idea behind the following.

Lemma 1.21. *If $a_n \to A$, then $\{a_n\}$ is bounded.*

Proof. This is Exercise 1.26. □

Example 1.22. We saw in Example 1.20 that the sequence $\{a_n\} = \{\frac{n^2}{n+1}\} = \{\frac{n}{1+1/n}\}$ is not bounded. So applying Lemma 1.21 in its contrapositive form, $\{a_n\}$ diverges. We might say that $\{a_n\}$ **diverges to** $+\infty$. ◇

The converse of Lemma 1.21 *does not* hold: The sequence $\{a_n\} = \left\{(-1)^{n+1}\right\}$ is bounded, but as we saw in Example 1.14, it diverges.

Example 1.23. Suppose that $a_n \to A$. We show that $a_n^2 \to A^2$. But here we forgo the formal definition (i.e., the ε and the N); this is usually done by people with some experience in real analysis. By the triangle inequality,

$$\left|a_n^2 - A^2\right| = |a_n + A|\,|a_n - A| \le (|a_n| + |A|)\,|a_n - A|.$$

Now by Lemma 1.21, $\{a_n\}$ is bounded because it converges. That is, there is $M > 0$ such that

$$|a_n| \le M \quad \text{for } n = 1, 2, 3, \ldots.$$

Therefore

$$(|a_n| + |A|)\,|a_n - A| \le (M + |A|)\,|a_n - A| \quad \text{for } n = 1, 2, 3, \ldots.$$

And again, since $\{a_n\}$ converges,

$$(M + |A|)\,|a_n - A| \to 0.$$

Therefore $\left|a_n^2 - A^2\right| \to 0$, and so $a_n^2 \to A^2$. ◇

We close this section by stating four more useful facts which again, one uses routinely without explicit mention. We assume that the reader can prove these, or is quite comfortable in accepting them. We leave their proofs as exercises.

Lemma 1.24. *Let $\alpha, \beta \in \mathbf{R}$. If $a_n \to A$ and $b_n \to B$, then $\alpha a_n + \beta b_n \to \alpha A + \beta B$. In particular, $a_n + b_n \to A + B$ and $a_n - b_n \to A - B$.*

Proof. This is Exercise 1.28. □

Lemma 1.25. *If $a_n \to A$ and $b_n \to B$, then $a_n b_n \to AB$.*

Proof. This is Exercise 1.29. □

Lemma 1.26. *If $a_n \to A$ and $b_n \to B$, with $b_n \neq 0$ for all n and $B \neq 0$, then $\frac{a_n}{b_n} \to \frac{A}{B}$.*

Proof. This is Exercise 1.30. □

Lemma 1.27. *If $a_n \to A$ and $a_n \geq 0$, then $A \geq 0$. Consequently (upon consideration of $b_n - a_n$), if $a_n \to A$ and $b_n \to B$, with $a_n \leq b_n$, then $A \leq B$.*

Proof. This is Exercise 1.31. □

These four lemmas say, respectively, that convergent sequences respect linear combinations, products, quotients, and nonstrict inequalities. *Nonstrict* because even if $a_n > 0$ in Lemma 1.27, we can still only conclude that $A \geq 0$. For example, $1/n > 0$, but $\lim_{n\to\infty} 1/n = 0$.

Example 1.28. As indicated, the proof of Lemma 1.25 is the content of Exercise 1.29. But here is another approach. In Example 1.23 we showed that if $c_n \to C$ then $c_n^2 \to C^2$. Now it is easily verified that

$$a_n b_n = \frac{1}{4}\left((a_n + b_n)^2 - (a_n - b_n)^2\right).$$

So if $a_n \to A$ and $b_n \to B$ then $a_n b_n \to AB$, by applying Lemma 1.24, in various combinations, to the right-hand side. ◇

1.5 The Increasing Bounded Sequence Property

We have seen that $\mathbf{Q}$ is not closed under the operation of taking the limit of a sequence which is increasing and bounded above. (Again, such a sequence may well have a limit, but this limit may not be in $\mathbf{Q}$—as is the case with $\sqrt{2}$.) So appending to $\mathbf{Q}$ all such limits, we get the set real numbers $\mathbf{R}$:

$$\mathbf{R} = \mathbf{Q} \cup \{\text{limits of all sequences from } \mathbf{Q} \text{ which are increasing \& bounded above}\}.$$

Example 1.29. The irrational number $\sqrt{2}$ is the limit of the sequence of rational numbers

$$\{a_n\} = \{\, 1.4,\ 1.41,\ 1.414,\ 1.4142,\ 1.41421,\ 1.414213,\ 1.4142135,\ \ldots\},$$

which is increasing and bounded above. Therefore $\sqrt{2} \in \mathbf{R}$. ◇

Example 1.30. The number $x = 1.234567891011121314\ldots$ is irrational because its decimal expansion continues endlessly, with no repeating pattern. This number is the limit of the sequence of rational numbers

$$\{a_n\} = \{1,\ 1.2,\ 1.23,\ 1.234,\ 1.2345,\ 1.23456,\ 1.234567,\ \ldots\}.$$

And this sequence is increasing and bounded above (by 2, say). So $x \in \mathbf{R}$. ◇

Remark 1.31. The two sequences in Examples 1.29 and 1.30, which converge to $\sqrt{2}$ and to x respectively, are not unique. The reader should think of other sequences $\{a_n\}$ which are increasing and bounded above, for which $a_n \to \sqrt{2}$, and $a_n \to x$. ◦

The following is a very important example of a sequence which is increasing and bounded above.

Example 1.32. Using an idea from [12], we show that sequence

$$\left\{\left(1 + \frac{1}{n}\right)^n\right\}$$

of rational numbers is increasing and bounded above. As such, it converges to some real number. For $n \in \mathbf{N}$ and $a \neq b$, the following identity can be found by doing long division on the left-hand side, or simply verified by cross multiplication:

$$\frac{b^{n+1} - a^{n+1}}{b - a} = b^n + ab^{n-1} + a^2b^{n-2} + \cdots + a^{n-2}b^2 + a^{n-1}b + a^n.$$

There are $n + 1$ terms on the right-hand side so for $0 \le a < b$,

$$\frac{b^{n+1} - a^{n+1}}{b - a} < (n + 1)b^n.$$

This inequality is easily rearranged to get

$$b^n\big[\,(n + 1)a - nb\,\big] < a^{n+1}.$$

Now setting $a = 1 + \frac{1}{n+1}$ and $b = 1 + \frac{1}{n}$ we get

$$\left(1 + \frac{1}{n}\right)^n < \left(1 + \frac{1}{n + 1}\right)^{n+1},$$

and so $\{(1+\frac{1}{n})^n\}$ is an increasing sequence. Setting instead $a = 1$ and $b = 1+\frac{1}{2n}$, we get

$$\left(1+\frac{1}{2n}\right)^n < 2, \quad \text{and so} \quad \left(1+\frac{1}{2n}\right)^{2n} < 4.$$

But since $\{(1+\frac{1}{n})^n\}$ is increasing, $(1+\frac{1}{n})^n < (1+\frac{1}{2n})^{2n}$. Therefore $(1+\frac{1}{n})^n < 4$, and so $\{(1+\frac{1}{n})^n\}$ is bounded above. ◇

The real number to which $\{(1+\frac{1}{n})^n\}$ converges is denoted by e. The symbol e is used in honor of the great Swiss mathematician Leonhard Euler (1701–1783); it is often called **Euler's number**. We shall prove in Sect. 8.4 that e is irrational. Approximately, e = 2.718281828459 ⋯.

Remark 1.33. Saying that e is irrational is the same as saying that e is not the solution to any equation $ax + b = 0$, where a and b are integers. Notice that $ax+b = 0$ is a polynomial equation of degree 1. The French mathematician Charles Hermite (1822–1901) proved in 1873 that e is not a solution to *any* polynomial equation of *any* degree with integer coefficients. That is, e is a **transcendental number**. So even by somehow attaching to **Q** all nth roots of all rational numbers, or even all linear combinations of these, we would still not obtain all of the real numbers because e would remain excluded. ○

The following theorem contains a fundamental property of the real numbers. We shall appeal to it many times. It says that **R** *is* closed under the operation of taking the limit of a sequence which is increasing and bounded above.

Theorem 1.34. (The Increasing Bounded Sequence Property of **R**.) *Any sequence of real numbers which is increasing and bounded above converges to a real number.*

Proof. Let $\{a_n\}$ be a sequence of real numbers which is increasing and bounded above. If each $a_n \in \mathbf{Q}$ then $a_n \to A \in \mathbf{R}$, exactly by our definition of **R**, and we are finished. Otherwise, we consider a related sequence $\{b_n\}$ defined by

$$b_n = a_n, \quad \text{but truncated after the } n\text{th decimal place.}$$

Then $\{b_n\}$ is increasing and bounded above and each $b_n \in \mathbf{Q}$, and so we must have $b_n \to B$, for some $B \in \mathbf{R}$. Now by the triangle inequality,

$$\begin{aligned} |a_n - B| &= |a_n - b_n + b_n - B| \le |a_n - b_n| + |b_n - B| \\ &\le \frac{1}{10^n} + |b_n - B|. \end{aligned}$$

Therefore, since $\frac{1}{10^n} \to 0$ and $b_n \to B$, we see that $a_n \to B$ also, as desired. □

Remark 1.35. There's that trick again: subtracting b_n and adding b_n, then using the triangle inequality. ○

Example 1.36. Consider the sequence $\{a_n\}$ of real numbers defined recursively via $a_1 = 1$, and

$$a_{n+1} = \sqrt{1 + a_n} \quad \text{for } n = 1, 2, 3, \ldots .$$

The first few terms of this sequence are as follows.

$$\{a_n\} = \left\{1,\ \sqrt{2},\ \sqrt{1+\sqrt{2}},\ \sqrt{1+\sqrt{1+\sqrt{2}}},\ \sqrt{1+\sqrt{1+\sqrt{1+\sqrt{2}}}}, \ldots\right\}$$

$$\cong \{1,\ 1.414214,\ 1.553774,\ 1.598053,\ 1.611848, \ldots\}.$$

This is a sequence of *real numbers* which is increasing and bounded above. (We leave the verification of this for Exercise 1.40.) As such, by the Increasing Bounded Sequence Property (Theorem 1.34), $\{a_n\}$ converges to some real number φ. To find φ, notice that since $a_n \to \varphi$ and $a_{n+1} = \sqrt{1 + a_n}$, we must have $\varphi = \sqrt{1 + \varphi}$. Then squaring both sides and using the quadratic formula gives $\varphi = \frac{1+\sqrt{5}}{2} \cong 1.618$. ◇

Remark 1.37. The number φ is called the **golden mean**. It is irrational. But it is clearly not transcendental because as we saw, it is the root of a quadratic equation with integer coefficients. There is some debate among historians of mathematics as to whether $\sqrt{2}$ or φ was the first-ever irrational number to be discovered [23]. ○

Example 1.38. The ubiquitous number π is the ratio of the circumference to the diameter of any circle. The reader is surely familiar with the formula $A = \pi r^2$, where A is the area of a circle with radius r. By approximating the area of a circle of radius $r = 1$ with the area of an inscribed equilateral triangle, square, regular pentagon, regular hexagon etc., we see that π is the limit of an increasing sequence of real numbers which is bounded above. As such, π is a real number. We shall prove in Sect. 12.2 that π is irrational. (So, in particular, $\pi \neq 22/7$!!) ◇

Remark 1.39. The German mathematician F. Lindemann (1852–1939) proved in 1882 that π is in fact transcendental. So again, even by somehow attaching to $\mathbf{Q}$ all nth roots of all rational numbers, or even all linear combinations of these, we would still not obtain all of the real numbers—π would remain excluded. ○

Remark 1.40. One doesn't normally worry too much about such things, but all of this gives meaning to arithmetic in $\mathbf{R}$. For example, consider $\sqrt{2} + \mathrm{e}$. Each of $\sqrt{2}$ and e is the limit of a sequence of rational numbers which is increasing and bounded above, say $a_n \to \sqrt{2}$, and $b_n \to \mathrm{e}$. Then each $a_n + b_n$ is a rational number, and $\sqrt{2}+\mathrm{e}$ is the limit of $\{a_n + b_n\}$, a sequence of rational numbers which is increasing and bounded above. And one can verify (but not easily) that this limit is independent of the specific choice of the sequences $\{a_n\}$ and $\{b_n\}$ of rationals, as long as each is increasing and bounded above, with $a_n \to \sqrt{2}$ and $b_n \to \mathrm{e}$ respectively. ○

1.6 The Nested Interval Property

If $\{a_n\}$ converges to A, then $\{-a_n\}$ clearly converges to $-A$. Therefore, by the Increasing Bounded Sequence Property of **R** (Theorem 1.34), every sequence $\{a_n\}$ of real numbers which is decreasing and bounded below also converges to a real number. This leads to another very important property of **R**, as follows.

Theorem 1.41. (Nested Interval Property of **R**) *For any collection of* ***nested intervals*** $[a_1, b_1] \supseteq [a_2, b_2] \supseteq [a_3, b_3] \supseteq \cdots$ *with the property that* $b_n - a_n \to 0$, *there is a unique point which belongs to each interval.*

Proof. The sequence $\{a_n\}$ is increasing and bounded above (by b_1) and so by the Increasing Bounded Sequence Property (Theorem 1.34), it converges to some $A \in \mathbf{R}$. The sequence $\{b_n\}$ is decreasing and bounded below (by a_1) and so it converges to some $B \in \mathbf{R}$. We must then have $a_n \le A \le B \le b_n$ for $n = 1, 2, 3, \ldots$. But we cannot have $A < B$ because $b_n - a_n \to 0$. Therefore $A = B$, and this real number belongs to each interval $[a_n, b_n]$, as desired. □

Example 1.42. Again, the decimal expansion for $\sqrt{2}$ begins $1.41421356\cdots$. The number $\sqrt{2}$ is the only point that belongs to each of the nested intervals:

$$[1.4,\ 1.5] \supseteq [1.41,\ 1.42] \supseteq [1.414,\ 1.415] \supseteq [1.4142,\ 1.4143] \supseteq \cdots . \qquad \diamond$$

Example 1.43. We showed in Example 1.32 that the sequence $\left\{(1+\frac{1}{n})^n\right\}$ is increasing and bounded above. So it has a limit, which is denoted by e (Euler's number). In a similar way, which we leave for Exercise 1.39, it happens that the sequence $\left\{(1+\frac{1}{n})^{n+1}\right\}$ is decreasing and bounded below. Now observe that

$$\left(1+\frac{1}{n}\right)^{n+1} - \left(1+\frac{1}{n}\right)^{n} = \left(1+\frac{1}{n}\right)^{n}\left(\left(1+\frac{1}{n}\right)-1\right) = \left(1+\frac{1}{n}\right)^{n}\frac{1}{n} \to 0.$$

Therefore, by the Nested Interval Property of **R** (Theorem 1.41), the collection of nested intervals

$$\left[\left(1+\tfrac{1}{n}\right)^n,\ \left(1+\tfrac{1}{n}\right)^{n+1}\right], \quad \text{where } n = 1, 2, 3, \ldots$$

contains a single point, which must be e. Taking $n = 5{,}000$, for example, gives $2.7180 \le \mathrm{e} \le 2.7186$. ◇

The basic string of inequalities which comes from Examples 1.32 and 1.43 is

$$\left(1+\frac{1}{n}\right)^{n} < \mathrm{e} < \left(1+\frac{1}{n}\right)^{n+1}.$$

We shall revisit these inequalities many times.

Example 1.44. By approximating the area of a circle of radius 1 by areas of inscribed regular polygons with increasing areas, we described π as the limit of an increasing sequence of real numbers which is bounded above. By also approximating the area of the same circle by areas of *circumscribed* regular polygons with decreasing areas, we can describe π as the single point which belongs to a sequence of nested intervals. ◇

Remark 1.45. Around 250 B.C., Archimedes used 96-sided inscribed and circumscribed polygons to obtain the rather impressive estimates

$$3.14085 \cong \frac{223}{71} < \pi < \frac{22}{7} \cong 3.14286. \qquad \circ$$

Finally, we point out that the three *collections* of nested intervals in Examples 1.42–1.44, which yielded the numbers $\sqrt{2}$, e, and π, are not unique.

Exercises

1.1. Let $x \in \mathbf{R}$. **(a)** Find all real numbers y for which $|x+y| = |x| + |y|$.
(b) Find all real numbers y for which $\big||x| - |y|\big| = |x-y|$.

1.2. **(a)** Show that

$$|x| = \begin{cases} x & \text{if } x \geq 0 \\ -x & \text{if } x < 0. \end{cases}$$

(b) Use the observation $-|x| \leq x \leq |x|$ (and $-|y| \leq y \leq |y|$) to prove the triangle inequality. This is the proof in most books.
(c) We proved the reverse triangle inequality using the triangle inequality. Prove the reverse triangle inequality using the definition of absolute value.

1.3. **(a)** Prove that if $x, y, z \in \mathbf{R}$, then $|x+y+z| \leq |x| + |y| + |z|$.
(b) Prove that if $x_1, x_2, \ldots, x_n \in \mathbf{R}$, then $|x_1 + x_2 + \ldots + x_n| \leq |x_1| + |x_2| + \cdots + |x_n|$.

1.4. [2] Show that if $|x+y| = |x| + |y|$, then $|ux + vy| = u|x| + v|y|$ for all $u, v \geq 0$.

1.5. Let $x < y$. **(a)** Show that $x < \frac{x+y}{2} < y$.
(b) For $p, q > 0$, show that $x < \frac{px+qy}{p+q} < y$. (The quotient $\frac{px+qy}{p+q}$ is a *weighted average* of x and y.)

1.6. [5] Let $a > b \geq 0$ and let $x \in (0,1)$. Show that

$$\left(1 + x^b\right)^a > \left(1 + x^a\right)^b.$$

1.7. Let $a, b \in \mathbf{R}$. Show that

$$\max\{a,b\} = \frac{a+b+|a-b|}{2} \quad \text{and} \quad \min\{a,b\} = \frac{a+b-|a-b|}{2}.$$

1.8. Let $a, b \in \mathbf{R}$. Prove that $|a| + |b| \le |a+b| + |a-b|$. When does equality occur?

1.9. Let $a, b \in \mathbf{R}$. Prove that $\frac{|a+b|}{1+|a+b|} \le \frac{|a|}{1+|a|} + \frac{|b|}{1+|b|}$. When does equality occur?

1.10. [8] Let $x \in \mathbf{R}$. Prove that $|1+x| \le |1+x|^2 + |x|$. When does equality occur?

1.11. [14] Fill in the details of the following proof of a sharpened version of the triangle inequality: *For nonzero* $x_1, x_2, \ldots, x_n \in \mathbf{R}$,

$$\left|\sum_{j=1}^{n} x_j\right| + \left(n - \sum_{j=1}^{n} \frac{x_j}{|x_j|}\right) \min_{1\le j\le n} |x_j| \le \sum_{j=1}^{n} |x_j|.$$

(a) Let $|x_k| = \min_{1\le j\le n} |x_j|$ and $K = \{j : 1 \le j \le n \text{ and } |x_j| \ne |x_k|\}$. Verify that

$$\left|\sum_{j=1}^{n} \frac{x_j}{|x_j|}\right| = \left|\sum_{j=1}^{n} \frac{x_j}{|x_k|} - \sum_{j\in K}\left(\frac{1}{|x_k|} - \frac{1}{|x_j|}\right) x_j\right|.$$

(b) Apply the reverse triangle inequality. This result is improved somewhat in [18].

1.12. Here is a simple but useful fact, which is really just a consequence of algebraic manipulations. We shall refer to it in a number of subsequent exercises in this book, but not in the text proper. Show that the area A of a triangle T with vertices (x_1, y_1), (x_2, y_2), and (x_3, y_3) is given by

$$A = \frac{1}{2}\left|x_1(y_2 - y_3) + x_3(y_1 - y_2) + x_2(y_3 - y_1)\right|.$$

(If the vertices of the triangle are arranged counterclockwise then $A > 0$ without the absolute value signs.) Hint: In **Fig. 1.3**, use the fact that $Area(T) = Area(\text{Trapezoid } ABQP) + Area(\text{Trapezoid } BCRQ) - Area(\text{Trapezoid } ACRP)$. Note: Readers who know some Linear Algebra might recognize that

$$A = \frac{1}{2}\left|\det\begin{bmatrix} x_1 & x_2 \\ y_1 & y_2 \end{bmatrix} + \det\begin{bmatrix} x_2 & x_3 \\ y_2 & y_3 \end{bmatrix} + \det\begin{bmatrix} x_3 & x_1 \\ y_3 & y_1 \end{bmatrix}\right|$$

$$= \frac{1}{2}\left|\det\begin{bmatrix} 1 & 1 & 1 \\ x_1 & x_2 & x_2 \\ y_1 & y_2 & y_3 \end{bmatrix}\right|.$$

Fig. 1.3 For Exercise 1.12

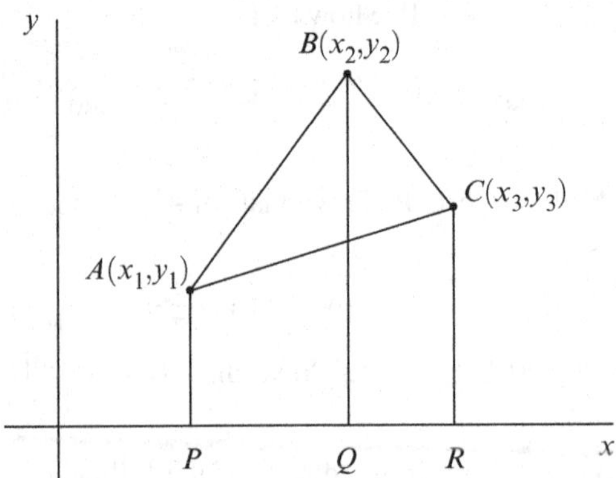

Here, "det" is short for *determinant*. See [4] for some interesting extensions of this formula.

1.13. Fill in the details of the standard textbook proof that $\sqrt{2}$ is irrational. (This proof was known to Euclid (~300 B.C.) and was probably known even to Aristotle (384–322 B.C.)) Assume that $\sqrt{2} = p/q$ is rational, so that $2 = p^2/q^2$. Then $2q^2 = p^2$ must be even. Therefore p is even. Therefore q^2 even and so q is even. So from each of p and q we may cancel a factor of 2. Repeat.

1.14. [6] Fill in the details of the following very slick proof that $\sqrt{2}$ is irrational, due to American mathematician Ivan Niven (1915–1999). If $\sqrt{2}$ is rational then there is a smallest positive integer b such that $b\sqrt{2}$ is an integer. Then $b\sqrt{2} - b$ is a smaller positive integer. Now consider $(b\sqrt{2} - b)\sqrt{2}$.

1.15. **(a)** Prove that $\sqrt{3}$ is irrational. **(b)** Prove that $\sqrt[3]{11}$ irrational. **(c)** Prove that $\sqrt[7]{45}$ irrational. **(d)** Describe how to prove that the nth root of any natural number which is not itself an nth power, is irrational.

1.16. [16] Fill in the details of another proof that the square root of any natural number that is not itself a perfect square is irrational: Let $\sqrt{n} = p/q$ where p and q are positive integers, and have no common factors. Then p^2 and q also have no common factors. But then $p^2/q = p\sqrt{n} = qn$, which is an integer.

1.17. [3] Fill in the details of another proof that the square root of any natural number that is not itself a perfect square is irrational: Suppose that $\sqrt{n} = p/q$ where p and q are positive integers, and have no common factors. Then $\sqrt{n} = nq/p$ also, but this is not in lowest terms. Therefore p is an integer multiple of q. Therefore n is a perfect square, a contradiction.

1.18. We haven't officially met logarithms yet. Still, prove that $\log_{10}(2)$ is irrational. ($\log_{10}(2)$ is that real number x for which $10^x = 2$.)

1.19. **(a)** Write $0.823, 0.455$, and $-0.9999\ldots$ as fractions.

(b) Multiply $x = 0.\overline{2}$ by 10 and subtract x from the result. Then solve for x to write x as a fraction.
(c) Multiply $y = 0.\overline{91}$ by 100 and subtract y from the result. Then solve for y to write y as a fraction.
(d) Write $0.\overline{237}$ and $6.457\overline{132}$ as fractions.
(e) Describe how to prove that any decimal which eventually repeats is a rational number. We'll see another way to do some of this in Sect. 2.1; see also Exercise 2.4.

1.20. Show by doing long division that **(a)** $1/3 = 0.\overline{3}$ (and conclude that $0.\overline{9} = 1$), **(b)** $\frac{1}{11} = 0.\overline{09}$, and **(c)** $\frac{22}{7} = 3.\overline{142857}$. **(d)** Describe how to prove that any rational number is either a terminating or repeating decimal. **(e)** Let $x = p/q$ be a rational number. What is the longest that the repeating string in its decimal expansion could possibly be? (The length of the repeating string in the decimal expansions in each of $1/7$ and $1/97$, for example, is maximal.)

1.21. **(a)** Is it true that the sum of two rational numbers is rational? Explain.
(b) How about the sum of a rational number and an irrational number? Explain.
(c) How about the sum of two irrational numbers? Explain.

1.22. Prove Lemma 1.12 another way: Looking for a contradiction, assume that $A_1 \neq A_2$, say $A_1 < A_2$. Then let $\varepsilon = (A_2 - A_1)/2$.

1.23. Four of the following seven sequences converge. Decide which four they are, then prove that each of them converges. **(a)** $\{\frac{1}{n^{3/2}}\}$, **(b)** $\{\cos(\frac{n\pi}{2}) + \frac{1}{n}\}$, **(c)** $\{n^{(-1)^n}\}$, **(d)** $\{\sqrt{n+1} - \sqrt{n}\}$, **(e)** $\{\frac{1}{5n-61}\}$, **(f)** $\left\{\frac{n^2+3}{n^2+2n-1}\right\}$, **(g)** $\left\{\frac{1}{\sqrt{n+1}-\sqrt{n}}\right\}$.

1.24. Use the reverse triangle inequality to prove that if $a_n \to A$ then $|a_n| \to |A|$.

1.25. **(a)** Prove that if $a_n \to A$ then for any $\varepsilon > 0$ there exists $N > 0$ such that all of the terms of $\{a_n\}$ belong to the interval $(A - \varepsilon, A + \varepsilon)$, except possibly $a_1, a_2, \ldots, a_{N-1}, a_N$.
(b) Use (a) to prove the following. If $a_n \to A$ with $A > 0$ then there exists $m > 0$ and $N > 0$ such that $a_n \geq m$ for $n > N$.

1.26. Prove Lemma 1.21. Suggestion: Use Exercise 1.25(a).

1.27. **(a)** Prove that if $a_n \to 0$ and $\{b_n\}$ is a bounded sequence, then $a_n b_n \to 0$.
(b) Show that $\frac{(-1)^n+1/2}{n} \to 0$.
(c) Show that $\frac{\sin(n)}{\sqrt{n}+3} \to 0$.
(d) Show that $\frac{n\cos(n!)}{n^2+100} \to 0$.

1.28. Prove Lemma 1.24. Hint: Use

$$|(\alpha a_n + \beta b_n) - (\alpha A + \beta B)| \leq |\alpha a_n - \alpha A| + |\beta b_n - \beta B| = |\alpha|\,|a_n - A| + |\beta|\,|b_n - B|.$$

1.29. Prove Lemma 1.25. Hint: Observe that

$$\begin{aligned}|a_n b_n - AB| &= |a_n b_n - a_n B + a_n B - AB| \\ &= |a_n(b_n - B) + B(a_n - A)| \le |a_n|\,|b_n - B| + |B|\,|a_n - A|,\end{aligned}$$

then use Lemma 1.21.

1.30. Prove Lemma 1.26. Hint: Observe that

$$\left|\frac{a_n}{b_n} - \frac{A}{B}\right| = \left|\frac{a_n B - Ab_n}{b_n B}\right| = \left|\frac{a_n B - AB + AB - Ab_n}{b_n B}\right| = \left|\frac{B(a_n - A) + A(B - b_n)}{b_n B}\right|,$$

then use the triangle inequality and Exercise 1.25(b).

1.31. **(a)** Prove the first part of Lemma 1.27. Hint: Assume that $A < 0$ to get a contradiction.

(b) Explain how (a) implies the second part of Lemma 1.27: if $a_n \to A$ and $b_n \to B$, with $a_n \le b_n$, then $A \le B$.

1.32. Prove that if $a_n \to A$, with $a_n \ge 0$ (so that $A \ge 0$ too) then $\sqrt{a_n} \to \sqrt{A}$. Hint: First dispense with the case $A = 0$. Then for $A \ne 0$,

$$\left|\sqrt{a_n} - \sqrt{A}\right| = \left|\frac{a_n - A}{\sqrt{a_n} + \sqrt{A}}\right| \le \left|\frac{a_n - A}{\sqrt{A}}\right|.$$

1.33. Suppose that $a_n \to A$. For $n = 1, 2, 3, \ldots$, set $b_n = (a_1 + a_2 + \cdots + a_n)/n$. Show that $b_n \to A$. Is the converse true? Explain.

1.34. Associate with each of the four real numbers $x = 3.6912151821242730\ldots$, $x = 0.\overline{3}$, $x = 0.1002000300004\ldots$, and $x = 10.567$ an increasing sequence which converges to x.

1.35. (e.g., [1, 11, 13, 15])

(a) Prove that there exist irrational numbers a and b such that a^b is rational. Hint: Begin by considering $\sqrt{2}^{\sqrt{2}}$.

(b) Prove that there exist irrational numbers a and b such that a^b is irrational. Hint: Begin by considering $\sqrt{2}^{\sqrt{2}+1}$. See also Exercise 1.36.

1.36. [20] Here is a *constructive* approach to Exercise 1.35.

(a) We haven't officially met logarithms yet. Still, prove that $\log_2(3)$ is irrational. ($\log_2(3)$ is that real number x for which $2^x = 3$.)

(b) Verify that $\sqrt{2}^{2\log_2(3)}$ is rational. So there are irrational numbers a and b such that a^b is rational.

(c) Verify that $\sqrt{2}^{\log_2(3)}$ is irrational. So there are irrational numbers a and b such that a^b is irrational.

1.37. [19] Consider the area between the circumscribed and inscribed circles of a regular n sided polygon with side lengths 1. Show that this area is independent of n and find the area.

1.38. **(a)** Show that between any two real numbers there are infinitely many rational numbers.
(b) Show that between any two real numbers there are infinitely many irrational numbers.

1.39. [12] Show that the sequence $\{(1 + \frac{1}{n})^{n+1}\}$ is decreasing and bounded below (and hence converges), as follows. **(a)** Show that for $0 \le a < b$,

$$a^n(b-a) < \frac{b^{n+1} - a^{n+1}}{n+1}.$$

(b) Now set $a = 1 + \frac{1}{n+1}$ and $b = 1 + \frac{1}{n}$.

1.40. Consider the sequence $\{a_n\} \subset \mathbf{R}$ defined recursively via $a_1 = 1$, and $a_{n+1} = \sqrt{1 + a_n}$ for $n = 1, 2, 3, \ldots$. Use the Increasing Bounded Sequence Property (Theorem 1.34) and mathematical induction to show that $\{a_n\}$ converges to a real number φ. Show that $\varphi = (1 + \sqrt{5})/2$, the golden mean. Show that φ is irrational.

1.41. For each of the four real numbers $x = 3.69121518\ldots$, $x = 0.1002000300004\ldots$, $x = 0.\overline{3}$, and $x = 10.567$, construct a sequence of nested intervals $[a_1, b_1] \supseteq [a_2, b_2] \supseteq \ldots$, with $b_n - a_n \to 0$, such that each interval contains x.

1.42. Consider the sequence defined by $a_0 = 1$, and $a_{n+1} = \frac{1}{1+a_n}$ for $n = 0, 1, 2, 3, \ldots$.

(a) Show that $\{[a_{2n}, a_{2n+1}]\}$ is a collection of nested intervals, with $a_{2n+1} - a_{2n} \to 0$.
(b) Show that the point given by the Nested Interval Property (Theorem 1.41) is the golden mean $\varphi = (1 + \sqrt{5})/2$.

1.43. A set A is **countable** if there is a one-to-one onto function $\sigma : \mathbf{N} \to A$. (So all of its elements can be listed off: $\sigma(1), \sigma(2), \sigma(3), \ldots$.)

(a) Show that $\mathbf{Z}$ is countable.
(b) Show that $\{x \in \mathbf{Q} : 0 < x < 1\}$ is countable.
(c) Show that $\{x \in \mathbf{R} : 0 < x < 1\}$ is **uncountable**, that is, is not countable.
(d) Show that $\mathbf{Q}$ is countable. (A *formula* for a one-to-one onto $\sigma : \mathbf{Z} \to \mathbf{Q}$ can be found in [9]. For a very slick proof that $\mathbf{Q}$ is countable, see [7] or [22].)

1.44. [17] (If you did Exercise 1.43.) Fill in the details of the following proof that the set of algebraic numbers—that is, the set of all roots of all polynomials of any degree, with integer coefficients—is countable. This amazing fact was discovered in 1871 by the great German mathematician Georg Cantor (1845–1918). But first:

(a) Show that a quick consequence of Cantor's discovery is that $\mathbf{Q}$ is countable.
(b) Consider the polynomial equation

$$p(x) = a_n x^n + a_{n-1}x^{n-1} + \ldots + a_1 x + a_0 = 0,$$

where the a_j's are integers. This equation has at most n solutions. We may assume that $a_n \geq 1$. How?

(c) Define the *index* of any such polynomial p as

$$index(p) = |a_n| + |a_{n-1}| + \ldots + |a_1| + |a_0|.$$

Show, for example, that there is only one such polynomial with index 2. There are four such polynomials with index 3. There are 11 such polynomials with index 4. Argue that there are only finitely many polynomials with a given index.

(d) Now show that the set of algebraic numbers is countable.

(e) Show that the set of transcendental numbers is uncountable.

References

1. Ash, J.M., Tan, Y.: A rational number of the form a^a with a irrational. Math. Gaz. **96**, 106–108 (2012)
2. Baker, J.A: Isometries in normed spaces. Am. Math. Mon. **78**, 655–658 (1971)
3. Berresford, G.C.: A *simpler* proof of a well-known fact. Am. Math. Mon. **115**, 524 (2008)
4. Braden, B.: The surveyor's area formula. Coll. Math. J. **17**, 326–337 (1986)
5. Braken, P., Ackerman, S., Flanigan, F.J., Sanders, M., Tesman, B.: Problem 594. Coll. Math. J. **29**, 70–71 (1998)
6. Bumcrot, R.J.: Irrationality made easy. Coll. Math. J. **17**, 243 (1986)
7. Campbell, S.L.: Countability of sets. Am. Math. Mon. **93**, 480–481 (1986)
8. Eustice, D., Meyers, L.F.: Problem Q663. Math. Mag. **52**, 317, 323 (1979)
9. Freilich, G.: A denumerability formula for the rationals. Am. Math. Mon. **72**, 1013–1014 (1965)
10. Hopkinson, J: Further evidence that $\sqrt{2}$ is irrational. Math. Gaz. **59**, 275 (1975)
11. Jarden, D.: A simple proof that a power of an irrational number to an irrational exponent may be rational. Scr. Math. **19**, 229 (1953)
12. Johnsonbaugh, R.F.: Another proof of an estimate for e. Am. Math. Mon. **81**, 1011–1012 (1974)
13. Jones, J.P., Toporowski, S.: Irrational numbers. Am. Math. Mon. **80**, 423–424 (1973)
14. Kato, M., Saito, K.S., Tamura, T.: Sharp traingle inequality and its reverse in Banach spaces. Math. Ineq. Appl. **10**, 451–460 (2007)
15. Körner, T.W.: A Companion to Analysis. American Mathematical Society, Providence (2004)
16. Lewin, M.: An even shorter proof that $\sqrt{n}$ is either irrational or integral. Math. Gaz. **60**, 295 (1976)
17. Niven, I.M.: Numbers Rational and Irrational. Mathematical Association of America, Washington, DC (1961)
18. Pečarić, J., Rajić, R.: The Dunkl-Williams inequality with n elements in normed linear spaces. Math. Ineq. Appl. **10**, 461–470 (2007)
19. Rebman, K.: Problem Q778. Math. Mag. **64**, 198, 206 (1991)
20. Sasane, S.: Irrational numbers redux. Am. Math. Mon. **119**, 380 (2012)
21. Schielack Jr., V.P.: A quick counting proof of the irrationality of $\sqrt[n]{k}$. Math. Mag. **68**, 386 (1995)
22. Touhey, P.: Countability via bases other than 10. Coll. Math. J. **27**, 382–384 (1996)
23. von Fritz, K.: The discovery of incommensurability by Hippasus of Metapontum. Ann. Math. **46**, 242–264 (1945)

Chapter 2
Famous Inequalities

I speak not as desiring more, but rather wishing a more strict restraint.

—Isabella, in *Measure for Measure,* by William Shakespeare

In this chapter we meet three very important inequalities: Bernoulli's Inequality, the Arithmetic Mean–Geometric Mean Inequality, and the Cauchy–Schwarz Inequality. At first we consider only pre-calculus versions of these inequalities, but we shall soon see that a thorough study of inequalities cannot be undertaken without calculus. And really, calculus cannot be thoroughly understood without some knowledge of inequalities. We define Euler's number e by a more systematic method than that of Example 1.32. We'll see that this method engenders many fine extensions.

2.1 Bernoulli's Inequality and Euler's Number e

The following is a very useful little inequality. It is named for the Swiss mathematician Johann Bernoulli (1667–1748).

Lemma 2.1. (Bernoulli's Inequality) *Let* $n = 1, 2, 3, \ldots$. *Then for* $x > -1$,

$$(1+x)^n \geq 1 + nx\,.$$

Proof. If $n = 1$, the inequality holds, with equality. For $n = 2$,

$$(1+x)^2 = 1 + 2x + x^2 \;\geq\; 1 + 2x.$$

For $n = 3$, we use the $n = 2$ case:

$$(1+x)^3 = (1+x)(1+x)^2 \;\geq\; (1+x)\,(1+2x) = 1 + 3x + 2x^2 \;\geq\; 1 + 3x.$$

P.R. Mercer, *More Calculus of a Single Variable*, Undergraduate Texts in Mathematics, DOI 10.1007/978-1-4939-1926-0_2

For $n = 4$, we use the $n = 3$ case:

$$(1+x)^4 = (1+x)(1+x)^3 \ \geq\ (1+x)\,(1+3x) = 1+4x+4x^2 \ \geq\ 1+4x.$$

Etcetera: We could clearly continue this procedure up to any positive integer n, and so our proof is complete. □

Example 2.2. We show that for $|x| < 1$, the sequence $\{x^n\}$ converges to 0. First, for $0 < x < 1$ we can write $x = \frac{1}{1+q}$, for some $q > 0$. Then by Bernoulli's Inequality (Lemma 2.1),

$$(1+q)^n \ \geq\ 1+nq,$$

so that

$$0 < x^n = \frac{1}{(1+q)^n} \leq \frac{1}{1+nq}.$$

Letting $n \to \infty$, the result follows. For $-1 < x < 0$, we simply replace x with $-x$ above. (The case $x = 0$ is trivial.) ◇

Exercise 2.1 contains the fact that for any $x > 0$, $\left\{x^{1/n}\right\}$ converges to 1.

Example 2.3. The series

$$1 + x + x^2 + x^3 + \cdots = \sum_{k=0}^{\infty} x^k$$

is called a **geometric series**. In it, each term x^k after the first is the *geometric mean* of the term just before it and the term just after it: $x^k = \sqrt{x^{k-1}x^{k+1}}$. Here we find a formula for the sum of a geometric series, when it exists. For $x \neq 1$ the following identity can be found by doing long division on the right-hand side, or simply verified by cross multiplication:

$$\sum_{k=0}^{n} x^k = 1 + x + x^2 + \cdots + x^n = \frac{1-x^{n+1}}{1-x}.$$

So if $|x| < 1$ then by Example 2.2, the sequence of *partial sums* $\{S_n\} = \left\{\sum_{k=0}^{n} x^k\right\}$ converges to $\frac{1}{1-x}$. Therefore,

$$\sum_{k=0}^{\infty} x^k = \frac{1}{1-x} \qquad \text{for } |x| < 1.$$ ◇

Example 2.4. We write $x = 0.611111\ldots$ as a fraction. Observe that

$$10x = 6 + \sum_{k=1}^{\infty}\left(\frac{1}{10}\right)^k = 5 + \sum_{k=0}^{\infty}\left(\frac{1}{10}\right)^k.$$

The series here is a geometric series, so by Example 2.3,

$$10x = 5 + \frac{1}{1 - 1/10} = \frac{55}{9}.$$

Therefore, $x = 11/18$. ◇

Example 2.5. [2] Here we show that the sequence $\{n^{1/n}\}$ converges to 1. Setting $x = 1/\sqrt{n}$ in Bernoulli's Inequality (Lemma 2.1) we get

$$\left(1 + \frac{1}{\sqrt{n}}\right)^n \geq 1 + \sqrt{n} > \sqrt{n}.$$

Therefore

$$\left(1 + \frac{1}{\sqrt{n}}\right)^2 > \left(\sqrt{n}\right)^{2/n} = n^{1/n} \geq 1,$$

and the result follows upon letting $n \to \infty$. ◇

Example 2.6. [12,16,55] Using Bernoulli's Inequality (Lemma 2.1) we show again (cf. Example 1.32) that $\left\{\left(1 + \frac{1}{n}\right)^n\right\}$ converges. We have seen that the number to which this sequence converges is Euler's number e. First, observe that

$$\frac{(1 + \frac{1}{n+1})^{n+1}}{(1 + \frac{1}{n})^n} = \left(1 + \frac{1}{n}\right)\left(\frac{1 + \frac{1}{n+1}}{1 + \frac{1}{n}}\right)^{n+1} = \left(1 + \frac{1}{n}\right)\left(1 - \frac{1}{(n+1)^2}\right)^{n+1}.$$

Then applying Bernoulli's Inequality,

$$\left(1 + \frac{1}{n}\right)\left(1 - \frac{1}{(n+1)^2}\right)^{n+1} \geq \left(1 + \frac{1}{n}\right)\left(1 - \frac{1}{n+1}\right) = 1.$$

Therefore

$$\left(1 + \frac{1}{n+1}\right)^{n+1} \geq \left(1 + \frac{1}{n}\right)^n$$

and so $\left\{\left(1 + \frac{1}{n}\right)^n\right\}$ is increasing. In a very similar way, which we leave for Exercise 2.6 (see also Example 1.43 and Exercise 1.39), one can show that $\left\{\left(1 + \frac{1}{n}\right)^{n+1}\right\}$ is decreasing. Then we have

$$\left(1 + \frac{1}{n}\right)^n \leq \left(1 + \frac{1}{n}\right)^{n+1} \leq \left(1 + \frac{1}{1}\right)^{1+1} = 4,$$

and so $\left\{\left(1 + \frac{1}{n}\right)^n\right\}$ is also bounded above. Therefore, by the Increasing Bounded Sequence Property (Theorem 1.34), this sequence has a limit: Euler's number, e. ◇

Again we emphasize that the basic string of inequalities here, is:

$$\left(1+\frac{1}{n}\right)^n < e < \left(1+\frac{1}{n}\right)^{n+1} \qquad \text{for } n = 1, 2, 3, \cdots .$$

2.2 The AGM Inequality

We begin with a simple yet incredibly useful fact. It turns out to be a special case of the main result of this section (Theorem 2.10).

Lemma 2.7. *Let a and b be positive real numbers. Then*

$$\sqrt{ab} \le \frac{a+b}{2}, \quad \textit{and equality occurs here} \Leftrightarrow a = b.$$

Proof. It is easily verified that

$$(a+b)^2 - 4ab = (a-b)^2 \ge 0.$$

Therefore,

$$(a+b)^2 \ge 4ab, \quad \text{or} \quad \frac{a+b}{2} \ge \sqrt{ab}\,.$$

Now if $a = b$, then clearly $\sqrt{ab} = \frac{a+b}{2}$. Conversely, if $\sqrt{ab} = \frac{a+b}{2}$ then in the first line of the proof we must have $(a-b)^2 = 0$, and so $a = b$. □

The average $A = \frac{a+b}{2}$ is known as the *Arithmetic Mean* of a and b. The quantity $G = \sqrt{ab}$ is known as their *Geometric Mean*. A rather satisfying Proof Without Words for Lemma 2.7, which also suggests why $\sqrt{ab}$ is called the Geometric Mean, is shown **Fig. 2.1.** See also Exercise 2.19

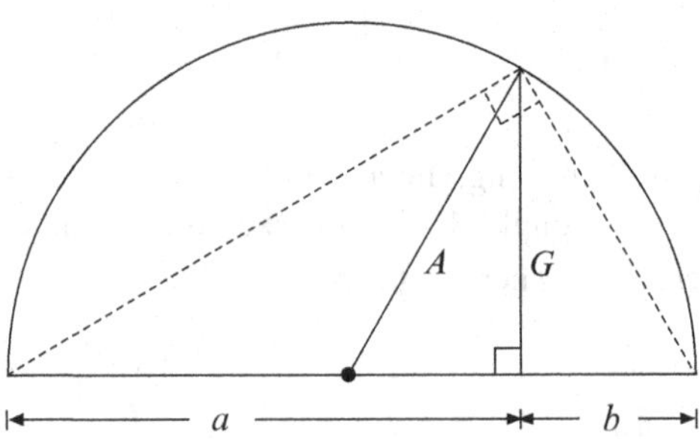

Fig. 2.1 $G = \sqrt{ab} \le A = \frac{a+b}{2}$

Example 2.8. Suppose we use a balance to determine the mass of an object. We place the object on the left side of the balance and a known mass on the right side, to obtain a measurement a. Then we place the object on the right side of the balance and a known mass on the left side, to obtain a measurement b. By the principle of the lever (or more generally, the principle of moments) the true mass of the object is the Geometric Mean $\sqrt{ab}$. (See also Exercise 2.15.) ◇

Sometimes the most important feature of an inequality is the case in which equality occurs, as the following example illustrates.

Example 2.9. A rectangle with side lengths a and b has perimeter $P = 2a + 2b$ and area $T = ab$. Lemma 2.7 reads $ab \leq (\frac{a+b}{2})^2$, or $T \leq (P/4)^2$. So a rectangle with given perimeter has greatest area when $a = b$, i.e., when the rectangle is a square. Likewise, a rectangle with given area has least perimeter when $a = b$, again when the rectangle is a square. In either case, $T = (P/4)^2$. ◇

The most natural extension of Lemma 2.7 is to allow n positive numbers instead of just two. But we need to know what would be meant by *Arithmetic Mean* and *Geometric Mean* in this case. These turn out to be exactly as one might expect, as follows.

Let $a_1, a_2, \ldots, a_n$ be real numbers. Their **Arithmetic Mean** is given by

$$A = \frac{a_1 + a_2 + \cdots + a_n}{n} = \frac{1}{n}\sum_{j=1}^{n} a_j \,.$$

If these numbers are also nonnegative, then their **Geometric Mean** given by

$$G = \left((a_1)(a_2)\cdots(a_n)\right)^{1/n} = \left(\prod_{j=1}^{n} a_j\right)^{1/n} .$$

A number $M = M(a_1, a_2, \ldots, a_n)$ which depends on $a_1, a_2, \ldots, a_n$ is called a **mean** simply if it satisfies

$$\min_{1 \leq j \leq n} \{a_j\} \leq M \leq \max_{1 \leq j \leq n} \{a_j\}.$$

However, for practical purposes one often desires other properties, like (i) having $M(a_1, a_2, \ldots, a_n)$ independent of the order in which the numbers $a_1, a_2, \ldots, a_n$ are arranged, and (ii) having $M(ta_1, ta_2, \ldots, ta_n) = tM(a_1, a_2, \ldots, a_n)$ for any $t \geq 0$. The reader should agree that A and G each satisfy (i) and (ii).

The Arithmetic Mean–Geometric Mean Inequality below, or what we shall call the **AGM Inequality** for short, extends Lemma 2.7 to n numbers. This inequality is of fundamental importance in mathematical analysis. The great French mathematician Augustin Cauchy (1789–1857) was the first to prove it, in 1821. We provide his proof at the end of this section. (The Scottish mathematician Colin Maclaurin (1698–1746) had an earlier proof, around 1729, which wasn't quite complete.)

The list of mathematicians who have offered proofs of the AGM Inequality over the years is impressive. It includes Liouville, Hurwitz, Steffensen, Bohr, Riesz, Sturm, Rado, Hardy, Littlewood, and Polya. (See, e.g., [5,9,18,32,49]; the book [9] contains over 75 proofs.) Below we provide the clever 1976 proof given by K.M. Chong [10].

Theorem 2.10. (AGM Inequality) *Let $a_1, a_2, \ldots, a_n$ be n positive real numbers, where $n \geq 2$. Then*

$$G \leq A,$$

and equality occurs here $\Leftrightarrow$ $a_1 = a_2 = \cdots = a_n$.

Proof. If $n = 2$ then the result is simply Lemma 2.7, so we consider $n \geq 3$. By rearranging the a_j's if necessary, we may suppose that $a_1 \leq a_2 \leq \cdots \leq a_{n-1} \leq a_n$. Then $0 < a_1 \leq A \leq a_n$, and so

$$A(a_1 + a_n - A) - a_1 a_n = (a_1 - A)(A - a_n) \geq 0.$$

That is,

$$a_1 + a_n - A \geq \frac{a_1 a_n}{A}. \tag{2.1}$$

Take $n = 3$ here, and notice that the Arithmetic Mean of the *two* numbers a_2 and $a_1 + a_3 - A$ is A. Now we apply Lemma 2.7 to *these two numbers*, along with (2.1) to get

$$A^2 \geq a_2(a_1 + a_3 - A) \geq a_2 \frac{a_1 a_3}{A}.$$

That is,

$$A^3 \geq a_1 a_2 a_3.$$

Now take $n = 4$. The Arithmetic Mean of the *three* numbers a_2, a_3 and $a_1 + a_4 - A$ is again A, so by what we have just shown applied to *these three numbers*, along with (2.1),

$$A^3 \geq a_2 a_3 (a_1 + a_4 - A) \geq a_2 a_3 \frac{a_1 a_4}{A}.$$

That is,

$$A^4 \geq a_1 a_2 a_3 a_4.$$

Clearly we could continue this procedure indefinitely, showing that $A \geq G$ for any positive integer n, and so we have proved the main part of the theorem. Now to address the equality conditions. If $a_1 = a_2 = \cdots = a_n$, then it is easily verified that

$G = A$. Conversely, if $A = G$ for some particular n, then in the argument above, $(a_1 - A)(A - a_n) = 0$ so that $a_1 = a_n = A$, and therefore $a_1 = a_2 = \cdots = a_n = A$. □

Example 2.11. Suppose that an investment returns 10 % in the first year, 50 % in the second year, and 30 % in the third year. Using the Geometric Mean

$$[(1.1)(1.5)(1.3)]^{1/3} \cong 1.289,$$

the average rate of return over the 3 years is just under 29 %. The Arithmetic Mean gives an overestimate of the average rate of return, at 30 %. ◇

Example 2.12. We saw in Example 2.9 that a rectangle with given perimeter has greatest area when the rectangle is a square, and that a rectangle with given area has least perimeter when the rectangle is a square. Likewise, using the AGM Inequality (Theorem 2.10), a box (even in n dimensions) with given surface area has greatest volume when the box is a cube, and a box (even in n dimensions) with given volume has least surface area when the box is a cube. ◇

Example 2.13. Named for Heron of Alexandria (c. 10–70 AD), **Heron's formula** gives the area T of a triangle in terms of its three side lengths a, b, c and perimeter P, as follows:

$$16T^2 = P(P - 2a)(P - 2b)(P - 2c).$$

So if we apply the AGM Inequality (Theorem 2.10) to the three numbers $P - 2a$, $P - 2b$ and $P - 2c$, we obtain

$$16T^2 \le P\left(\frac{(P-2a)+(P-2b)+(P-2c)}{3}\right)^3, \quad \text{or}$$

$$T \le \frac{P^2}{12\sqrt{3}}.$$

Therefore, for a triangle with fixed perimeter P, its area T is largest possible when $P - 2a = P - 2b = P - 2c$. This is precisely when $a = b = c$, that is, when the triangle is equilateral. Likewise a triangle with fixed area T has least perimeter P when it is an equilateral triangle. In either case, $T = P^2/(12\sqrt{3})$. ◇

Remark 2.14. We saw in Example 2.13 that for a triangle, we have $T \le P^2/(12\sqrt{3})$. In Example 2.9, we saw that for a rectangle, $T \le P^2/16$. This latter inequality persists for all quadrilaterals having area T and perimeter P. See Exercise 2.29. These inequalities are called **isoperimetric inequalities**. The isoperimetric inequality for an n-sided polygon is

$$T \le \frac{P^2}{4n \tan(\pi/n)},$$

and equality holds if and only if the polygon is regular. The isoperimetric inequality for *any* plane figure with area T and perimeter P is

$$T \leq \frac{P^2}{4\pi}.$$

The famous **isoperimetric problem** was to prove that equality holds here if and only if the plane figure is a circle. The solution of the isoperimetric problem takes up an important and interesting episode in the history of mathematics [37]. For a polished modern solution, see [25]. The reader might find it somewhat comforting that

$$\lim_{n\to\infty}\left[n \tan\left(\frac{\pi}{n}\right)\right] = \pi.$$

This can be verified quite easily (see Exercise 5.46) using *L'Hospital's Rule*, which we meet in Sect. 5.3. ○

Example 2.15. [26] We show that Bernoulli's Inequality (Lemma 2.1)

$$(1+x)^n \geq 1+nx$$

for $x > -1$ and $n = 1, 2, 3, \ldots$ follows from the AGM Inequality (Theorem 2.10). First, if $-1 < x \leq -1/n$, then $1+nx \leq 0 < 1+x$, and so $1+nx < (1+x)^n$. Therefore we assume that $x > -1/n$. We write

$$1+x = \frac{1+nx+(n-1)}{n} = \frac{1+nx+1+1+\cdots+1}{n},$$

where there are $n-1$ 1's to the right of nx in the second numerator. Then applying the AGM Inequality (Theorem 2.10) to the n positive numbers $1+nx,\ 1,\ 1,\ \ldots,\ 1$, we get

$$(1+x)^n = \left(\frac{1+nx+1+1+\cdots+1}{n}\right)^n \geq (1+nx)(1)(1)\cdots(1) = 1+nx,$$

as desired. ◇

Conversely, it happens that the AGM Inequality (Theorem 2.10) follows from Bernoulli's Inequality (Lemma 2.1), and so the two are equivalent. We leave the verification of this for Exercise 2.10.

Example 2.16. For n positive numbers $a_1, a_2, \ldots, a_n$, their **Harmonic Mean** is given by

$$H = \left(\frac{1/a_1 + 1/a_2 + \cdots + 1/a_n}{n}\right)^{-1}.$$

Replacing a_j with $1/a_j$ in the AGM Inequality (Theorem 2.10) we get

$$H \leq G.$$

Then in $H \le G \le A$, the outside inequality can be rewritten rather nicely as

$$\sum_{j=1}^{n} a_j \sum_{j=1}^{n} \frac{1}{a_j} \ge n^2. \tag{2.2}$$

Again, equality occurs here if and only if $a_1 = a_2 = \cdots = a_n$. The Harmonic Mean of two numbers $a, b > 0$ is simply

$$H = \frac{2ab}{a+b}.$$

In this case, (2.2) reads

$$(a+b)\left(\frac{1}{a} + \frac{1}{b}\right) \ge 4,$$

and equality occurs here if and only if $a = b$. ◇

To close this section, we supply Cauchy's brilliant 1821 proof of the AGM Inequality (Theorem 2.10) but without addressing the equality conditions—these we leave for Exercise 2.35. The pattern of argument here is powerful and has since been used by mathematicians in many other contexts. (We shall see it applied in one other context in Sect. 8.3.)

Proof. Again, if $n = 2$, this is simply Lemma 2.7. If $n = 4$, we use Lemma 2.7 twice:

$$\begin{aligned}
(a_1 \cdot a_2 \cdot a_3 \cdot a_4)^{1/4} &= \left((a_1 \cdot a_2)^{1/2}\right)^{1/2} \left((a_3 \cdot a_4)^{1/2}\right)^{1/2} \\
&\le \left(\frac{1}{2}(a_1 + a_2)\right)^{1/2} \cdot \left(\frac{1}{2}(a_3 + a_4)\right)^{1/2} \\
&\le \frac{1}{2}\left(\frac{1}{2}(a_1 + a_2) + \frac{1}{2}(a_3 + a_4)\right) \\
&= \frac{1}{4}(a_1 + a_2 + a_3 + a_4).
\end{aligned}$$

If $n = 8$, we use Lemma 2.7 then the $n = 4$ case:

$$\begin{aligned}
&(a_1 \cdot a_2 \cdot a_3 \cdot a_4 \cdot a_5 \cdot a_6 \cdot a_7 \cdot a_8)^{1/8} \\
&= \left((a_1 \cdot a_2 \cdot a_3 \cdot a_4)^{1/4}\right)^{1/2} \left((a_5 \cdot a_6 \cdot a_7 \cdot a_8)^{1/4}\right)^{1/2} \\
&\le \left(\frac{1}{4}(a_1 + a_2 + a_3 + a_4)\right)^{1/2} \cdot \left(\frac{1}{4}(a_5 + a_6 + a_7 + a_8)\right)^{1/2}
\end{aligned}$$

$$\leq \frac{1}{2}\left(\frac{1}{4}(a_1 + a_2 + a_3 + a_4) + \frac{1}{4}(a_5 + a_6 + a_7 + a_8)\right)$$

$$= \frac{1}{8}(a_1 + a_2 + a_3 + a_4 + a_5 + a_6 + a_7 + a_8) .$$

Clearly we could continue this procedure indefinitely, and so we may assume that we have proved that $G \leq A$ for any n of the form $n = 2^m$. For any (other) n, we choose m so large that $2^m > n$. Now,

$$\frac{a_1 + a_2 + \cdots + a_n + (2^m - n)A}{2^m} = A.$$

The numerator of the left-hand side here has 2^m members in the sum and so we can apply what we have proved so far to see that

$$\left(a_1 \cdot a_2 \cdots a_n \cdot A^{(2^m - n)}\right)^{1/2^m} \leq A.$$

That is,

$$a_1 \cdot a_2 \cdots a_n \cdot A^{2^m - n} \leq A^{2^m} \quad \Rightarrow \quad G^n \leq A^n .$$

□

Remark 2.17. Extending Lemma 2.7 to $n = 4, 8, 16, \ldots$ as above is not too hard, just a bit messy. Cauchy's genius lies in being able to extending the result to *any* n. With this in mind we mention that T. Harriet proved the AGM Inequality (Theorem 2.10) for $n = 3$ around 1,600 [39]. No small feat for the time. ∘

2.3 The Cauchy–Schwarz Inequality

Let $a_1, a_2, \ldots, a_n$ and $b_1, b_2, \ldots, b_n$ be real numbers. The Cauchy–Schwarz Inequality provides an upper bound for the sum of products $\sum_{j=1}^{n} a_j b_j$. The proof we provide below uses Lemma 2.7.

Theorem 2.18. (Cauchy–Schwarz Inequality) *Let* $a_1, a_2, \ldots, a_n$ *and* $b_1, b_2, \ldots, b_n$ *be real numbers. Then*

$$\left(\sum_{j=1}^{n} a_j b_j\right)^2 \leq \sum_{j=1}^{n} a_j^2 \sum_{j=1}^{n} b_j^2 .$$

Proof. If $\sum_{j=1}^{n} a_j^2 = 0$ or $\sum_{j=1}^{n} b_j^2 = 0$ then the inequality holds, with equality. Otherwise, we set $a_j = \frac{a_j^2}{\sum_{k=1}^{n} a_k^2}$ and $b_j = \frac{b_j^2}{\sum_{k=1}^{n} b_k^2}$ in Lemma 2.7 to obtain

$$\frac{a_j}{\sqrt{\sum_{k=1}^{n} a_k^2}} \frac{b_j}{\sqrt{\sum_{k=1}^{n} b_k^2}} \leq \frac{1}{2}\left(\frac{a_j^2}{\sum_{k=1}^{n} a_k^2} + \frac{b_j^2}{\sum_{k=1}^{n} b_k^2}\right).$$

Then summing from $j = 1$ to n we get

$$\frac{\sum_{j=1}^{n} a_j b_j}{\sqrt{\sum_{k=1}^{n} a_k^2}\sqrt{\sum_{k=1}^{n} b_k^2}} \leq \frac{1}{2}\left(\frac{\sum_{j=1}^{n} a_j^2}{\sum_{k=1}^{n} a_k^2} + \frac{\sum_{j=1}^{n} b_j^2}{\sum_{k=1}^{n} b_k^2}\right) = \frac{1}{2}(1+1) = 1,$$

which is really what we wanted to show. □

Example 2.19. The **Root Mean Square** of the real numbers $a_1, a_2, \ldots, a_n$ is:

$$R = \sqrt{\frac{1}{n}\sum_{j=1}^{n} a_j^2}.$$

For example, suppose that three squares with side lengths a_1, a_2 and a_3 have average area T. Then the single square with area T is the one with side length R. The reader should verify that R is a *mean*. We have seen that $G \leq A$. The Cauchy–Schwarz Inequality (Theorem 2.18) shows that $A \leq R$, on taking $b_1 = b_2 = \cdots = b_n = 1/n$. ◇

Remark 2.20. Readers who know some linear algebra might recognize the Cauchy–Schwarz Inequality in the following form. For two vectors $\mathbf{u} = (a_1, a_2, \ldots, a_n)$ and $\mathbf{v} = (b_1, b_2, \ldots, b_n)$ in $\mathbf{R}^n$, their **dot product** is given by $\mathbf{u} \cdot \mathbf{v} = \sum_{j=1}^{n} a_j b_j$, and the **length** of $\mathbf{u}$ is given by $\|\mathbf{u}\| = \sqrt{\mathbf{u} \cdot \mathbf{u}}$. Then the Cauchy–Schwarz Inequality reads

$$|\mathbf{u} \cdot \mathbf{v}| \leq \|\mathbf{u}\| \, \|\mathbf{v}\|.$$

(See [48], for example, for a proof of the Cauchy–Schwarz Inequality in this context.) This says that for non-zero vectors $\mathbf{u}$ and $\mathbf{v}$ we have

$$-1 \leq \frac{\mathbf{u} \cdot \mathbf{v}}{\|\mathbf{u}\| \, \|\mathbf{v}\|} \leq 1,$$

so we may *define* the **angle** θ between such vectors in $\mathbf{R}^n$ as that $\theta \in [0, \pi]$ for which

$$\cos(\theta) = \frac{\mathbf{u} \cdot \mathbf{v}}{\|\mathbf{u}\| \, \|\mathbf{v}\|}.$$

Moreover,

$$\|\mathbf{u} + \mathbf{v}\|^2 = (\mathbf{u} + \mathbf{v}) \cdot (\mathbf{u} + \mathbf{v}) = \|\mathbf{u}\|^2 + 2\mathbf{u} \cdot \mathbf{v} + \|\mathbf{v}\|^2 \leq \|\mathbf{u}\|^2 + 2 \, \|\mathbf{u}\| \, \|\mathbf{v}\| + \|\mathbf{v}\|^2,$$

by the Cauchy–Schwarz Inequality. This last piece equals $\left(\|\mathbf{u}\| + \|\mathbf{v}\| \right)^2$, and so we have the **triangle inequality** in $\mathbf{R}^n$:

$$\|\mathbf{u} + \mathbf{v}\| \leq \|\mathbf{u}\| + \|\mathbf{v}\| \, .$$

So the Cauchy–Schwarz Inequality is fundamental for working in $\mathbf{R}^n$. And, in $\mathbf{R}^n$ it is evident why the triangle inequality is so named—see **Fig. 2.2.** ○

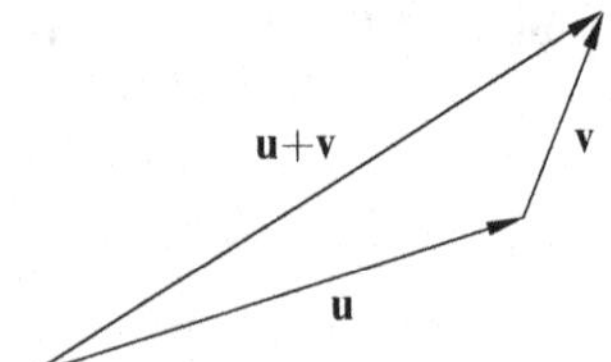

Fig. 2.2 The Triangle Inequality $\|\mathbf{u} + \mathbf{v}\| \leq \|\mathbf{u}\| + \|\mathbf{v}\|$

There are many other proofs of the Cauchy–Schwarz Inequality, a few of which we explore in the exercises. However, we would be remiss if we did not supply what is essentially H. Schwarz's (1843–1921) own ingenious proof, as follows. (See also Exercises 2.38 and 2.41.) For any real number t,

$$\sum_{j=1}^{n} (ta_j + b_j)^2 \geq 0.$$

That is,

$$t^2 \sum_{j=1}^{n} a_j^2 + 2t \sum_{j=1}^{n} a_j b_j + \sum_{j=1}^{n} b_j^2 \geq 0.$$

Now the left-hand side is a quadratic in the variable t and since it is ≥ 0, it must have either *no* real root or *one* real root. (It cannot have two distinct real roots.) Therefore its discriminant $B^2 - 4AC$ must be ≤ 0. That is,

$$\left(2\sum_{j=1}^{n} a_j b_j\right)^2 - 4\sum_{j=1}^{n} a_j^2 \sum_{j=1}^{n} b_j^2 \leq 0.$$

Rearranging this inequality yields the desired result. Pretty slick.

Exercises

2.1. Show that for $x > 0$, the sequence $\{x^{1/n}\}$ converges to 1.
Hint: Write $x^{1/n} = \left(1 + n\frac{x-1}{n}\right)^{1/n}$.

2.2. Show that Bernoulli's Inequality (Lemma 2.1), implies Lemma 2.7:

$$\sqrt{ab} \leq \frac{a+b}{2}.$$

Hint: Assume that $a \leq A = (a+b)/2$, take $n = 2$, and $x = A/a - 1$.

2.3. [35] Show that Bernoulli's Inequality (Lemma 2.1), holds also for $x \in [-2, -1]$.

2.4. **(a)** Write $0.\overline{237}$ and $6.45\overline{7132}$ as fractions. **(b)** Describe how to write any repeating decimal as a fraction.

2.5. [2] Take $x = -1/n^2$ in Bernoulli's Inequality (Lemma 2.1) to show that the sequence $\{(1 + \frac{1}{n})^n\}$ is increasing.

2.6. Show that $\left\{ \left(1 + \frac{1}{n}\right)^{n+1} \right\}$ is decreasing, as follows.

(a) Verify that

$$\frac{(1 + \frac{1}{n+1})^{n+2}}{(1 + \frac{1}{n})^{n+1}} = \left(1 + \frac{1}{n+1}\right)\left(1 - \frac{1}{(n+1)^2}\right)^{n+1}.$$

(b) Apply Bernoulli's Inequality (Lemma 2.1) to get

$$\left(1 + \frac{1}{n+1}\right)\left(1 - \frac{1}{(n+1)^2}\right)^{n+1} \leq \left(1 - \frac{1}{(n+1)^4}\right)^{n+1} < 1.$$

2.7. [47] Find the least positive integer N such that for all $n > N$,

$$\left(\frac{n^{n+1}}{(n+1)^n}\right)^n < n! < \left(\frac{n^{n+1}}{(n+1)^n}\right)^{n+1}.$$

2.8. (e.g., [24, 34, 52, 54]) Show that $x = \left(1 + \frac{1}{u}\right)^u$ and $y = \left(1 + \frac{1}{u}\right)^{u+1}$ (and vice-versa) are solutions to the equation $x^y = y^x$, for $x, y > 0$. (Taking $u = 1, 2, 3, \ldots$ in fact yields *all* nontrivial (i.e., $x \neq y$) rational solutions to this equation.)

2.9. [28] Denote by $\lfloor x \rfloor$ the greatest integer not exceeding x. This function is often called the *Floor function*. Prove that for $x \geq 1$,

$$\left(1 + \frac{x}{\lfloor x \rfloor}\right)^{\lfloor x \rfloor} \leq 2^x \leq \left(1 + \frac{x}{\lfloor x + 1 \rfloor}\right)^{\lfloor x+1 \rfloor}.$$

2.10. [26] In Example 2.15 we showed that the AGM Inequality implies Bernoulli's Inequality. Show that Bernoulli's Inequality implies the AGM Inequality.

2.11. Explain how the inequality $(a + b)^2 - 4ab = (a - b)^2 \geq 0$ in the proof of Lemma 2.7 relates to **Fig. 2.3.**

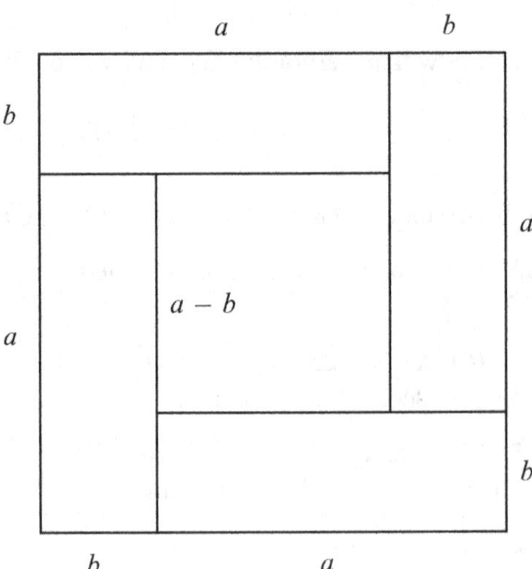

Fig. 2.3 For Exercise 2.11

2.12. **(a)** Fill in the details of another proof of Lemma 2.7, as follows. Let $a, b \geq 0$. Since $(t - \sqrt{a})(t - \sqrt{b})$ has real zeros, conclude that $\sqrt{ab} \leq \frac{a+b}{2}$.

(b) Let $a, c > 0$. Show that if $|b| > a + c$ then $ax^2 + bx + c$ has two (distinct) real roots.

2.13. [19] In **Fig. 2.4**, ABCD is a trapezoid with AB parallel to DC, and EF is parallel to each of these. Show that m is a weighted average of a and b. That is, $m = \frac{pb+qa}{p+q}$, for some $p, q > 0$.

2.14. **(a)** Show that for $x > 0$, we have $x + 1/x \geq 2$, with equality if and only if $x = 1$.

(b) Conclude (even though we have not yet officially met the exponential function) that

$$\cosh(x) = \frac{e^x + e^{-x}}{2} \geq 1,$$

with equality if and only if $x = 0$.

(c) [31] Show that for $x > 0$,

$$\frac{x^n}{1 + x + x^2 + \cdots + x^{2n}} \leq \frac{1}{2n+1}.$$

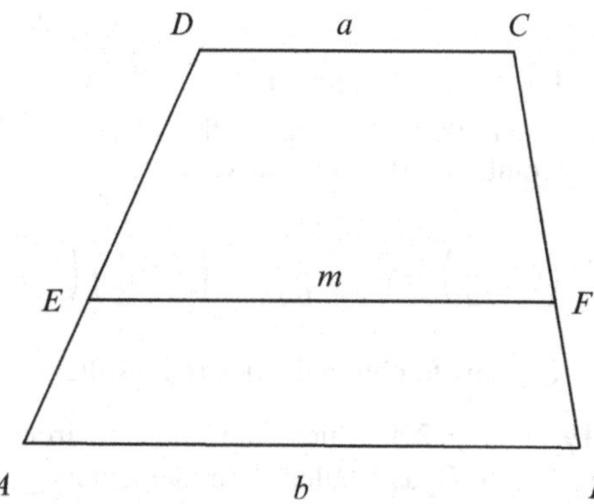

Fig. 2.4 For Exercise 2.13

2.15. [51] The bank in your town sells British pounds at the rate 1£ = \$$S$ and buys them at the rate 1£ = \$$B$. You and your friend want to exchange dollars and pounds between the two of you, at a rate that is fair to both. Show that the fair exchange rate is 1£ = $\sqrt{SB}$, the Geometric Mean of S and B.

2.16. [13] Let $H \leq G \leq A$ denote respectively the Harmonic, Geometric and Arithmetic Means of two positive numbers.

(a) Show that H, G, and A are the side lengths of a triangle if and only if

$$\frac{3-\sqrt{5}}{2} < \frac{A}{H} < \frac{3+\sqrt{5}}{2}.$$

(b) Show that H, G, and A are the side lengths of a right triangle if and only if A/H is the *golden mean*:

$$\frac{A}{H} = \frac{1+\sqrt{5}}{2}.$$

2.17. [56]

(a) Prove that for any natural number n, we have $\sum_{k=1}^{n} k = \frac{n(n+1)}{2}$.

(b) For $n = 1$, it is clear that $n! = \left(\frac{n+1}{2}\right)^n$. Use (a) and the AGM Inequality (Theorem 2.10) to prove that for integers $n \geq 2$,

$$n! < \left(\tfrac{n+1}{2}\right)^n.$$

2.18. [31] Let $a, b > 0$, with $a + b = 1$. Show, as follows, that

$$\left(a + \frac{1}{a}\right)^2 + \left(b + \frac{1}{b}\right)^2 \geq \frac{25}{2}.$$

(a) For such a, b, show that $\frac{1}{ab} \geq 4$.
(b) Use Lemma 2.7 to show that $(a+1/a)^2 + (b+1/b)^2 \geq 2\,(a+1/a)\,(b + 1/b)$.
(c) Combine (a) and (b) to show that

$$\left(a + \frac{1}{a}\right)^2 + \left(b + \frac{1}{b}\right)^2 - 2\left(a + \frac{1}{a}\right)\left(b + \frac{1}{b}\right) \geq (1+4)^2 - \left(a + \frac{1}{a}\right)^2 - \left(b + \frac{1}{b}\right)^2.$$

(d) Use this to obtain the desired result.

2.19. In **Fig. 2.5**, which shows a semicircle with diameter $a + b$, we can see that $A > G > H$ as labeled. Use elementary geometry to show that A, G, and H are respectively the Arithmetic, Geometric and Harmonic Means of a and b.

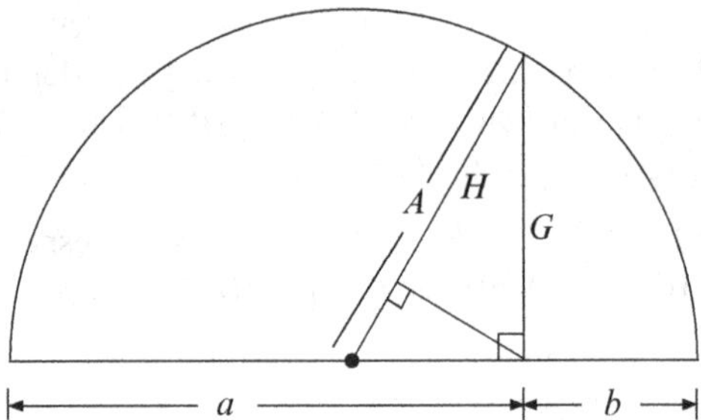

Fig. 2.5 For Exercise 2.19

2.20. (a) Suppose that a car travels at a miles per hour from point A to point B, then returns at b miles per hour. Show that the average speed for the trip is the Harmonic Mean of a and b.
(b) Show that $G - H < A - G$, where H, G and A the Harmonic, Geometric, and Arithmetic Means of two numbers $a, b > 0$.
(c) For $a, b > 0$, **Heron's Mean**, named for Heron of Alexandria (c. 10–70 AD), is

$$\hat{H} = \frac{a + \sqrt{ab} + b}{3}.$$

Show that if $a \neq b$, then $G < \hat{H} < A$, where G and A are the Geometric and Arithmetic means of a and b.

2.21. Let $a_1, a_2, \ldots, a_n > 0$. We saw in (2.2) that

$$\sum_{j=1}^{n} a_j \sum_{j=1}^{n} \frac{1}{a_j} \geq n^2.$$

Apply this to the three numbers $a+b, a+c$, and $b+c$ to obtain **Nesbitt's Inequality**:

$$\frac{a}{b+c} + \frac{b}{a+c} + \frac{c}{a+b} \geq \frac{3}{2}.$$

2.22. Denote by H, G and A the Harmonic, Geometric, and Arithmetic Means of two numbers $a, b > 0$. We have seen that $H \leq G \leq A$.

(a) Show that $A - H = (b-a)/2$.
(b) Let $a_1 = H(a,b)$ and $b_1 = A(a,b)$. Then for $n = 1, 2, 3, \ldots$, let

$$a_{n+1} = H(a_n, b_n) \quad \text{and} \quad b_{n+1} = A(a_n, b_n).$$

Show that $\{[a_n, b_n]\}$ is a sequence of nested intervals, with $b_n - a_n \to 0$. Conclude by the Nested Interval Property (Theorem 1.41) that there is c belonging to each of these intervals.
(c) Show that $\sqrt{a_n b_n} = \sqrt{ab} = G$ for all n to conclude that $c = G$. (For example, if $a = 1$ and $b = 2$, then $\{a_n\}$ is an increasing sequence of rational numbers which converges to $\sqrt{2}$.)

2.23. [43] In Example 2.5 we used Bernoulli's Inequality (Lemma 2.1) to show that $\sqrt[n]{n} \to 1$ as $n \to \infty$. Prove this using the AGM Inequality (Theorem 2.10), by setting $a_1 = a_2 = \cdots = a_{n-1}$ and $a_n = \sqrt{n}$.

2.24. Show that

$$\lim_{n\to\infty} \frac{2+4+6+\cdots+(2n)}{1+3+5+\cdots+(2n-1)} = \mathrm{e}.$$

2.25. [30]

(a) In Example 2.6 we used Bernoulli's Inequality (Lemma 2.1) to show that $\left\{\left(1+\frac{1}{n}\right)^n\right\}$ is an increasing sequence. Show this using the AGM Inequality (Theorem 2.10). Hint: Consider the $n+1$ numbers $1, \frac{n+1}{n}, \frac{n+1}{n}, \ldots, \frac{n+1}{n}$.
(b) Show that $\left\{\left(1+\frac{1}{n}\right)^{n+1}\right\}$ is a decreasing sequence using the AGM Inequality (Theorem 2.10). Hint: Consider the $n+2$ numbers $1, \frac{n}{n+1}, \frac{n}{n+1}, \ldots, \frac{n}{n+1}$, apply the AGM Inequality, then take reciprocals.
(c) Use the AGM Inequality (Theorem 2.10) to show that $\left\{\left(1-\frac{1}{n}\right)^{-n}\right\}$ is a decreasing sequence (for $n = 2, 3, \ldots$).
Hint: Consider the $n+1$ numbers $1, 1-\frac{1}{n}, 1-\frac{1}{n}, \ldots, 1-\frac{1}{n}$.

Note: Many variations of Exercise 2.25 have been discovered and rediscovered over the years (e.g., [15, 22, 27, 29–31, 41, 57]). Other approaches can be found in [3, 17, 42, 44].

2.26. [59] Apply the AGM Inequality (Theorem 2.10) to the $n+k$ numbers

$$\left(1+\frac{1}{n}\right), \left(1+\frac{1}{n}\right), \cdots, \left(1+\frac{1}{n}\right), \frac{k-1}{k}, \frac{k-1}{k}, \cdots, \frac{k-1}{k}$$

to show that

$$\left(1+\frac{1}{n}\right)^n < \left(\frac{k}{k-1}\right)^k.$$

So, for example, taking $k = 6$, we may conclude that $e \leq (\frac{6}{5})^6 \cong 2.986 < 3$.

2.27. [22] Use $(1 + 1/n)^n < e$ and induction to show that $(n/e)^n < n!$.

2.28. [32, 49] Let

$$p(x) = x^n + a_{n-1}x^{n-1} + \cdots + a_1 x + a_0$$

be a polynomial with roots $x_1, x_2, \ldots, x_n$.

(a) Show that $a_0 = (-1)^n x_1 x_2 \cdots x_n$.

(b) Show that $a_{n-1} = -\sum_{j=1}^{n} x_j$.

(c) Show that $a_1 = (-1)^{n-1}\left[x_2x_3\cdots x_n + x_1x_3\cdots x_n + \cdots + x_1x_2\ldots x_{n-1}\right]$.

(d) Show that if all of the roots are positive, then $a_1 a_{n-1}/a_0 \geq n^2$.

2.29. Suppose a quadrilateral has side lengths $a, b, c, d > 0$ and denote by s its semi perimeter: $s = (a + b + c + d)/2$. **Bretschneider's formula** says that the area of the quadrilateral is given by

$$A = \sqrt{(s-a)(s-b)(s-c)(s-d) - abcd\cos^2(\theta)},$$

where θ is half of the sum of any pair of opposite angles. If the quadrilateral can be inscribed in a circle then elementary geometry shows that $\theta = \pi/2$ and we get **Brahmagupta's formula**

$$A = \sqrt{(s-a)(s-b)(s-c)(s-d)}.$$

(And if $d = 0$ then the quadrilateral is in fact a triangle and we get Heron's formula.) Show that among all quadrilaterals with a given perimeter, the square has the largest area.

2.30. In our proof (that is, K.M. Chong's) of the AGM Inequality (Theorem 2.10) we focused on A and used the inequality (2.1). Fill in the details of the following proof, which focuses instead on G.

(a) Show (assuming again $a_1 \leq a_2 \leq \cdots \leq a_n$) that

$$a_1 + a_n - G \geq \frac{a_1 a_n}{G}.$$

(b) Use this to prove the AGM Inequality.

2.31. Fill in the details of H. Dorrie's beautiful 1921 proof of the AGM Inequality (Theorem 2.10), as follows. (This proof was rediscovered by P.P. Korovkin in 1952 [22] and again by G. Ehlers in 1954 [5].) Lemma 2.7 is the case $n = 2$, so we proceed by induction, assuming that the result is true for $n - 1$ numbers. What we want to show is that $\sum_{j=1}^{n} a_j \geq nG$.

(a) Argue that since $G^n = \prod_{j=1}^{n} a_j$, at least one a_j must be $\leq G$, and some other a_j must be $\geq G$. So we may assume that $a_1 \leq G$ and $a_2 \geq G$.
(b) Show that $a_1 \leq G$ and $a_2 \geq G$ imply that

$$a_1 + a_2 \geq G + \frac{a_1 a_2}{G}, \quad \text{and so} \quad \sum_{j=1}^{n} a_j \geq G + \frac{a_1 a_2}{G} + \sum_{j=3}^{n} a_j.$$

(c) Now apply the assumed result to the $n - 1$ numbers $\frac{a_1 a_2}{G}, a_3, a_4, \ldots, a_n$.

2.32. [7,50] Let $a_1, a_2, \ldots, a_n$ be nonnegative real numbers. Show that

$$1 + \prod_{j=1}^{n} a_j^{1/n} \leq \prod_{j=1}^{n} (1 + a_j)^{1/n}.$$

Hint: Consider the left-hand side divided by the right-hand side, apply the AGM Inequality (Theorem 2.10), then tidy up.

2.33. [11] Let A, B and C be the interior angles of a triangle. Show that

$$\sin(A) + \sin(B) + \sin(C) \leq \frac{3\sqrt{3}}{2}.$$

Hint:
$A + B + C = \pi \Rightarrow \sin(A) + \sin(B) + \sin(C) = 4\cos(A/2)\cos(B/2)\cos(C/2)$.

2.34. [6] Prove that

$$\left(\sqrt{2} - 1\right)\left(\sqrt[3]{6} - \sqrt{2}\right)\cdots\left(\sqrt[n+1]{(n+1)!} - \sqrt[n]{n!}\right) < \frac{n!}{(n+1)^n}.$$

2.35. Analyze Cauchy's proof of the AGM Inequality (Theorem 2.10) given at the end of Sect. 2.2 to obtain necessary and sufficient conditions for equality.

2.36. In Cauchy's proof of the AGM Inequality (Theorem 2.10) given at the end of Sect. 2.2, we focused on A and had (for $2^m > n$):

$$A = \frac{a_1 + a_2 + \cdots + a_n + (2^m - n)A}{2^m}.$$

Then we applied the result for the 2^m case. For a proof which focuses instead on G, verify (for $2^m > n$) that

$$G^{2^m} = a_1 \cdot a_2 \cdots a_n \cdot G^{2^m - n},$$

then apply the result for the 2^m case.

2.37. We used Lemma 2.7 to prove the Cauchy–Schwarz Inequality (Theorem 2.18). Then we used the Cauchy–Schwarz Inequality to show that $A \leq R$. Show that $A \leq R$ using Lemma 2.7 directly. When does equality hold?

2.38. Fill in the details of the following proof of the Cauchy–Schwarz Inequality (Theorem 2.18), which is very similar to Schwarz's.

(a) Dispense with the case $a_1 = a_2 = \cdots = a_n = 0$.

(b) Expand the sum in the expression $0 \leq \sum_{j=1}^{n} (ta_j + b_j)^2$.

(c) Set $t = -\sum_{j=1}^{n} a_j b_j / \sum_{j=1}^{n} a_j^2$.

(This is the t at which the quadratic $\sum_{j=1}^{n} (ta_j + b_j)^2$ attains its minimum.)

2.39. Fill in the details of another proof of the Cauchy–Schwarz Inequality (Theorem 2.18), as follows.

(a) Replace a with a^2 and b with b^2 in Lemma 2.7 to get $ab \leq \frac{1}{2}a^2 + \frac{1}{2}b^2$.

(b) Write $ab = \sqrt{t}a\frac{1}{\sqrt{t}}b$ in (a) to show that for numbers a, b and any $t > 0$,

$$ab \leq \frac{t}{2}a^2 + \frac{1}{2t}b^2.$$

(c) Now write $a_j b_j = \sqrt{t}a_j \frac{1}{\sqrt{t}}b_j$ then sum from $j = 1$ to n to get

$$\sum_{j=1}^{n} a_j b_j \leq \frac{t}{2}\sum_{j=1}^{n} a_j^2 + \frac{1}{2t}\sum_{j=1}^{n} b_j^2.$$

(d) Dispense with the case $a_1 = a_2 = \cdots = a_n = 0$, then set

$$t = \left(\sum_{j=1}^{n} b_j^2\right)^{1/2} \Big/ \left(\sum_{j=1}^{n} a_j^2\right)^{1/2},$$

then simplify. (This is the t at which $\frac{t}{2}\sum_{j=1}^{n} a_j^2 + \frac{1}{2t}\sum_{j=1}^{n} b_j^2$ attains its minimum.)

2.40. Apply Schwarz's idea, as in his proof of the Cauchy–Schwarz Inequality (Theorem 2.18), to $\sum_{j=1}^{n}(a_j + t)^2$. What do you get? Can you prove whatever you got using the Cauchy–Schwarz Inequality?

2.41. [53] Fill in the details of the following proof of the Cauchy–Schwarz Inequality (Theorem 2.18), which is quite possibly just as slick as Schwarz's. Observe that

$$\frac{\sum_{j=1}^{n} a_j b_j}{\sqrt{\sum_{j=1}^{n} a_j^2}\sqrt{\sum_{j=1}^{n} b_j^2}} = 1 - \frac{1}{2}\sum_{j=1}^{n}\left(\frac{a_j}{\sqrt{\sum_{j=1}^{n} a_j^2}} - \frac{b_j}{\sqrt{\sum_{j=1}^{n} b_j^2}}\right)^2.$$

2.42. Fill in the details of the following (ostensibly) different proof of the Cauchy–Schwarz Inequality (Theorem 2.18).

(a) Replace a with a^2 and b with b^2 in Lemma 2.7 to get $ab \le \frac{1}{2}a^2 + \frac{1}{2}b^2$.
(b) Dispense with the cases $a_1 = a_2 = \cdots = a_n = 0$ or $b_1 = b_2 = \cdots = b_n = 0$.
(c) Set $a = a_j\left(\sum_{j=1}^{n} a_j^2\right)^{-1/2}$ and $b = b_j\left(\sum_{j=1}^{n} b_j^2\right)^{-1/2}$ in (a), then sum from 1 to n.

2.43. Find necessary and sufficient conditions for equality to hold in the Cauchy–Schwarz Inequality (Theorem 2.18).

2.44. [33] Let $a_1, a_2, \ldots, a_n$ be positive real numbers and let $r \le n$ be a positive integer. Set

$$A_1 = \frac{1}{r}\sum_{j=1}^{r} a_j, \quad A = \frac{1}{n}\sum_{j=1}^{n} a_j, \quad \text{and} \quad \sigma^2 = \frac{1}{n}\sum_{j=1}^{n}(a_j - A)^2.$$

The number σ^2 is called the **variance** of $a_1, a_2, \ldots, a_n$. Show that

$$r(A_1 - A)^2 \le (n - r)\sigma^2.$$

2.45. [58] Let $x_1, x_2, \ldots, x_n \in \mathbf{R}$. Show that

$$\frac{x_1}{1+x_1^2} + \frac{x_2}{1+x_1^2+x_2^2} + \cdots + \frac{x_n}{1+x_1^2+x_2^2+\cdots+x_n^2} \le \sqrt{n}.$$

2.46. [14] Show that

$$\left(\sum_{k=1}^{n} \sqrt{\frac{k-\sqrt{k^2-1}}{\sqrt{k(k+1)}}} \right)^2 \le n\sqrt{\frac{n}{n+1}}.$$

2.47. [23,38] For n data points $(a_1, b_1), \ldots, (a_n, b_n)$,

$$A = \frac{1}{n}\sum_{j=1}^{n} a_j, \quad \text{and} \quad B = \frac{1}{n}\sum_{j=1}^{n} b_j,$$

Pearson's coefficient of linear correlation is

$$\rho = \frac{\sum_{j=1}^{n} (a_j - A)(b_j - B)}{\sqrt{\sum_{j=1}^{n} (a_j - A)^2}\sqrt{\sum_{j=1}^{n} (b_j - B)^2}}.$$

Clearly the Cauchy–Schwarz Inequality (Theorem 2.18) implies that $|\rho| \le 1$. Show that $|\rho| \le 1$ implies the Cauchy–Schwarz Inequality. (For readers who know a little linear algebra, [21] contains a neat relationship between ρ and something called the *Gram determinant*.)

2.48. [1,50]

(a) Show that for $x_1, x_2, \ldots, x_n \in \mathbf{R}$ and $y_1, y_2, \ldots, y_n > 0$,

$$\frac{(x_1+x_2+\cdots+x_n)^2}{y_1+y_2+\cdots+y_n} \le \frac{x_1^2}{y_1} + \frac{x_2^2}{y_2} + \cdots + \frac{x_n^2}{y_n}.$$

(b) Set $x_j = a_j b_j$ and $y_j = b_j^2$ to obtain the Cauchy–Schwarz Inequality (Theorem 2.18).

(c) Let $a, b, c > 0$. Use (a) to obtain **Nesbitt's Inequality**:

$$\frac{a}{b+c} + \frac{b}{a+c} + \frac{c}{a+b} \ge \frac{3}{2}.$$

(d) Use (a) to show that for $a, b > 0$,

$$a^4 + b^4 \ge \tfrac{1}{8}(a^4 + b^4).$$

2.49. [8,46] Let $a, b, c > 0$.

(a) Show that if

$$a\cos^2(x) + b\sin^2(x) < c\,,$$

then

$$\sqrt{a}\cos^2(x) + \sqrt{b}\sin^2(x) < \sqrt{c}.$$

(b) Show that

$$(abc)^{2/3} \leq \frac{ab+ac+bc}{3} \leq \left(\frac{a+b+c}{3}\right)^2.$$

2.50. [20] We saw in Remark 2.14 that the isoperimetric inequality for an n-sided polygon with area T and perimeter P is

$$T \leq \frac{P^2}{4n\tan(\pi/n)}.$$

Show that if $a_1, a_2, \ldots, a_n$ are the side lengths of an n-sided polygon, then

$$\sum_{j=1}^{n} a_j^2 \geq 4T\tan(\pi/n).$$

2.51. Let $a_1, a_2, \ldots, a_n$, and $b_1, b_2, \ldots, b_n$ be real numbers, with $\sum_{j=1}^{n} b_j = 0$. Show that

$$\left(\sum_{j=1}^{n} a_j b_j\right)^2 \leq \left(\sum_{j=1}^{n} a_j^2 - \left(\sum_{j=1}^{n} a_j\right)^2\right)\sum_{j=1}^{n} b_j^2.$$

2.52. cf. [36] Let $a_1, a_2, \ldots, a_n$, and $b_1, b_2, \ldots, b_n$ be real numbers with $0 < a \leq a_j \leq A$ and $0 < b \leq b_j \leq B$. Fill in the following details to obtain a **reversed version of the Cauchy–Schwarz Inequality**:

$$\frac{\sum_{j=1}^{n} a_j^2 \sum_{j=1}^{n} b_j^2}{\left(\sum_{j=1}^{n} a_j b_j\right)^2} \leq \frac{1}{4}\left(\sqrt{\frac{AB}{ab}} + \sqrt{\frac{ab}{AB}}\right).$$

(a) Verify that $\left(\frac{a_j}{b_j} - \frac{a}{B}\right)\left(\frac{b_j}{a_j} - \frac{A}{b}\right) \le 0.$
(b) Use this to verify that

$$a_j^2 + \frac{aA}{bB}b_j^2 \le \left(\frac{a}{B} + \frac{A}{b}\right)a_j b_j.$$

(c) Write

$$\left(\sum_{j=1}^{n} a_j^2\right)^{1/2}\left(\sum_{j=1}^{n} b_j^2\right)^{1/2} = \sqrt{\frac{Bb}{Aa}}\left(\sum_{j=1}^{n} a_j^2\right)^{1/2}\left(\sum_{j=1}^{n} \frac{Aa}{Bb}b_j^2\right)^{1/2},$$

then use Lemma 2.7 and (b) to obtain the desired result.

2.53. Use the Cauchy–Schwarz Inequality (Theorem 2.18) to prove **Minkowski's Inequality**: *Let* $a_1, \ldots, a_n, b_1, \ldots, b_n \in \mathbf{R}$. *Then*

$$\left(\sum_{j=1}^{n} \left(a_j + b_j\right)^2\right)^{1/2} \le \left(\sum_{j=1}^{n} a_j^2\right)^{1/2} + \left(\sum_{j=1}^{n} b_j^2\right)^{1/2}.$$

(Notice that if $n = 1$ this is simply the triangle inequality $|a + b| \le |a| + |b|$.) Hint: Write $\left(a_j + b_j\right)^2 = a_j\left(a_j + b_j\right) + b_j\left(a_j + b_j\right)$, then sum, then apply the Cauchy–Schwarz Inequality (Theorem 2.18) to each piece.

2.54. [45] The Cauchy–Schwarz Inequality (Theorem 2.18) gives an upper bound for $\sum_{j=1}^{n} a_j b_j$. Under certain circumstances, a lower bound is given by **Chebyshev's Inequality:** *Let* $\{a_1, a_2, \ldots, a_n\}$ *and* $\{b_1, b_2, \ldots, b_n\}$ *be sequences of real numbers, with either both increasing or both decreasing. Then*

$$\frac{1}{n}\sum_{j=1}^{n} a_j \cdot \frac{1}{n}\sum_{j=1}^{n} b_j \le \frac{1}{n}\sum_{j=1}^{n} a_j b_j.$$

And the inequality is reversed if the sequences have opposite monotonicity. Fill in the details of the following proof of Chebyshev's Inequality, for the $a_j's$ and $b_j's$ both increasing. (The other case is handled similarly.) First, let $A = \frac{1}{n}\sum_{j=1}^{n} a_j$.

(a) Show that there is k between 1 and n such that

$$a_1 \le a_2 \le \cdots \le a_k \le A \le a_{k+1} \le \cdots \le a_n.$$

(b) Conclude that

$$\left(a_j - A\right)\left(b_j - b_k\right) \ge 0 \text{ for } j = 1, 2, \ldots, n,$$

and therefore

$$\frac{1}{n}\sum_{j=1}^{n}\left(a_j - A\right)\left(b_j - b_k\right) \geq 0.$$

(c) Expand then simplify the sum on the left hand side in (b).

2.55. [45] (If you did Exercise 2.54.) **(a)** Use **Chebyshev's Inequality** to prove that for $0 \leq a_1 \leq a_2 \leq \cdots \leq a_n$ and $n \geq 1$,

$$\left(\frac{1}{n}\sum_{j=1}^{n} a_j\right)^n \leq \frac{1}{n}\sum_{j=1}^{n} a_j^n.$$

(b) Let a, b, c be the side lengths of a triangle with area T and perimeter P. Show that $a^3 + b^3 + c^3 \geq \frac{4\sqrt{3}}{3} PT$ and that $a^4 + b^4 + c^4 \geq 16T^2$.

2.56. [40] (If you did Exercise 2.54.) Use **Chebyshev's Inequality** and the AGM Inequality (Theorem 2.10) to prove that for $0 < a_1 \leq a_2 \leq \cdots \leq a_n$,

$$\sum_{j=1}^{n} a_j^{n+1} \geq a_1 a_2 \cdots a_n \sum_{j=1}^{n} a_j.$$

2.57. For $a_1, a_2, \ldots, a_n \in \mathbf{R}$, their **variance** is the number $\sigma^2 = \frac{1}{n}\sum_{j=1}^{n}(a_j - A)^2$, where $A = \frac{1}{n}\sum_{j=1}^{n} a_j$ is their Arithmetic Mean. Suppose that $m \leq a_j \leq M$ for all j.

(a) Verify that

$$\frac{1}{n}\sum_{j=1}^{n}(a_j - A)^2 = (M - A)\left(A - m\right) - \frac{1}{n}\sum_{j=1}^{n}(M - a_j)(a_j - m),$$

in order to conclude that $\frac{1}{n}\sum_{j=1}^{n}(a_j - A)^2 \leq (M - A)\left(A - m\right)$. (This inequality was obtained differently, and generalized considerably, in [4].)

(b) Show that this inequality is better than, that is, is a *refinement* of **Popoviciu's Inequality**:

$$\frac{1}{n}\sum_{j=1}^{n}(a_j - A)^2 \leq \frac{1}{4}(M - m)^2.$$

Hint: Show that the quadratic $(Q-x)(x-q)$ is maximized when $x = \frac{1}{2}(Q+q)$.

2.58. [32]

(a) Extend Exercise 2.57 to prove **Grüss's Inequality**: *Let* $a_1, a_2, \ldots, a_n$, *and* $b_1, b_2, \ldots, b_n$ *be real numbers, with* $m \leq a_j \leq M$ *and* $\gamma \leq b_j \leq \Gamma$. *Then*

$$\left| \frac{1}{n}\sum_{j=1}^{n} a_j b_j - \frac{1}{n}\sum_{j=1}^{n} a_j \cdot \frac{1}{n}\sum_{j=1}^{n} b_j \right| \leq \frac{1}{4}(M-m)(\Gamma-\gamma).$$

Hint: Let $A = \frac{1}{n}\sum\limits_{j=1}^{n} a_j$, $B = \frac{1}{n}\sum\limits_{j=1}^{n} b_j$ and begin by applying the Cauchy–Schwarz Inequality (Theorem 2.18) to

$$\left(\frac{1}{n}\sum_{j=1}^{n} (a_j - A)(b_j - B) \right)^2.$$

(b) Show by providing an example (take $a_j = b_j$ for simplicity) that the constant $1/4$ in Grüss's Inequality cannot be replaced by any smaller number. That is, the $1/4$ is *sharp*.

References

1. Andreescu, T., Enescu, B: Mathematical Olympiad Treasures. Birkhauser, Boston (2003)
2. Armitage, D.H.: Two Applications of Bernoulli's Inequality. Math. Gaz. **66**, 309–310 (1982)
3. Barnes, C.W.: Euler's constant and e. Am. Math. Mon. **91**, 428–430 (1984)
4. Bhatia, R., Davis, C.: A better bound on the variance. Am. Math. Mon. **107**, 353–357 (2000)
5. Beckenbach, E.F., Bellman, R.: Inequalities. Springer, Berlin (1965)
6. Bencze, M., Howard, J.: Problem 1353. Math. Mag. **74**, 275–276 (1991)
7. Borwein, J.M., Bailey, D.H., Girgensohn, R., Experimentation in Mathematics: Computational Paths to Discovery. A.K. Peters, Natick (2004)
8. Brenner, J.L., Tom, W.W.: Problem 184. Coll. Math. J. **13**, 210–211 (1982)
9. Bullen, P.S.: Handbook of Means and their Inequalities. Kluwer Academic, Dordrecht/Boston (2003)
10. Chong, K.M.: An inductive proof of the A.M. – G.M. inequality. Am. Math. Mon. **83**, 369 (1976)
11. Curry, T.R., Just, W., Schaumberger, N.: Problem E1634. Am. Math. Mon. **71**, 915–916 (1964)
12. Darst, R.B: Simple proofs of two estimates for e. Am. Math. Mon. **80**, 194 (1973)
13. Di Domenico, A.: The golden ratio – the right triangle – and the arithmetic, geometric, and harmonic means. Math. Gaz. **89**, 261 (2005)
14. Díaz-Barrero, J.L., Gibergans-Báguena, J.: Problem 903. Coll. Math. J. **41**, 244–245 (2010)
15. Forder, H.G.: Two inequalities. Math. Gaz. **16**, 267–268 (1932)
16. Friesecke, G., Wehrstedt, J.C.: An elementary proof of the Gregory-Mengoli-Mercator formula. Math. Intell. **28**, 4–5 (2006)
17. Goodman, T.N.T.: Maximum products and $\lim(1 + 1/n)^n = \mathrm{e}$. Am. Math. Mon. **93**, 638–640 (1986)

18. Hardy, G.H., Littlewood, J.E., Polya, G.: Inequalities. Cambridge University Press, Cambridge/New York (1967)
19. Hoehn, L.: A geometrical interpretation of the weighted mean. Coll. Math. J. **15**, 135–139 (1984)
20. Just, W., Schaumberger, N., Marsh, D.C.B.: Problem E1634. Am. Math. Mon. **71**, 796 (1964)
21. Kass, S.: An eigenvalue characterization of the correlation coefficient. Am. Math. Mon. **96**, 910–911 (1989)
22. Korovkin, P.P.: Inequalities. Pergamon Press, Oxford (1961) (Originally published as Neravenstva, Gostekhizdat, Moscow, 1952.)
23. Kundert, K.R.: Correlation – a vector approach. Coll. Math. J. **11**, 52 (1980)
24. Kupitz, Y.S., Martini, H.: On the equation $x^y = y^x$. Elem. Math. **55**, 95–101 (2000)
25. Lax, P.D.: A short path to the shortest path. Am. Math. Mon. **102**, 158–159 (1995)
26. Maligranda, L.: The AM-GM inequality is equivalent to the Bernoulli inequality. Math. Intell. **35**, 1–2 (2012)
27. Mahmoud. M., Edwards, P.: Some inequalities and sequences converging to e. Int. J. Math. Ed. Sci. Tech. **30**, 430–434 (1999)
28. Marjanovic, M., Spital, S.: Problem E1950. Am. Math. Mon. **75**, 543–544 (1968)
29. Melzak, Z.A.: On the exponential function. Am. Math. Mon. **82**, 842–844 (1975)
30. Mendelsohn, N.S.: An application of a famous inequality. Am. Math. Mon. **58**, 563 (1951)
31. Mitrinovic, D.S: Elementary Inequalities. P. Noordhoff, Groningen (1964)
32. Mitrinovic, D.S.: Analytic Inequalities. Springer, Berlin/New York (1970)
33. Murty, V.N., Dana, J.L: Problem 224. Coll. Math. J. **15**, 73–74 (1984)
34. Nakhli, F.: A transcendental sequence and a problem on commutativity of exponentiation of real numbers. Math. Mag. **62**, 185–190 (1980)
35. Persky, R.L.: Extending Bernoulli's inequality. Coll. Math. J. **25**, 230 (1994)
36. Poly, G., Szegő, G.: Problems and Theorems in Analysis I. Springer, Berlin/Heidelberg/New York (2004)
37. Rademacher, H., Toeplitz, O.: The Enjoyment of Math. Princeton Univ Press, Princeton (1966)
38. Rose, D.: The Pearson and Cauchy-Schwarz inequalities. Coll. Math. J. **39**, 64 (2008)
39. Roy, R.: Sources in the Development of Mathematics. Cambridge University Press, Cambridge/New York (2011)
40. Sadoveanu, I., Vowe, M., Wagner, R.J.: Problem 458. Coll. Math. J. **23**, 344–345 (1992)
41. Sándor, J.: Monotonic convergence to e via the arithmetic-geometric mean. Math. Mag. **80**, 228–229 (2007)
42. Schaumberger, N.: Another application of the mean value theorem. Coll. Math. J. **10**, 114–115 (1979)
43. Schaumberger, N.: Another look at $x^{1/x}$. Coll. Math. J. **15**, 249–250 (1984)
44. Schaumberger, N.: Alternate approaches to two familiar results. Coll. Math. J. **15**, 422–423 (1984)
45. Schaumberger, N.: Another proof of Chebyshev's inequality. Coll. Math. J. **20**, 141–142 (1989)
46. Schaumberger, N., Eberhart, H.O.: Problem 134. Coll. Math. J. **11**, 212–213 (1980)
47. Schaumberger, N., Johnsonbaugh, R.: Problem 70. Coll. Math. J. **8**, 295 (1977)
48. Slay, J.C., Solomon, J.L.: A note on the Cauchy-Schwarz inequality. Coll. Math. J. **10**, 280–281 (1979)
49. Steele, J.M: The Cauchy-Schwarz Master Class. Mathematical Association of America/Cambridge University Press, Washington DC – Cambridge/New York (2004)
50. Steele, J.M.: (2014) http://www-stat.wharton.upenn.edu/~steele
51. Schwenk, A.J.: Selection of a fair currency exchange rate. Coll. Math. J. **13**, 154–155 (1982)
52. Sved, M.: On the rational solutions of $x^y = y^x$. Math. Mag. **63**, 30–33 (1990)
53. Szüsz, P.: Math Bite. Math. Mag. **68**, 97 (1995)
54. Weaver, C.S.: Rational solutions of $x^y = y^x$. In: Page, W. (ed.) Two-Year College Mathematics Readings, pp. 115–117. Mathematical Association of America, Washington DC (1981)
55. Wiener, J.: Bernoulli's inequality and the number e. Coll. Math. J. **16**, 399–400 (1985)

56. Wilker, J.B.: Stirling ideas for freshman calculus. Math. Mag. **57**, 209–214 (1984)
57. Yang, H., Yang, H.: The arithmetic-geometric mean inequality and the constant e. Math. Mag. **74**, 321–323 (2001)
58. Yi, H., Anderson, K.F.: Problem 11430. Am. Math. Mon. **118**, 88 (2011)
59. Young, R.M.: Another quick proof that $e < 3$. Math. Gaz. **97**, 333–334 (2013)

Chapter 3
Continuous Functions

It is easy to be brave from a safe distance.

—Aesop

Let $I \subseteq \mathbf{R}$ be an interval—open, closed, or otherwise. For the sake of simplicity, but without great loss, we mainly consider functions $f : I \to \mathbf{R}$. Roughly speaking, if f is continuous then $f(x)$ is close to $f(x_0)$ whenever $x \in I$ is close to $x_0 \in I$. Many functions which arise naturally in applications are continuous on some interval I. We shall see that continuous functions have very nice properties. The two big theorems in the world of continuous functions are the Intermediate Value Theorem and the Extreme Value Theorem. We prove these using *bisection algorithms*.

3.1 Basic Properties

Let I be an interval (open, closed, or otherwise) and let $f : I \to \mathbf{R}$. We say that f is **continuous on** I if f is continuous at *every* $x_0 \in I$. That is, for every $x_0 \in I$ and for any sequence $\{x_n\}$ in I for which $x_n \to x_0$,

$$\lim_{n\to\infty} f(x_n) = f(\lim_{n\to\infty} x_n) = f(x_0).$$

So the operation defined by f and the operation of taking the limit can be interchanged. More precisely: For any sequence $\{x_n\}$ in I for which $x_n \to x_0 \in I$, and for any $\varepsilon > 0$, there is a number N such that $|f(x_n) - f(x_0)| < \varepsilon$ for $n > N$.

If a function is continuous on I then its graph has no jumps nor breaks on I. (So one cannot really know that a particular function has a graph with no jumps nor breaks until it has been verified that the function is continuous.) In **Fig. 3.1**, the graphed function is continuous on (a, b), except at two points.

Example 3.1. We can rely very heavily on what we know about sequences to prove things about continuous functions. For example, if f and g are each continuous at x_0 then so is $f + g$. Here's why: If $x_n \to x_0$ then $f(x_n) \to f(x_0)$ and $g(x_n) \to g(x_0)$,

P.R. Mercer, *More Calculus of a Single Variable*, Undergraduate Texts in Mathematics, DOI 10.1007/978-1-4939-1926-0_3

because f and g are continuous. Therefore $f(x_n) + g(x_n) \to f(x_0) + g(x_0)$, because we know this about sequences (Lemma 1.24). So $f + g$ is continuous at x_0. ◇

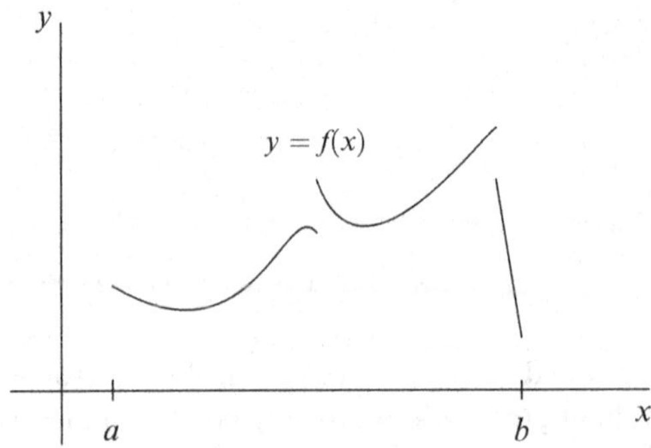

Fig. 3.1 A function continuous on (a, b), except at two points

Example 3.2. For any $y_0 \in \mathbf{R}$, the function

$$f(x) = \begin{cases} \dfrac{x}{|x|} & \text{if } x \neq 0 \\ y_0 & \text{if } x = 0 \end{cases}$$

is continuous on $(-\infty, 0)$ and on $(0, +\infty)$, but f is not continuous at $x_0 = 0$: Consider $x_n = (-1)^n/n$. Then $x_n \to 0$, yet $\{f(x_n)\} = \{-1, 1, -1, 1, \ldots\}$, which diverges (Example 1.14). So we *do not* have $f(x_n) \to f(0)$. See **Fig. 3.2.** ◇

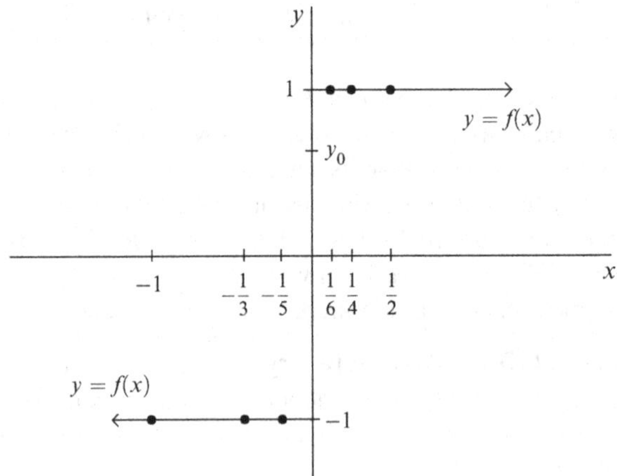

Fig. 3.2 For Example 3.2. Here, $x_n = (-1)^n/n \to 0$ and $\{f(x_n)\} = \{-1, 1, -1, 1, \ldots\}$, which diverges. So we do not have $f(x_n) \to f(0)$

Example 3.3. For any $y_0 \neq 1$, the function

$$f(x) = \begin{cases} 1 & \text{if } x \neq 0 \\ y_0 & \text{if } x = 0 \end{cases}$$

is continuous on $(-\infty, 0)$ and on $(0, +\infty)$, but f is not continuous at $x_0 = 0$: Any sequence $\{x_n\}$ for which $x_n \to 0$ (and $x_n \neq 0$) has $|f(x_n) - f(0)| = |1 - y_0|$ and so we *do not* have $f(x_n) \to f(0)$. See **Fig. 3.3.** ◇

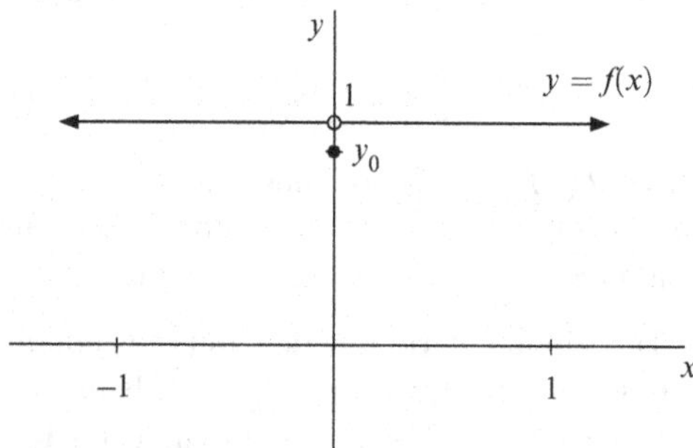

Fig. 3.3 For Example 3.3. Here, $x_n \to 0$ can be arbitrarily close to 0, but with $f(x_n)$ always being a fixed positive distance from $f(0)$. So $f(x)$ does not get close to $f(0)$ as x gets close to 0

Roughly, the idea is that for a continuous function f defined on I, if $x \in I$ is close to $x_0 \in I$ then $f(x)$ is close to $f(x_0)$. (The formal definition is needed to make precise the two instances of the word *close*.) The following useful result illustrates this idea very nicely.

Lemma 3.4. *Let f be continuous on $[a, b]$, with $f(x_0) \neq 0$ for some $x_0 \in [a, b]$. Then there is a closed interval $J \subset [a, b]$ containing x_0 such that $f(x) \neq 0$ for every $x \in J$.*

Proof. For $n = 1, 2, 3 \ldots$, let J_n be any closed interval of length $\frac{(b-a)}{n}$ which contains x_0. If the conclusion of the lemma is not true, then there is a point $x_n \in J_n$ such that $f(x_n) = 0$. Now $\frac{(b-a)}{n} \to 0$ and so $x_n \to x_0$, and since f is continuous we must have $f(x_n) \to f(x_0)$. Finally, $f(x_n) = 0$ implies that $f(x_0) = 0$, a contradiction. □

We assume that the reader has some familiarity with continuous functions. We cite the following simple facts which are inherited from Lemmas 1.24–1.26. We shall use these facts freely, often without explicit mention. Their proofs are left as Exercises 3.2, 3.4 and 3.6, respectively. *If f and g are each continuous functions on I, then so are $\alpha f + \beta g$ (for any $\alpha, \beta \in \mathbf{R}$), $f \cdot g$, and f/g (as long as $g \neq 0$ on I).*

The reader should agree that it is immediate from the definition, that the functions $f(x) = 1$ and $g(x) = x$ are continuous on $\mathbf{R}$. Therefore, by the first two facts from the previous paragraph, any polynomial is continuous on **R**. And by the third fact, any rational function (a polynomial divided by a polynomial) is continuous wherever it is defined—that is, wherever its denominator is not zero.

We can add many more functions to our collection of continuous functions using the fact that a composition of continuous functions is a continuous function. More precisely: *Let* $g : J \to I$ *and* $f : I \to \mathbf{R}$ *be continuous functions. Then the composition* $f \circ g : J \to \mathbf{R}$ *defined by* $(f \circ g)(x) = f(g(x))$ *is a continuous function.* Here's why: If $x_n \to x_0$ then $g(x_n) \to g(x_0)$, because g is continuous. And then $f(g(x_n)) \to f(g(x_0))$, because f is continuous.

Example 3.5. We show that if g is continuous, then $|g|$ is continuous. Let $y_0 \in \mathbf{R}$. If $y_n \to y_0$ then by the reverse triangle inequality,

$$\big| |y_n| - |y_0| \big| \le |y_n - y_0| \to 0,$$

and so $f(y) = |y|$ is continuous at y_0. Now if g is continuous at x_0 and $g(x_0) = y_0$ then, being a composition of continuous functions, $f(g(x)) = |g(x)|$ is also continuous at x_0. That is, $|g|$ is continuous if g is continuous. ◇

The trigonometric function $\sin(x)$ is continuous on **R**. We leave the verification of this claim for Exercise 3.8. Then, being a composition of continuous functions, $\cos(x) = \sin(\pi/2 - x)$ is continuous on **R**. Then, being quotients of continuous functions, $\tan(x)$, $\csc(x)$, $\sec(x)$, and $\cot(x)$ are continuous wherever they are defined, i.e., wherever their denominators are not zero.

One can define the exponential function $f(x) = e^x$ for $x \in \mathbf{R}$ and then after some justification, name $f^{-1}(x) = \ln(x)$ as its inverse (for $x > 0$). Alternatively, one can define the natural logarithmic function $f(x) = \ln(x)$ for $x > 0$ and then after some justification, name $f^{-1}(x) = e^x$ as its inverse (for $x \in \mathbf{R}$). We shall say more about each of these approaches, in Chaps. 6 and 10 respectively. Either way, e^x is continuous on $(-\infty, +\infty)$, and $\ln(x)$ is continuous on $(0, +\infty)$. We assume that the reader is comfortable in accepting these two claims, even though we postpone their proper verification. Graphs of e^x and $\ln(x)$ are shown in **Fig. 3.4**.

Example 3.6. $f(x) = \frac{\cos(x)}{x^2+1} + e^{\cos(x^3)} + x\ln(\sin(x) + 2)$ is continuous for $x \in \mathbf{R}$. ◇

3.2 Bolzano's Theorem

The following important result is named for Italian mathematician Bernhard Bolzano (1781–1848). For its statement, we use the fact that two real numbers A and B have opposite signs if and only if $AB < 0$.

Theorem 3.7. (Bolzano's Theorem) *Let* f *be a continuous function on* $[a, b]$ *with* $f(a)f(b) < 0$. *Then there is at least one* $c \in (a, b)$ *for which* $f(c) = 0$.

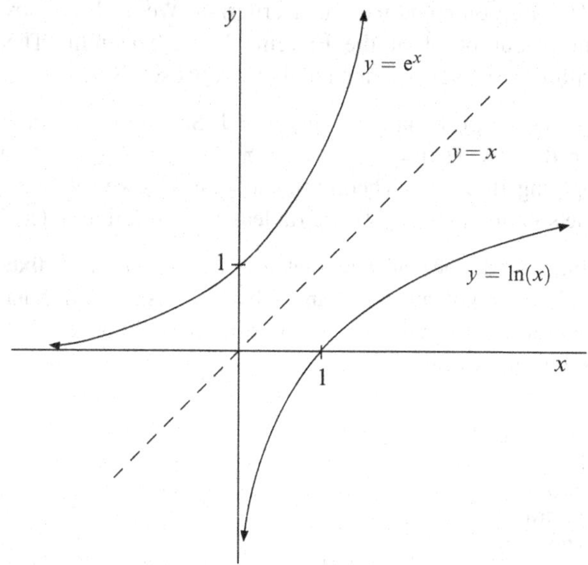

Fig. 3.4 The graphs of $y = e^x$ and its inverse, $y = \ln(x)$. Each is the graph of the other, reflected the line $y = x$

Proof. We employ a bisection algorithm. Write $[a, b] = [a_0, b_0]$, let $c_0 = \frac{a_0+b_0}{2}$, the midpoint of $[a_0, b_0]$, and bisect $[a_0, b_0]$ into intervals $[a_0, c_0]$ and $[c_0, b_0]$. Now if $f(c_0) = 0$ then we are done—that is, $c = c_0$ (and we count ourselves very fortunate). Otherwise, since f changes sign on $[a_0, b_0]$, it must change sign on either $[a_0, c_0]$ or on $[c_0, b_0]$ (or on both). Keep an interval on which f changes sign, rename it $[a_1, b_1]$ and discard the other. Now we continue this process. That is, for $n = 1, 2, 3, \ldots$ do the following:

(∗) Let $c_n = \frac{a_n+b_n}{2}$.
If $f(c_n) = 0$ then we are done—that is, $c = c_n$.
If $f(a_n)f(c_n) < 0$ then set $a_{n+1} = a_n$ and $b_{n+1} = c_n$, and go back to (∗).
If $f(c_n)f(b_n) < 0$ then set $a_{n+1} = c_n$ and $b_{n+1} = b_n$, and go back to (∗).

Then $[a, b] \supseteq [a_1, b_1] \supseteq [a_2, b_2] \supseteq [a_3, b_3] \supseteq \ldots$ is a sequence of nested intervals with $b_n - a_n = \frac{b-a}{2^n} \to 0$. So by the Nested Interval Property of **R** (Theorem 1.41), there is a unique point c belonging to each interval. Now $a_n \to c$ and $b_n \to c$ and f is continuous, so we must therefore have $f(a_n) \to f(c)$ and $f(b_n) \to f(c)$. Now we observe that $f(a_n)f(b_n) < 0$ after each pass through the algorithm, and so we must have $f(c)^2 \leq 0$ (by Lemma 1.27). This is only possible if $f(c) = 0$, as desired. □

See **Fig. 3.5** for an illustration of Bolzano's Theorem (Theorem 3.7). For its proof, we employed what is known as a *bisection algorithm.* At each step of such an algorithm an interval is bisected, then one of the halves is kept (and the other discarded), based on some particular criterion. We shall employ a bisection algorithm again in our proof of the Extreme Value Theorem (Theorem 3.23). Bisection algorithms are also used in a number of the exercises.

Example 3.8. Consider the equation $x\sin(x) = 1$. Set $f(x) = x\sin(x) - 1$, which is continuous on **R**. Now $f(0) = -1 < 0$, $f(\pi/2) = \pi/2 - 1 > 0$, and $f(\pi) = -1 < 0$, so applying Bolzano's Theorem (Theorem 3.7) we see that the equation has (at least) one solution in $(0, \pi/2)$ and (at least) one solution in $(\pi/2, \pi)$. ◇

Let f be a function defined on I and let $p \in I$. Then f **has a fixed point** p if $f(p) = p$. That is, the point p is not changed by f—it is *fixed*. If f has fixed point p, then the function $f(x) - x$ has a zero at $x = p$, and conversely. This simple observation can be very useful.

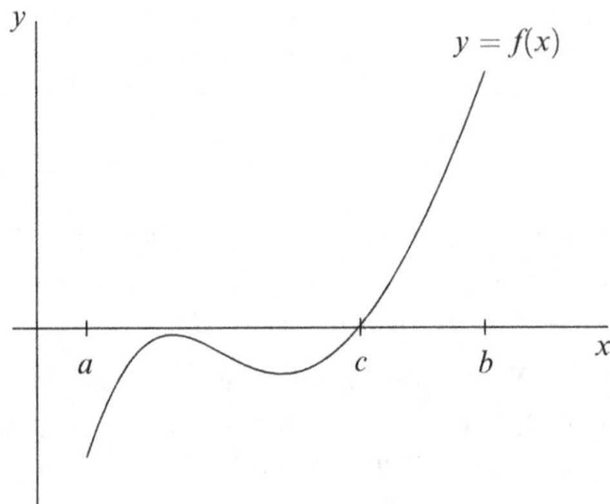

Fig. 3.5 Bolzano's Theorem (Theorem 3.7): The continuous function f goes from negative to positive, so its graph must cross the x-axis

Example 3.9. Let $f(x) = x^2 - x - 1/\sqrt{x}$. Then $p > 0$ is a zero of f if and only if p is a fixed point of $F(x) = x^2 - 1/\sqrt{x}$; $p > 0$ is a zero of f if and only if p is a fixed point of $H(x) = 1 + 1/x^{3/2}$; $p > 0$ is a zero of f if and only if p is a fixed point of $G(x) = \sqrt{x + 1/\sqrt{x}}$. ◇

The following result shows that if the graph of a continuous function f is entirely contained within the rectangle $[a,b]\times[a,b]$, then f must have a fixed point in $[a,b]$. That is, the graph must intersect the line $y = x$ at least once. See **Fig. 3.6.**

Lemma 3.10. (Fixed Point Lemma) *Let $f : [a,b] \to [a,b]$ be continuous. Then f has at least one fixed point in $[a,b]$.*

Proof. If $f(a) = a$ then a is a fixed point, or if $f(b) = b$ then b is a fixed point. So we may assume that $f(a) \neq a$ and $f(b) \neq b$. Let $g(x) = f(x) - x$. Then g is continuous on $[a,b]$, with $g(a) = f(a) - a > 0$ and $g(b) = f(b) - b < 0$. That is, $g(a)g(b) < 0$. So we apply Bolzano's Theorem (Theorem 3.7) to g to see that there is $p \in (a,b)$ for which $g(p) = 0$. That is, $g(p) = f(p) - p = 0$, or $f(p) = p$, as desired. □

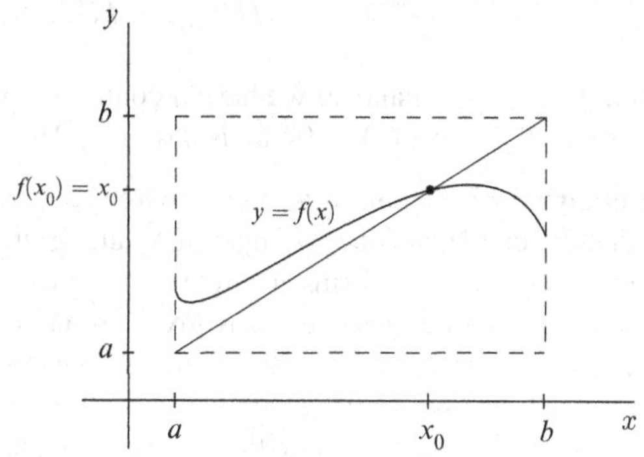

Fig. 3.6 A continuous $f : [a, b] \to [a, b]$ has a fixed point: $f(x_0) = x_0$

Remark 3.11. The Fixed Point Lemma (Lemma 3.10) is a special case of Brouwer's Theorem, due to Dutch mathematician L.E.J. Brouwer (1881–1966), which holds in much more generality. Here is an amusing instance of the theorem being applied in two dimensions. A map of Wyoming say, can be regarded as a function from Wyoming to a large piece of paper: Each actual point in Wyoming is mapped by the function to a dot on the paper which represents that point. Then placing the map flat and wholly within Wyoming (anywhere on the ground, say) can be regarded as a mapping from Wyoming to a subset of Wyoming. Brouwer's Theorem says that there must be a dot on the map which sits exactly over the actual point in Wyoming which the dot represents. (See also Exercise 3.20, and the reader might consult [3, 4, 9] for other amusing examples.) ∘

3.3 The Universal Chord Theorem

A function f defined on I has a **horizontal chord** if $f(a) = f(b)$ for some $a < b \in I$. The length of this horizontal chord is then $b - a$. A continuous function need not, of course, have any horizontal chords ($f(x) = x$, for example). The result below shows however, that if a continuous function happens to have a horizontal chord of length $b - a$, then it must also have a horizontal chord of length $(b - a)/2$. See **Fig. 3.7**. We state and prove this result on $[0, 1]$ instead of $[a, b]$, only for the sake of simplicity; the $[a, b]$ case is left for Exercise 3.25. (See also Exercise 3.26.)

Lemma 3.12. (Half-Chord Lemma) *Let f be continuous on $[0, 1]$, with $f(0) = f(1)$. Then there is $c \in [0, 1/2]$ such that $f(c + 1/2) = f(c)$.*

Proof. Define the function g on $[0, 1/2]$ via $g(x) = f(x + 1/2) - f(x)$. Then g is continuous on $[0, 1/2]$. We want to show that g has a zero in $[0, 1/2]$. If g does not have a zero in $[0, 1/2]$ then, by Bolzano's Theorem (Theorem 3.7), g is either always positive or always negative on $[0, 1/2]$. Say it's positive; if it's negative we would consider $-g$. Then $f(x) < f(x + 1/2)$ on $[0, 1/2]$. Setting $x = 0$ and then $x = 1/2$, we get

$$f(0) < f(1/2) < f(1/2 + 1/2) = f(1).$$

But $f(0) = f(1)$ and so we have a contradiction. Therefore g indeed has a zero c in $[0, 1/2]$. Then $g(c) = 0$ yields $f(c + 1/2) = f(c)$, as desired. □

Remark 3.13. Think of a loop of wire shaped like a circle, heated in any manner whatsoever. Then since temperature along the wire is a continuous function, and since the wire forms a circle (i.e., $f(0) = f(1)$), the Half-Chord Lemma (Lemma 3.12) implies that at any given moment, there is a pair of opposite points on the wire having the same temperature. Notice that the wire doesn't really need to be a *circular* loop—we only need to have a well defined notion of *opposite*. For example, at any given moment in time there is a pair of antipodal points on the earth's equator having the same temperature, and another pair of antipodal points having the same wind speed (and another pair having the same atmospheric pressure, etc.). ○

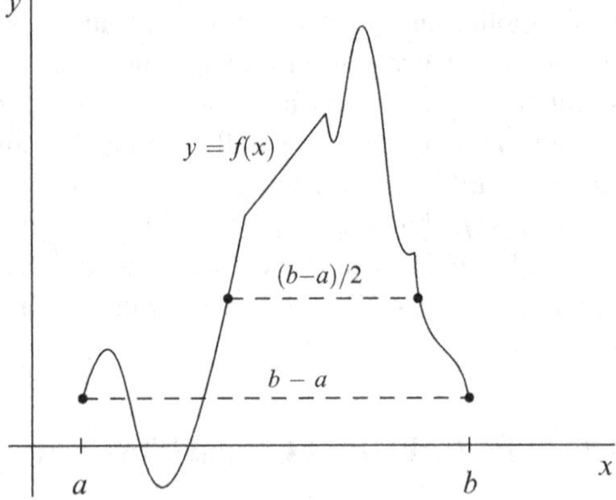

Fig. 3.7 The Half-Chord Lemma (Lemma 3.12) on $[a, b]$: The continuous function f has a horizontal chord of length $b - a$ so it must also have a horizontal chord of length $(b - a)/2$

Applying the Half-Chord Lemma (Lemma 3.12), or more precisely Exercise 3.25, over and over again, we can see that if a continuous function has a horizontal chord of length L, then it must also have horizontal chords of lengths $L/2^k$, for each $k \in \mathbf{N}$. But even more is true, as follows. Again, we take $[a, b] = [0, 1]$ for the sake of simplicity; the general $[a, b]$ case is left for Exercise 3.27. (See also Exercise 3.28.)

Theorem 3.14. (Universal Chord Theorem) *Let f be continuous on $[0, 1]$ with $f(0) = f(1)$, and let k be any positive integer. Then there is $c \in [0, 1 - 1/k]$ such that $f(c + 1/k) = f(c)$. That is, f has a horizontal chord of length $1/k$.*

Proof. The hypothesis of the theorem is $k = 1$, so we let $k \geq 2$. Consider the function $g(x) = f(x + 1/k) - f(x)$ on $[0, 1 - 1/k]$. Then g is continuous on $[0, 1 - 1/k]$. We want to show that g has a zero in $[0, 1 - 1/k]$. If not, then by

Bolzano's Theorem (Theorem 3.7), g is either always positive or always negative on $[0, 1 - 1/k]$. Let's say it's positive; if it's negative we could consider $-g$. Then $f(x) < f(x + 1/k)$ on $[0, 1 - 1/k]$. Setting $x = 0$, $x = 1/k$, $x = 2/k, \ldots,$ and $x = (k-1)/k$ we get

$$f(0) < f(1/k) < f(2/k) < f(3/k) < \cdots < f(k/k) = f(1).$$

But $f(0) = f(1)$ and so we have a contradiction. Therefore g indeed has a zero c in $[0, 1 - 1/k]$. Then $g(c) = 0$ yields $f(c + 1/k) = f(c)$, as desired. □

Remark 3.15. Think again of a circular wire of length L, heated in any manner whatsoever. The Universal Chord Theorem (Theorem 3.14) implies that for each $m \in \mathbf{N}$, there is a pair of points on the wire of distance L/m from each other (measured along the wire) which have the same temperature. ∘

We leave it for Exercise 3.29 to show that a continuous function f on $[0, 1]$ with $f(0) = f(1)$ need not have a horizontal chord of length t, if $t \in (1/2, 1)$. But more interesting is the fact that f need not have a horizontal chord of length $1/m$, if $m \neq 1, 2, 3, \ldots$. In this sense the Universal Chord Theorem (Theorem 3.14) is as good as it can be. This is demonstrated by the function

$$f(x) = x \sin^2(m\pi) - \sin^2(m\pi x).$$

Here, f is continuous on $[0, 1]$, with $f(0) = f(1) = 0$, yet one can check that

$$f(x + \tfrac{1}{m}) - f(x) = \tfrac{1}{m} \sin^2(m\pi) = 0$$

only if m is an integer. This is Exercise 3.30.

Remark 3.16. This section owes much to the excellent book [3]. From there: "Even though the Universal Chord Theorem was discovered by A.M. Ampere in 1806, it is commonly attributed to P. Levy, who rediscovered it in 1934, but also showed that it is optimal." Levy showed that it is optimal by using precisely the $f(x)$ from the above paragraph. ∘

3.4 The Intermediate Value Theorem

The Intermediate Value Theorem is arguably the most important theorem about continuous functions. It amounts to improving Bolzano's Theorem (Theorem 3.7) to allow $f(a)$ and $f(b)$ to be *any* two values—not just one negative and one positive. It is equivalent to Bolzano's Theorem but it gets used more often in this latter form. The theorem says that a continuous function on a closed interval $[a, b]$ attains *every* value between $f(a)$ and $f(b)$. See **Fig. 3.8**. This property is called the **Intermediate Value Property** on $[a, b]$.

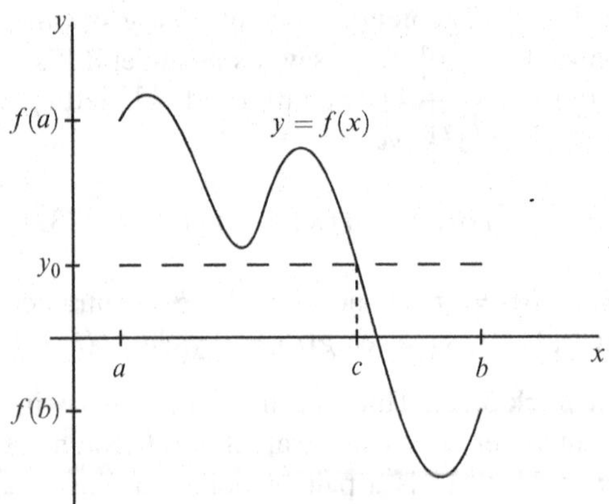

Fig. 3.8 The Intermediate Value Theorem (Theorem 3.17): There is $c \in [a,b]$ for which $f(c) = y_0$

Theorem 3.17. (Intermediate Value Theorem) *Let f be continuous on $[a,b]$ and let y_0 be any number between $f(a)$ and $f(b)$. Then there is at least one $c \in [a,b]$ for which $f(c) = y_0$.*

Proof. If $y_0 = f(a)$ or $y_0 = f(b)$ then we take $c = a$ or $c = b$ and we are done. So let y_0 be strictly between $f(a)$ and $f(b)$ and consider the function $g(x) = f(x) - y_0$. Then g is continuous on $[a,b]$. And since y_0 is strictly between $f(a)$ and $f(b)$ we have $g(a)g(b) = [f(a) - y_0][f(b) - y_0] < 0$. Applying Bolzano's Theorem (Theorem 3.7) to g we see that there is $c \in (a,b)$ for which $g(c) = 0$. That is, $f(c) = y_0$ as desired. □

Remark 3.18. The reader should agree, perhaps after making a sketch or two, that a function may have the Intermediate Value Property on some particular interval, yet not be continuous on that interval. Nevertheless, it might seem that the Intermediate Value Property should characterize continuous functions in the following way: If f satisfies the Intermediate Value Property on *every subinterval* of $[a,b]$, then f should be continuous on $[a,b]$. But this is not the case either, as the following function demonstrates. Let $\alpha \in \mathbf{R}$ and define

$$f(x) = \begin{cases} \sin\left(\frac{1}{x}\right) & \text{if } x \neq 0 \\ \alpha & \text{if } x = 0\,. \end{cases}$$

This function attains every value between -1 and $+1$ (infinitely many times) on any interval which contains $x = 0$, yet it is not continuous on any such interval, no matter what value is chosen for α. We leave the verification of this claim for Exercise 3.34. The graph of $y = f(x)$ is shown (very roughly) in **Fig. 3.9.** ∘

The following is a useful consequence of the Intermediate Value Theorem (Theorem 3.17). Among other things, it is the basis for some important results that we shall meet in Chap. 9.

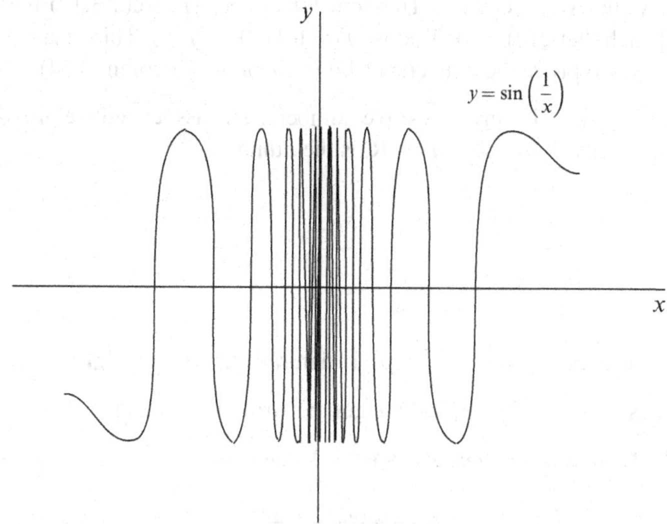

Fig. 3.9 The graph of $f(x) = \sin\left(\frac{1}{x}\right)$. This function has the Intermediate Value Property on any interval which contains $x = 0$, yet it is not continuous on any such interval

Theorem 3.19. (Average Value Theorem for Sums) *Let f be continuous on $[a, b]$, and let $x_1, x_2, \ldots, x_n \in [a, b]$. Then the average value of f, evaluated at $x_1, x_2, \ldots, x_n$, is attained. That is, there is $c \in [a, b]$ such that*

$$f(c) = \frac{1}{n}\sum_{j=1}^{n} f(x_j).$$

Proof. By suitably rearranging the x_j's, if necessary, we may assume that

$$f(x_1) \leq f(x_2) \leq \cdots \leq f(x_n).$$

And then since the average value (i.e., the Arithmetic Mean) is a *mean*, we have

$$f(x_1) \leq \frac{1}{n}\sum_{j=1}^{n} f(x_j) \leq f(x_n).$$

So by the Intermediate Value Theorem (Theorem 3.17), there is c between x_1 and x_n such that $f(c) = \frac{1}{n}\sum_{j=1}^{n} f(x_j)$, as desired. □

Example 3.20. Here is another way to prove the Half-Chord Lemma (Lemma 3.12). Still, consider function $g(x) = f(x + 1/2) - f(x)$ on $[0, 1/2]$. However, notice that $g(0) = -g(1/2)$, and so

$$\frac{1}{2}\left(g(0) + g(\tfrac{1}{2})\right) = 0.$$

Therefore, by the Average Value Theorem for Sums (Theorem 3.19) there exists $c \in [0, 1/2]$ such that $g(c) = 0$. That is, $f(c + 1/2) = f(c)$. This idea is extended in Exercise 3.28 to prove the Universal Chord Theorem (Theorem 3.14). ◇

Let $p_1, p_2, \ldots, p_n$ be any n positive numbers. The associated **weighted Arithmetic Mean**, of any $a_1, a_2, \ldots, a_n \in \mathbf{R}$, is the number

$$\frac{\sum_{j=1}^{n} p_j a_j}{\sum_{k=1}^{n} p_k}.$$

In this context, each $w_j = p_j / \sum_{k=1}^{n} p_k$ is naturally called a **weight**. (Taking each $p_j = 1$ makes each $w_j = 1/n$, and we get the Arithmetic Mean.)

Example 3.21. In the weighted Arithmetic Mean

$$\frac{2a_1+7a_2+a_3+5a_4+6a_5}{21}$$

we have $w_1 = 2/21$, $w_2 = 1/3$, $w_3 = 1/21$, $w_4 = 5/21$, and $w_5 = 2/7$. Then

$$\frac{2a_1+7a_2+a_3+5a_4+6a_5}{21} = \tfrac{2}{21}a_1 + \tfrac{1}{3}a_2 + \tfrac{1}{21}a_3 + \tfrac{5}{21}a_4 + \tfrac{2}{7}a_5 = \sum_{j=1}^{5} w_j a_j. \qquad \diamond$$

It is easily verified that the weighted Arithmetic Mean is indeed a *mean* and so the following result holds in very much the same way as the Average Value Theorem for Sums (Theorem 3.19).

Theorem 3.22. (Mean Value Theorem for Sums) *Let f be continuous on $[a, b]$, let $x_1, x_2, \ldots, x_n \in [a, b]$, and let $p_1, p_2, \ldots, p_n$ be any n positive numbers. Then the weighted Arithmetic Mean of f, evaluated at $x_1, x_2, \ldots, x_n$, is attained. That is, there is $c \in [a, b]$ such that*

$$f(c) = \frac{\sum_{j=1}^{n} p_j f(x_j)}{\sum_{j=1}^{n} p_j}.$$

Proof. This is Exercise 3.35. □

We point out that the conclusion of the Average Value Theorem for Sums (Theorem 3.19) reads

$$f(c) = \frac{1}{n}\sum_{j=1}^{n} f(x_j) = \frac{\sum_{j=1}^{n} 1 f(x_j)}{\sum_{j=1}^{n} 1}.$$

Replacing the $1's$ on the right-hand side with $p_j > 0$ yields the Mean Value Theorem for Sums (Theorem 3.22).

3.5 The Extreme Value Theorem

The following theorem is attributed to the great German mathematician Karl Weierstrass (1815–1897). If the Intermediate Theorem (Theorem 3.17) is not the most important theorem about continuous functions, then this one surely is. The theorem says that a continuous function on a closed interval *attains* smallest and largest values; see **Fig. 3.10**. Our proof follows [6], but see also [1,4].

Theorem 3.23. (Extreme Value Theorem) *Let f be continuous on $[a,b]$. Then there are numbers $x_m, x_M \in [a,b]$ such that*

$$f(x_m) \le f(x) \le f(x_M) \quad \textit{for every } x \in [a,b].$$

Proof. We prove the part of the theorem pertaining to x_M. The part pertaining to x_m is proved very similarly; we leave it as Exercise 3.38. We employ another bisection algorithm. Write $[a,b] = [a_0,b_0]$, let $c_0 = \frac{a_0+b_0}{2}$, the midpoint of $[a_0,b_0]$, and bisect $[a_0,b_0]$ into intervals $[a_0,c_0]$ and $[c_0,b_0]$. If there is a point $t \in [a_0,c_0]$ such that $f(t) \ge f(x)$ for each $x \in [c_0,b_0]$ then we keep $[a_0,c_0]$, rename it $[a_1,b_1]$, and discard $[c_0,b_0]$. Otherwise we keep $[c_0,b_0]$, rename it $[a_1,b_1]$, and discard $[a_0,c_0]$. We continue this process. That is, for $n = 1,2,3,\ldots$ we do the following:

(∗) Let $c_n = \frac{a_n+b_n}{2}$.
If there is $t \in [a_n,c_n]$ such that $f(t) \ge f(x)$ for each $x \in [c_n,b_n]$,
then set $a_{n+1} = a_n$ and $b_{n+1} = c_n$, and go back to (∗).
Otherwise, set $a_{n+1} = c_n$ and $b_{n+1} = b_n$, and go back to (∗).

Then $[a,b] \supset [a_1,b_1] \supseteq [a_2,b_2] \supseteq [a_3,b_3] \supseteq \ldots$ is a sequence of nested intervals with $b_n - a_n = \frac{b-a}{2^n} \to 0$. As such there is a point c belonging to each interval, by the Nested Interval Property of **R** (Theorem 1.41). We claim that $f(c) \ge f(x)$ for all $x \in [a,b]$.

If not (looking for a contradiction), there is $u \in [a,b]$ such that $f(u) > f(c)$. Then the function $g(x) = f(u) - f(x)$ is continuous on $[a,b]$, and it is positive at c. Therefore by Lemma 3.4, g is positive on some closed interval J containing c.

For n large enough we shall have $[a_n, b_n] \subset J$, with $u \notin [a_n, b_n]$, and therefore $f(u) > f(x)$ for all $x \in [a_n, b_n]$. But by the choice of $[a_n, b_n]$ at each stage of the algorithm, there is $t \in [a_n, b_n]$ such that

$$f(t) \geq f(x) \text{ for all } x \notin [a_n, b_n].$$

So $f(u) > f(c)$ cannot occur. This is our contradiction, as desired. □

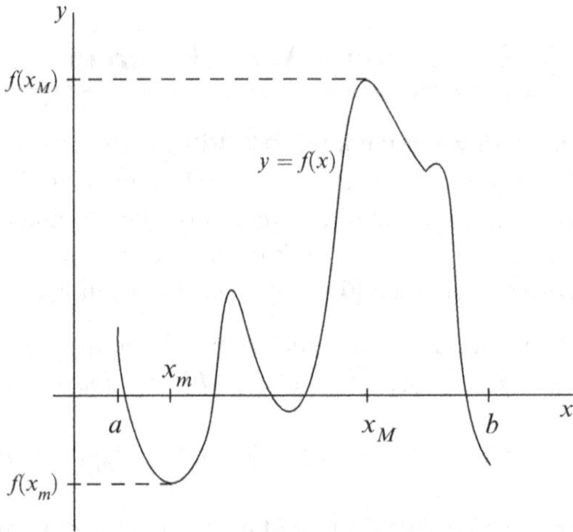

Fig. 3.10 The Extreme Value Theorem (Theorem 3.23): $f(x_m) \leq f(x) \leq f(x_M)$ for every $x \in [a, b]$

Again, the Extreme Value Theorem (Theorem 3.23) says that f *attains* smallest and largest values. In particular, a continuous function f on $[a, b]$ is necessarily **bounded**. That is, there exists a number M such that $|f(x)| \leq M$ for every $x \in [a, b]$. Indeed, taking any $M \geq \max\{|f(x_m)|, |f(x_M)|\}$ will suffice.

The Intermediate Value Theorem (Theorem 3.17) then implies that the function $f : [a, b] \to [f(x_m), f(x_M)]$ is **onto**: for each $y_0 \in [f(x_m), f(x_M)]$, there is $x_0 \in [a, b]$ such that $f(x_0) = y_0$. Therefore, the set $f([a, b])$ is *an interval.*

Exercises

3.1. Let f be a function defined on I.

(a) Suppose that for every $\varepsilon > 0$ there is $\delta > 0$ such that $|f(x) - f(x_0)| < \varepsilon$ whenever $x \in I$ and $|x - x_0| < \delta$. Show that f is continuous at x_0.

(b) Suppose that f is continuous at x_0. Show that for any $\varepsilon > 0$ there is $\delta > 0$ such that $|f(x) - f(x_0)| < \varepsilon$ whenever $x \in I$ and $|x - x_0| < \delta$.

3.2. Let $\alpha, \beta \in \mathbf{R}$. Use Lemma 1.24 to prove that if f and g are continuous on I, then $\alpha f + \beta g$ is continuous on I. (In particular, $f \pm g$ is continuous on I.)

3.3. Let f and g be continuous functions on I. Show that the functions

$$(f \wedge g)(x) = \min\{f(x), g(x)\} \quad \text{and} \quad (f \vee g)(x) = \max\{f(x), g(x)\}$$

are each continuous on I. Hint: Verify that $(f \wedge g)(x) = \frac{1}{2}(f(x) + g(x) - |f(x) - g(x)|)$ and $(f \vee g)(x) = \frac{1}{2}(f(x) + g(x) + |f(x) - g(x)|)$.

3.4. Use Lemma 1.25 to prove that if f and g are continuous on I, then the product $f \cdot g$ is continuous on I.

3.5. **(a)** Prove directly from the definition that if f is continuous, then so is f^2. Hint: Look at Example 1.23.
(b) Write $f \cdot g = \frac{1}{4}\left((f+g)^2 - (f-g)^2\right)$ to prove that if f and g are continuous, then $f \cdot g$ is continuous. Hint: Look at Example 1.28.

3.6. Use Lemma 1.26 to prove that if f and g are continuous at on I and $g(x) \neq 0$ for $x \in I$, then the quotient f/g is continuous on I.

3.7. Let f be continuous on I, with $f \geq 0$. Show that $\sqrt{f}$ is continuous on I.

3.8. In this exercise we show that $\sin(x)$ is continuous at every $x_0 \in \mathbf{R}$.

(a) Show (a picture will be helpful) that

$$0 \leq \sin(x) \leq x \quad \text{for } x \in [0, \pi/2].$$

(b) Use this and $\sin(-x) = -\sin(x)$ to show that $\lim_{x \to 0} \sin(x) = 0 = \sin(0)$. Conclude that $\sin(x)$ is continuous at $x_0 = 0$.
(c) Write $x = \frac{x+x_0}{2} + \frac{x-x_0}{2}$ and $x_0 = \frac{x+x_0}{2} - \frac{x-x_0}{2}$, then use the trigonometric identities

$$\sin(A \pm B) = \sin(A)\cos(B) \pm \cos(A)\sin(B)$$

to show that

$$\sin(x) - \sin(x_0) = 2\cos\left(\tfrac{x+x_0}{2}\right)\sin\left(\tfrac{x-x_0}{2}\right).$$

(d) Use (c) to show that $\sin(x)$ is continuous at every $x_0 \in \mathbf{R}$.

3.9. **(a)** Prove that the function $f(x) = x^2 - x - e^{-x}$ has a positive root.
(b) Prove that the equation $\cos(x) = x$ has a solution in $[0, \pi/2]$. Make a sketch.

3.10. Show that the equation $x^4 - x^2 - 2 = x$ has one negative solution and one positive solution. Draw a picture.

3.11. Prove that the equation $e^x = x^4$ has three solutions. Make a sketch.

3.12. Prove that a polynomial of odd degree has at least one root.

3.13. Consider the polynomial $p(x) = a_0 + a_1 x + a_2 x^2 + \cdots + a_n x^n$. Prove that if $a_0 a_n < 0$ then p has a positive root.

3.14. **(a)** Draw the graph of a continuous function $f : [a,b] \to [a,b]$ which has exactly two fixed points.

(b) Draw the graph of a continuous function $f : [a,b] \to [a,b]$ which has exactly three fixed points.

3.15. [2] Show that if $x = x_0$ is a solution to any of

$$\frac{2x^3 + 4x^2 - 5}{2} = x, \quad \sqrt{\frac{2x+5}{2x+4}} = x, \quad \text{or} \quad \sqrt[3]{\frac{2x + 5 - 4x^2}{2}} = x,$$

then x_0 is a solution to $2x^3 + 4x^2 - 2x - 5 = 0$.

3.16. **(a)** Show that $f(x) = \dfrac{1-x^2}{1+x^2}$ has a fixed point in $[0, 1]$.

(b) Sketch the graphs $y = f(x)$ and $y = x$ on $[0, 1]$.

(c) Show that $g(x) = \dfrac{1}{1+x}$ has a fixed point in $[0, 1]$. What is this fixed point?

(d) Sketch the graphs $y = g(x)$ and $y = x$ on $[0, 1]$.

3.17. **(a)** Show that $f(x) = \dfrac{1 + 2\cos(x)}{(\cos(x) + 2)^2}$ has a fixed point in $[0, \pi/2]$.

(b) Can you show that the fixed point is in fact in $[1/4, 1/2]$?

3.18. Let $f : [0, 1] \to [0, 1]$ be continuous.

(a) Show that there is $a \in [0, 1]$ such that $f(a) = a^2$.

(b) Show that there is $b \in [0, 1]$ such that $f(b) = \sqrt{b}$.

(c) Show that there is $c \in [0, 1]$ such that $f(c) = \sin(\pi c/2)$.

3.19. [7] Let f be continuous on $[-1, 1]$, with $f(-1) \geq -1$ and $f(1) \leq 1$. Show that f has a fixed point.

3.20. We saw in Remark 3.11 that the Fixed Point Lemma (Lemma 3.10) is a special case of Brouwer's Theorem, which holds in two dimensions. Use the fact that Brouwer's Theorem also holds in three dimensions to argue the following: After stirring a cup of coffee in any manner whatsoever and then letting it settle, there is a point in the coffee which ends up exactly where it began.

3.21. Suppose that one ride on a particular roller coaster lasts exactly 3 min. To keep people moving along, the amusement park staff runs a set of cars exactly 1.5 min after the previous set has left. Suppose that Hannah rides at the front of a set and Sarah rides at the front of the next set. Show that during Hannah's ride there is an instant at which she is at precisely the same elevation as Sarah.

3.22. [5] A snail begins to crawl up a stick at 6 am and reaches the top of the stick at noon. It spends the rest of the day and that night at the top. The next morning it leaves the top at 6 am and descends by the same route it used the day before, reaching the bottom at noon. Prove that there is a time between 6 am and noon at which the snail was at exactly the same spot on the stick on both days. Note: The snail may crawl at different speeds, rest, or even go backwards. Snails do that. (This problem has appeared in many places in many forms. It was originally posed by the American mathematician and science writer Martin Gardner (1914–2010).)

3.23. Suppose that one ride on a particular roller coaster lasts exactly 2 min and you take a ride. Show that there is a time interval 15 s in length after which your net change in elevation is zero.

3.24. Suppose that a roller ride coaster is 1.2 miles long and you take a ride. Show that there is a stretch of 0.24 miles after which your net change in elevation is zero.

3.25. Modify the proof of the Half-Chord Lemma (Lemma 3.12) to obtain a version for $[a, b]$, as follows. Let f be a continuous function on $[a, b]$, with $f(a) = f(b)$. Show that there is at least one $c \in [a, \frac{a+b}{2}]$ such that $f(c + \frac{b-a}{2}) = f(c)$. That is, f has a horizontal chord of length $(b - a)/2$.
Hint: Consider $g(x) = f(x + (b - a)/2) - f(x)$ on $[a, (a + b)/2]$.

3.26. **(a)** Prove the Half-Chord Lemma (Lemma 3.12) another way: Consider the function $g(x) = f(x+1/2) - f(x)$ on $[0, 1/2]$ and observe that $g(0)g(1/2) < 0$. Now apply Bolzano's Theorem (Theorem 3.7).

(b) Is this proof preferable to the one in the text? Why or why not?

(c) Modify the argument in (a) to prove the Half-Chord Lemma on $[a, b]$.

3.27. Modify the proof of the Universal Chord Theorem (Theorem 3.14) to obtain a version for $[a, b]$, as follows. Let f be a continuous function on $[a, b]$ with $f(a) = f(b)$, and let k be any positive integer. Show that there exists c in $[a, b - (b - a)/k]$ such that

$$f(c + (b - a)/k) = f(c).$$

That is, f has a horizontal chord of length $(b - a)/k$.
Hint: Consider $g(x) = f(x + (b - a)/k) - f(x)$ on $[a, b - (b - a)/k]$.

3.28. [3]

(a) Prove the Universal Chord Theorem (Theorem 3.14) another way, as follows. Let $g(x) = f(x + 1/k) - f(x)$. Verify that g continuous on $[0, 1 - 1/k]$ and that $\frac{1}{k} \sum_{j=1}^{k} g(\frac{j-1}{k}) = 0$. Apply the Average Value Theorem for Sums (Theorem 3.19).

(b) Modify the argument in (a) to prove the Universal Chord Theorem on $[a, b]$.

3.29. [3]

(a) Find an example which shows that even if f is continuous on $[0, 1]$ with $f(0) = f(1)$, f need not have a horizontal chord of length t, if $t \in (1/2, 1)$. A picture will suffice.

(b) Find an example which shows that even if f is continuous on $[0, 1]$ with $f(0) = f(1)$, f need not have a horizontal chord of length t, for $t \in (1/3, 1/2)$. A picture will suffice.

3.30. Verify the claim made prior to Remark 3.16, that P. Levy's example

$$f(x) = x\sin^2(m\pi) - \sin^2(m\pi x)$$

has $f(0) = f(1)$, yet has no horizontal chord of length $1/m$ unless m is a natural number. This shows that the Universal Chord Theorem (Theorem 3.14) is as good as it can be.

3.31. A snail begins to crawl up a stick at 6 am and reaches the top of the stick at noon. It spends the rest of the day and that night at the top. The next morning it leaves the top at 6 am and descends by the same route it used the day before, reaching the bottom at noon. Prove that there are two times, 21 h apart, at which the snail was at exactly the same spot on the stick. (The snail may crawl at different speeds, rest, or even go backwards. Snails do that.)

3.32. (a) Show that a 12-h clock that is stopped is correct twice a day.
(b) Show that the conclusion of the Intermediate Value Theorem (Theorem 3.17) no longer holds if f is not continuous on $[a, b]$.

3.33. Let $a > 0$. Show that the equation $x^4 - x^2 - x = a$ has one negative solution and one positive solution.

3.34. Let $y_0 \in \mathbf{R}$ and consider the function

$$f(x) = \begin{cases} \sin\left(\frac{1}{x}\right) & \text{if } x \neq 0 \\ y_0 & \text{if } x = 0\,. \end{cases}$$

Find two sequences $\{x_n\}$ and $\{y_n\}$ for which $x_n \to 0$ and $y_n \to 0$, and $f(x_n) \to A$, $f(y_n) \to B$, yet $A \neq B$. So f is not continuous at $x = 0$, even though f satisfies the Intermediate Value Property on any interval with contains 0. See [8] for an interesting classroom approach to this example.

3.35. (a) Let $a_1, a_2, \ldots, a_n \in \mathbf{R}$ and let $p_1, p_2, \ldots, p_n$ be n positive numbers. Show that

$$\min_{1\le j\le n}\{a_j\} \le \frac{\sum_{j=1}^{n} p_j a_j}{\sum_{j=1}^{n} p_j} \le \max_{1\le j\le n}\{a_j\}.$$

That is, the weighted Arithmetic Mean is indeed a *mean*.
(b) Prove the Mean Value Theorem for Sums (Theorem 3.22).

3.36. Let f be continuous on $[a,b]$ and let $x_1, x_2, \ldots, x_n \in [a,b]$. Prove that there is a number $c \in [a,b]$ at which the Root Mean Square of f evaluated at these points is attained. That is, prove that there is a number $c \in [a,b]$ such that

$$f(c) = \sqrt{\frac{1}{n}\left(f(x_1)^2 + f(x_2)^2 + \cdots + f(x_n)^2\right)}.$$

3.37. Let f be continuous on $[a,b]$ and let $x_1, x_2, \ldots, x_n \in [a,b]$. Prove that there is a number $c \in [a,b]$ at which the Geometric Mean of f evaluated at these points is attained. That is, prove that there is a number $c \in [a,b]$ such that

$$f(c) = \left(\prod_{j=1}^{n} f(x_j)\right)^{1/n} = \left(f(x_1)\cdot f(x_2)\cdots f(x_n)\right)^{1/n}.$$

3.38. Prove the part of the Extreme Value Theorem (Theorem 3.23) which pertains to x_m.

3.39. **(a)** Show that the conclusion of the Extreme Value Theorem (Theorem 3.23) is no longer true if f is not continuous on $[a,b]$.

(b) What happens if f is continuous, but on an interval that is not closed?

(c) Is the Extreme Value Theorem (Theorem 3.23) true if we assume only that f has the Intermediate Value Property on $[a,b]$?

References

1. Apostol, T.M.: Calculus, vol. 1. Blaisdell Publishing, New York (1961)
2. Bailey, D.F.: A historical survey of solution by functional iteration. Math. Mag. **62**, 155–166 (1989)
3. Boas Jr., R.P.: A Primer of Real Functions. Mathematical Association of America, Washington, DC (1981)
4. Courant, R., Robbins, H.: What Is Mathematics? Oxford University Press, London/New York (1941)
5. Gardner, M.: Mathematical games – a new collection of brain teasers. Sci. Am. **6**, 166–176 (1961)
6. Pennington, W.B.: Existence of a maximum of a continuous function. Am. Math. Mon. **67**, 892–893 (1960)
7. Reich, S.: A Poincaré type coincidence theorem. Am. Math. Mon. **81**, 52–53 (1974)
8. Schwenk, A.J.: Introduction to limits, or why can't we just trust the table? Coll. Math. J. **28**, 51 (1997)
9. Tanton, J.: A dozen areal maneuvers. Math. Horiz. **8**, 26–30,34 (2000)

Chapter 4
Differentiable Functions

Where the telescope ends the microscope begins, and who can say which has the wider vision?

—*Les Misérables*, by Victor Hugo

Again we denote by $I \subseteq \mathbf{R}$ a generic interval, and we mainly consider functions $f : I \to \mathbf{R}$. Roughly speaking, if a function has a derivative at $x \in I$ then it has a well-defined tangent line at $(x, f(x))$. Asking for a function to have a derivative is more than asking for it to be continuous. Still, many functions which arise naturally in applications do have a derivative for all x in some interval I. After defining the derivative, we remind the reader how to find derivatives of many different kinds of functions. And we shall see, for the sake of applications, that horizontal tangent lines are particularly desirable.

4.1 Basic Properties

Let I be an interval (open, closed, or otherwise) and let $f : I \to \mathbf{R}$. We say that f is **differentiable on** I if f is differentiable at *every* $x_0 \in I$. That is, for every $x_0 \in I$ and for any sequence $\{x_n\}$ in I for which $x_n \to x_0$, and $x_n \neq x_0$, it happens that

$$\lim_{n\to\infty} \frac{f(x_n) - f(x_0)}{x_n - x_0} \quad \text{exists,}$$

and depends only on x_0 (that is, not on $\{x_n\}$). When this is the case, we call the limit **the derivative of f at x_0** and we denote it by $f'(x_0)$.

To be more precise, if $f'(x_0)$ exists then for any sequence $\{x_n\} \subset I$ for which $x_n \to x_0$ and $x_n \neq x_0$, and for any $\varepsilon > 0$, there is a number N such that

$$\left| \frac{f(x_n) - f(x_0)}{x_n - x_0} - f'(x_0) \right| < \varepsilon \quad \text{for } n > N.$$

P.R. Mercer, *More Calculus of a Single Variable*, Undergraduate Texts in Mathematics, DOI 10.1007/978-1-4939-1926-0_4

The quotient

$$\frac{f(x_n) - f(x_0)}{x_n - x_0}$$

is the slope of the line through the points $(x_n, f(x_n))$ and $(x_0, f(x_0))$. So when it exists, the limit

$$f'(x_0) \quad \text{is the slope of the tangent line to} \quad y = f(x) \quad \text{at} \quad (x_0, f(x_0)).$$

The equation for this tangent line is

$$y - y_0 = f'(x_0)(x - x_0) \qquad (y_0 = f(x_0)).$$

Another way to think of $f'(x_0)$ is as follows: Continually zooming in on the graph of $y = f(x)$ in the vicinity of $(x_0, f(x_0))$, the graph looks more and more like a straight line and the slope of this straight line is $f'(x_0)$. See **Fig. 4.1.**

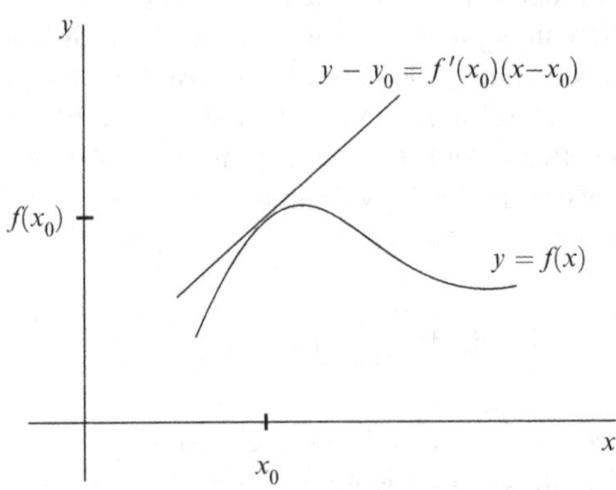

Fig. 4.1 The tangent line to $y = f(x)$ at $x = x_0$

Remark 4.1. A function may have a nonvertical tangent line at x_0 and not be differentiable at x_0, as the following example shows (for any $|x_0| < 1$).

$$f(x) = \begin{cases} \sqrt{(1 - x^2)} & \text{if } |x| \le 1 \text{ and } x \text{ is rational} \\ -\sqrt{(1 - x^2)} & \text{if } |x| < 1 \text{ and } x \text{ is irrational.} \end{cases}$$

This function is not continuous. For a *continuous* function, having a (nonvertical) tangent line and having a derivative are equivalent [27]. ○

Example 4.2. Let $f(x) \equiv C$ (constant) and let $x_n \to x_0$ with $x_n \neq x_0$. Then

$$\frac{f(x_n) - f(x_0)}{x_n - x_0} = \frac{C - C}{x_n - x_0} = 0.$$

Therefore f is differentiable on $\mathbf{R}$ and $f'(x_0) = 0$ for all $x_0 \in \mathbf{R}$. ◇

Example 4.3. Let $f(x) = x$ and let $x_n \to x_0$ with $x_n \neq x_0$. Then

$$\frac{f(x_n) - f(x_0)}{x_n - x_0} = \frac{x_n - x_0}{x_n - x_0} = 1.$$

Therefore f is differentiable on $\mathbf{R}$ and $f'(x_0) = 1$ for all $x_0 \in \mathbf{R}$. ◇

Example 4.4. Let $f(x) = x^2$ and let $x_n \to x_0$ with $x_n \neq x_0$. Then

$$\frac{f(x_n) - f(x_0)}{x_n - x_0} = \frac{x_n^2 - x_0^2}{x_n - x_0} = \frac{(x_n - x_0)(x_n + x_0)}{x_n - x_0} = x_n + x_0.$$

Therefore f is differentiable on $\mathbf{R}$ and $f'(x_0) = 2x_0$, for $x_0 \in \mathbf{R}$. For example (with $x_0 = 3$), the equation of the tangent line to $f(x) = x^2$ at $(3, f(3)) = (3, 9)$ is

$$y - 9 = 6(x - 3), \quad \text{that is,} \quad y = 6x - 9.$$ ◇

Example 4.5. Let $f(x) = |x|$ and consider the sequence $\{(-1)^n/n\}$, which converges to $x_0 = 0$. Then

$$\frac{f(x_n) - f(x_0)}{x_n - x_0} = \frac{|x_n| - |0|}{x_n - x_0} = \frac{1/n - 0}{(-1)^n/n - 0} = (-1)^n.$$

Now since the sequence $\{(-1)^n\}$ diverges, $f'(0)$ does not exist. Therefore f is not differentiable at $x = 0$. The problem here is that the graph of $f(x) = |x|$ has a cusp/corner at $(0, 0)$ and so there is no well-defined tangent line at $(0, 0)$. (However, f is differentiable on $(-\infty, 0) \cup (0, +\infty)$.) See **Fig. 4.2.** ◇

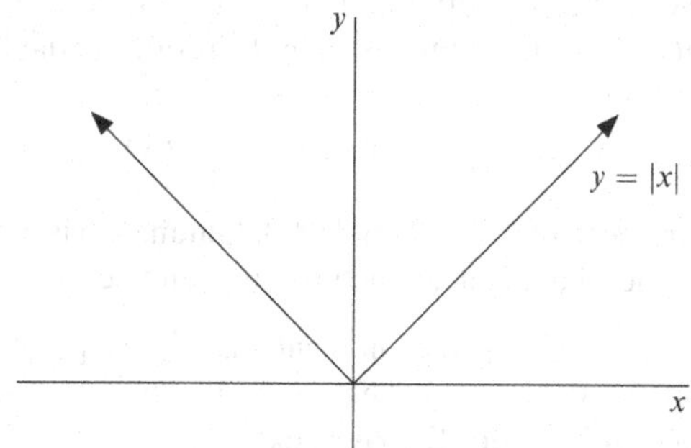

Fig. 4.2 For Examples 4.5 and 4.10: The graph of $y = |x|$. The derivative does not exist at $x_0 = 0$

Observe that if f is differentiable at x, then for any sequence $\{x_n\}$ with $x_n \to x$ (and $x_n \neq x$),

$$\lim_{n\to\infty} \frac{f(x_n) - f(x)}{x_n - x} \quad \text{exists.}$$

Clearly the denominator here $x_n - x \to 0$, so for the limit to exist it must be the case that the numerator $f(x_n) - f(x) \to 0$ also. That is, $f(x_n) \to f(x)$. This simple observation yields a connection between the differentiable functions and the continuous functions, as follows.

Lemma 4.6. *If f is differentiable at $x \in I$, then f is continuous at x.*

Proof. This is Exercise 4.1. □

Lemma 4.6 shows that the differentiable functions form a subset of the continuous functions. And it is a *proper* subset because $f(x) = |x|$ is continuous at $x = 0$ (Example 3.5) but as we saw in Example 4.5, it is not differentiable at $x = 0$. Still, many functions which arise naturally in applications are differentiable on some interval I.

Remark 4.7. It was long believed by mathematicians that a continuous function must be differentiable, except perhaps at some isolated points, just as $f(x) = |x|$ is differentiable everywhere except at $x = 0$. But around 1872, the German mathematician Karl Weierstrass (1815–1897) constructed a function which is continuous at each point of **R**, yet is differentiable at no point of **R**. Weierstrass's example has an important place in the history of mathematics. ○

Example 4.8. Suppose that h is differentiable at $x_0 \in \mathbf{R}$ and let $H(x) = h(x)^2$. Then for $x_n \to x_0$ with $x_n \neq x_0$,

$$\frac{H(x_n) - H(x_0)}{x_n - x_0} = \frac{h(x_n)^2 - h(x_0)^2}{x_n - x_0} = \frac{\big(h(x_n) + h(x_0)\big)\big(h(x_n) - h(x_0)\big)}{x_n - x_0}$$

$$= \big(h(x_n) + h(x_0)\big)\frac{h(x_n) - h(x_0)}{x_n - x_0}.$$

Now since h is differentiable at x_0, it is continuous at x_0, by Lemma 4.6. Therefore $h(x_n) \to h(x_0)$. Finally then, h^2 is differentiable on **R** and

$$(h^2)'(x_0) = 2h(x_0)h'(x_0) \text{ for } x_0 \in \mathbf{R}.$$ ◇

Remark 4.9. In Example 4.8, Lemma 4.6 is essential. Example 4.8 is a special case of the Chain Rule, which we meet in Sect. 4.2. ○

Suppose for the moment that $I = [a, b]$ is a *closed* interval and that f is differentiable on I. Then in particular, f is differentiable at $x_0 = a$. That is, the **derivative from the right** exists at $x_0 = a$. We denote this by $f'_R(a)$. Likewise, f is differentiable at $x_0 = b$. That is, the **derivative from the left** exists at $x_0 = b$. This is denoted by $f'_L(b)$.

Example 4.10. For $f(x) = |x|$, we have $f'_R(0) = 1$ and $f'_L(0) = -1$. Indeed, $f'(0)$ does not exist *because* these two limits are not equal. Again, see **Fig. 4.2.** ◇

If x_0 is not an endpoint of I, then for $|h|$ small enough, $x_0 + h \in I$. Therefore we may write

$$f'(x_0) = \lim_{h \to 0} \frac{f(x_0 + h) - f(x_0)}{h}.$$

When the derivative is to be thought of as a function, which is typically the case, the x_0 in $f'(x_0)$ is usually replaced simply by x (or by t, or by s, etc.).

It is customary to denote a small increment in x by Δx. Then for $y = f(x)$, the resulting increment in $f(x)$ is denoted by $\Delta y = f(x + \Delta x) - f(x)$. As such, we write

$$f'(x) = \lim_{\Delta x \to 0} \frac{f(x + \Delta x) - f(x)}{\Delta x} = \lim_{\Delta x \to 0} \frac{\Delta y}{\Delta x} = \frac{dy}{dx}.$$

Since $\frac{\Delta y}{\Delta x}$ is the *average* rate of change of f between x and $x + \Delta x$, this notation emphasizes the important fact that $f'(x) = \frac{dy}{dx}$ is the *instantaneous* rate of change of f with respect to x. Other notations are, depending on the context,

$$y' = f'(x) = \frac{dy}{dx} = \frac{d}{dx}y = \frac{d}{dx}f(x) = \frac{df}{dx}.$$

4.2 Differentiation Rules

In practice, appealing to the definition of the derivative is often unnecessary. Instead one uses various *differentiation rules*. Here we take a whirlwind tour of these rules.

The simplest differentiation rule—that derivatives respect linear combinations—is as follows. It is a direct consequence of Lemma 1.24.

Linear Combination Rule: *Let $\alpha, \beta \in \mathbf{R}$. If f and g are each differentiable for $x \in I$, then $\alpha f + \beta g$ is differentiable for $x \in I$, with*

$$\big(\alpha f(x) + \beta g(x)\big)' = \alpha f'(x) + \beta g'(x).$$

Proof. This is Exercise 4.2. □

For the case of a product of functions, the differentiation rule is not quite so straightforward.

Product Rule: *If f and g are each differentiable for $x \in I$ then the product $f \cdot g$ is differentiable for $x \in I$, with*

$$\big(f(x)g(x)\big)' = f(x)g'(x) + g(x)f'(x).$$

Proof. (cf. Example 1.28.) As in [9], we write

$$f(x)g(x) = \frac{1}{4}\left[(f(x)+g(x))^2 - (f(x)-g(x))^2\right].$$

Then by Example 4.8 (which uses Lemma 4.6) and the Linear Combination Rule,

$$(f(x)g(x))' = \frac{1}{4}[2(f(x)+g(x))(f'(x)+g'(x)) - 2(f(x)-g(x))(f'(x)-g'(x))].$$

Expanding the right-hand side, then some tidying, yields the desired result. □

The usual textbook proof of the Product Rule is the content of Exercise 4.3. In it, Lemma 4.6 is still indispensable.

Now we use the Product Rule to obtain derivatives of functions like x^3, x^4, x^5, etc. (See Exercises 4.7 and 4.8 for other methods.)

Power Rule for Positive Integer Powers: *Let n be a positive integer and for $x \in \mathbf{R}$, let $f(x) = x^n$. Then f is differentiable, and $f'(x) = nx^{n-1}$.*

Proof. We saw in Examples 9.18 and 9.19 that $(x)' = 1$ and $(x^2)' = 2x$. For x^3 we write $x^3 = x \cdot x^2$ and use the Product Rule:

$$(x^3)' = (x \cdot x^2)' = x(2x) + (1)x^2 = 3x^2.$$

For x^4 we do the same sort of thing:

$$(x^4)' = (x \cdot x^3)' = x(3x^2) + (1)x^3 = 4x^3.$$

Clearly we could continue this procedure indefinitely. So for *any* positive integer n,

$$(x^n)' = (x \cdot x^{n-1})' = x(n-1)x^{n-2} + (1)x^{n-1} = nx^{n-1},$$

as desired. □

We have seen that the function $f(x) \equiv 1$ is differentiable on $\mathbf{R}$ (with $f'(x) \equiv 0$). Therefore, by the Power Rule for Positive Integer Powers, and the Linear Combination Rule, any polynomial is differentiable on $\mathbf{R}$.

Assume that f and g are differentiable, and consider their quotient $h = f/g$ (wherever $g \neq 0$). As in [28], we write $hg = f$ and use the Product Rule to obtain

$$hg' + gh' = f'.$$

That is,

$$h' = \frac{f' - hg'}{g} = \frac{f' - \frac{f}{g}g'}{g} = \frac{gf' - fg'}{g^2}.$$

These manipulations *suggest* the following.

Quotient Rule: *Let f and g be differentiable on I. Then the quotient f/g is differentiable at $x \in I$* for which $g(x) \neq 0$, *and*

$$\left(\frac{f(x)}{g(x)}\right)' = \frac{g(x)f'(x) - f(x)g'(x)}{g(x)^2}.$$

Proof. This is Exercise 4.4 (or Exercise 4.5). □

The argument above only *suggests* the Quotient Rule, because it assumes that the derivative of f/g exists. In any proper proof, Lemma 4.6 is (again) essential. This argument is credited originally (editor's note in [9]) to Italian mathematician Maria Agnesi (1718–1799), who published a successful calculus textbook in 1748 [1, 10].

We have seen that any polynomial is differentiable on **R**. Then by the Quotient Rule any rational function (a polynomial divided by a polynomial) is differentiable on **R**, except wherever its denominator is zero.

Power Rule for All Integer Powers: *Let n be an integer and let $f(x) = x^n$ for $x \in \mathbf{R}$ (but $x \neq 0$ for $n < 0$). Then f is differentiable* on **R**, *and* $f'(x) = nx^{n-1}$.

Proof. We have already proved the result for $n = 0, 1, 2, \ldots$. For $n = -1, -2, -3, \ldots$ we use the result for $n = 1, 2, 3, \ldots$ and apply the Quotient Rule:

$$(x^n)' = \left(\frac{1}{x^{-n}}\right)' = \frac{x^{-n}(0) - (1)(-n)x^{-n-1}}{(x^{-n})^2} = \frac{nx^{-n-1}}{x^{-2n}} = nx^{n-1},$$

as desired. □

Thus far, we have seen how to obtain derivatives of linear combinations (including sums and differences), products, and quotients of functions. For compositions of functions, we use the Chain Rule below. Proofs of the Chain Rule are somewhat tricky so we supply one, which is motivated by [26]. (See also [5] and [23].) Another proof is outlined in Exercise 4.13.

Chain Rule: *Let $g : J \to I$ and $f : I \to$* **R**. *If g is differentiable on J, and f is differentiable on I, then their composition $f \circ g$ is differentiable on J, and*

$$(f \circ g)'(x) = f'(g(x))g'(x).$$

Proof. Let $x_0 \in J$ and let $\{x_n\}$ be a sequence in J, with $x_n \to x_0$ and $x_n \neq x_0$. If there exists N such that $g(x_n) \neq g(x_0)$ for $n > N$ then we may write (for such n):

$$\frac{f(g(x_n)) - f(g(x_0))}{x_n - x_0} = \frac{f(g(x_n)) - f(g(x_0))}{g(x_n) - g(x_0)} \frac{g(x_n) - g(x_0)}{x_n - x_0}.$$

Then as $n \to \infty$,

$$\frac{f(g(x_n)) - f(g(x_0))}{x_n - x_0} \to f'(g(x_0))g'(x_0),$$

and the theorem is proved. However, there are functions g for which there exists no such N as above. (See Exercise 4.11.) So if there is no such N, let a_1 be the x_j in $\{x_1, x_2, x_3, x_4, \ldots\}$ which has the smallest subscript, and for which $g(x_j) = g(x_0)$. Let a_2 be the x_j in $\{x_2, x_3, x_4, \ldots\}$ which has the smallest subscript, and for which $g(x_j) = g(x_0)$. Let a_3 be the x_j in $\{x_3, x_4, \ldots\}$ which has the smallest subscript, and for which $g(x_j) = g(x_0)$ etc. Then $\{a_n\}$ is a sequence in J, with $a_n \to x_0$ and $a_n \neq x_0$. Here we have

$$\frac{g(a_n) - g(x_0)}{a_n - x_0} = 0 \quad \text{for each } n.$$

Therefore $g'(x_0) = 0$. But we also have

$$\frac{f(g(a_n)) - f(g(x_0))}{a_n - x_0} = 0 \quad \text{for each } n,$$

and therefore

$$\frac{f(g(a_n)) - f(g(x_0))}{a_n - x_0} \to f'(g(x_0))g'(x_0),$$

as desired. □

In terms of instantaneous rates of change, the Chain Rule can be stated as follows. If f is a function of g, and g is a function of x, then ultimately f is a function of x. And if f and g are also differentiable, then

$$\frac{df}{dx} = \frac{df}{dg}\frac{dg}{dx}.$$

Indeed, if Fergus runs three times faster than Giuseppina and Giuseppina runs two times faster than Xavier, then Fergus runs six times faster than Xavier.

For $f(x) = x^{p/q}$ we write $f(x)^q = x^p$, then the Chain Rule and the Power Rule for Integer Powers can be used to obtain a Power Rule for Rational Powers, as stated below.

Power Rule for Rational Powers: *Let $p/q \in \mathbf{Q}$ be a rational number and for $x > 0$, let $f(x) = x^{p/q}$. Then f is differentiable for $x > 0$, and $f'(x) = \frac{p}{q}x^{\frac{p}{q}-1}$.*

Proof. This is Exercise 4.14. (See also Exercise 4.12.) □

4.3 Derivatives of Transcendental Functions

A *transcendental* function is a function that cannot be expressed as a finite combination of the operations of addition, subtraction, multiplication, division, raising to powers, and taking roots. That is, it *transcends* the basic algebraic operations.

The simplest examples of transcendental functions are the trigonometric functions and their inverses, and the exponential function e^x and its inverse the natural logarithmic function $\ln(x)$. It is beyond the scope of this book to show that these functions are indeed transcendental, so we shall have to be content with *transcendental* being simply a name.

We leave it for Exercise 4.15 to show that

$$\big(\sin(x)\big)' = \cos(x).$$

Here, x is in *radians*—otherwise this formula would not be quite so nice. Then since $\cos(x) = \sin(\frac{\pi}{2} - x)$, the Chain Rule gives:

$$\big(\cos(x)\big)' = -\cos(\tfrac{\pi}{2} - x) = -\sin(x).$$

The derivatives of the other four trigonometric functions can then be obtained using the Quotient Rule (Exercise 4.15). For example,

$$\big(\tan(x)\big)' = \sec^2(x).$$

For $x \in \mathbf{R}$, we denote by $\arctan(x)$ (often denoted by $\tan^{-1}(x)$ as well) the inverse of the function $\tan(x)$ on $(-\pi/2, \pi/2)$. That is, $\arctan(x)$ is the angle $\theta \in (-\pi/2, \pi/2)$ for which $\tan(\theta) = x$:

$$\begin{aligned} \arctan(\tan(\theta)) &= \theta \qquad \text{for } \theta \in (-\pi/2, \pi/2), \quad \text{and} \\ \tan(\arctan(x)) &= x \qquad \text{for } x \in \mathbf{R}. \end{aligned}$$

Differentiating the latter expression, the Chain Rule gives:

$$\sec^2(\arctan(x))(\arctan(x))' = 1.$$

Therefore

$$\big(\arctan(x)\big)' = \frac{1}{\sec^2(\arctan(x))} = \frac{1}{\tan^2(\arctan(x)) + 1} = \frac{1}{x^2 + 1}.$$

However, these manipulations only *suggest* the answer because they assume that the derivative of $\arctan(x)$ exists. One can show that it indeed exists using Exercise 4.16. But here, following [13], we do so more directly.

For $0 \le x < y$, set

$$\theta = \arctan(y) - \arctan(x).$$

Then since $\sin(\theta) \le \theta < \tan(\theta)$ for $0 \le \theta < \pi/2$,

$$\sin(\arctan(y) - \arctan(x)) \le \arctan(y) - \arctan(x) < \tan(\arctan(y) - \arctan(x)).$$

Applying the trigonometric identities

$$\sin(A - B) = \sin(A)\cos(B) - \cos(A)\sin(B) \quad \text{and}$$

$$\tan(A - B) = \frac{\tan(A) - \tan(B)}{1 + \tan(A)\tan(B)}$$

with $y = \tan(A)$ and $x = \tan(B)$ on the left-hand and right-hand sides respectively, we get

$$\frac{1}{\sqrt{1+y^2}\sqrt{1+x^2}} \leq \frac{\arctan(y) - \arctan(x)}{y - x} < \frac{1}{1+xy}.$$

Therefore, letting $y \to x$ (or $x \to y$), we get

$$\big(\arctan(x)\big)' = \big(\tan^{-1}(x)\big)' = \frac{1}{1+x^2}, \quad \text{for } x \geq 0.$$

Now since $\arctan(x)$ is an *odd* function: $\arctan(-x) = -\arctan(x)$ for all x, this is sufficient to show that the formula holds for all real x.

The exponential function e^x and its inverse function $\ln(x)$ are related by

$$\ln(e^x) = x \quad \text{for } x \in \mathbf{R}, \text{ and}$$
$$e^{\ln(x)} = x \quad \text{for } x > 0.$$

For the moment we assume that the reader is familiar with the basic properties of e^x and $\ln(x)$. One such property is

$$(e^x)' = e^x \quad \text{for } x \in \mathbf{R}.$$

Then differentiating $e^{\ln(x)} = x$ using the Chain Rule gives

$$e^{\ln(x)}(\ln(x))' = 1.$$

Therefore

$$\big(\ln(x)\big)' = \frac{1}{x} \quad \text{for } x > 0.$$

But again, these manipulations are only *suggestive* because in them, we have assumed that the derivative of $\ln(x)$ exists. One can show that it indeed exists using Exercise 4.16 but we shall do so more directly in Sect. 6.3—in a similar spirit to how we obtained the derivative of $\arctan(x)$ above.

Then the Chain Rule gives

$$\left(e^{\alpha(x)}\right)' = \alpha'(x)e^{\alpha(x)} \quad \text{for differentiable functions } \alpha \text{, and}$$

$$\left(\ln \beta(x)\right)' = \frac{\beta'(x)}{\beta(x)} \quad \text{for positive differentiable functions } \beta.$$

For $x > 0$, $r \in \mathbf{R}$ and $\alpha(x) = \ln(x^r) = r\ln(x)$, we obtain

$$\left(x^r\right)' = \left(e^{r\ln(x)}\right)' = \frac{r}{x}e^{r\ln(x)} = \frac{r}{x}x^r = rx^{r-1},$$

which we highlight as follows.

Power Rule for Real Powers: *Let $r \in \mathbf{R}$ and for $x > 0$, let $f(x) = x^r$. Then f is differentiable, with $f'(x) = rx^{r-1}$.*

Proof. As outlined above; a proper proof is postponed until Sect. 6.3. □

A similar trick can be used to find derivatives of functions for which the variable appears as part of an exponent. This is called **logarithmic differentiation**. Here are two more examples.

Example 4.11. Let $f(x) = 2^x$. We write this as $f(x) = e^{\ln(2)^x} = e^{x\ln(2)}$, then use the Chain Rule:

$$f'(x) = e^{x\ln(2)}\ln(2) = 2^x\ln(2). \quad \diamond$$

Example 4.12. For a more complicated example, let $g(x) = \left(1+x^2\right)^{\sin(x)}$. Here we write $g(x) = e^{\ln\left(1+x^2\right)^{\sin(x)}} = e^{\sin(x)\ln(1+x^2)}$. Then by the Chain Rule and the Product Rule,

$$g'(x) = \left(1+x^2\right)^{\sin(x)}\left(\sin(x)\frac{2x}{1+x^2} + \cos(x)\ln(1+x^2)\right). \quad \diamond$$

4.4 Fermat's Theorem and Applications

Surely the most practical application of the derivative is to find maximum and minimum values (these are called *extrema*) for various functions. To this end, the following result is very useful.

Theorem 4.13. (Fermat's Theorem) *Let f be defined on (a,b) and let $c \in (a,b)$ be such that $f(c) \geq f(x)$ for all $x \in (a,b)$. Then either $f'(c) = 0$ or $f'(c)$ does not exist.*

Proof. Let $\{x_n\}$ be any sequence in (a, b) with $x_n \to c$ (and $x_n \neq c$) so by hypothesis, $f(c) \geq f(x_n)$. Now whenever $c > x_n$, we have

$$\frac{f(x_n) - f(c)}{x_n - c} \geq 0 .$$

But whenever $c < x_n$, we have

$$\frac{f(x_n) - f(c)}{x_n - c} \leq 0 .$$

So if $f'(c)$ exists then we must have $f'(c) = 0$. (Otherwise, of course, $f'(c)$ does not exist.) □

This theorem is named for French mathematician Pierre de Fermat (1601–1665). Although his life predates the discovery of calculus proper, Fermat computed tangent lines and extrema for many families of curves.

The number $f(c)$ in this context is called a **local maximum** for f. It is *local* because f may attain larger values outside (a, b). Fermat's Theorem (Theorem 4.13) holds also for c yielding a **local minimum**: $f(c) \leq f(x)$ for all $x \in (a, b)$. We leave the verification of this claim for Exercise 4.24. The number $f(c)$ is called a **local extremum** if it is either a local maximum or a local minimum.

The number $f(c)$ is called an **absolute maximum** for f if $f(c) \geq f(x)$ for all x in the domain of f, and the number $f(c)$ is called an **absolute minimum** if $f(c) \leq f(x)$ for all x in the domain of f. The number $f(c)$ is called an **absolute extremum** if it is either an absolute maximum or an absolute minimum.

Any point c for which either $f'(c) = 0$ or $f'(c)$ does not exist is called a **critical point** for f. So Fermat's Theorem (Theorem 4.13) says, in short, that a local extremum for f must occur at a critical point for f. These are places at which f has either a horizontal tangent line or a cusp/corner. In either case, $f(c)$ is called a **critical value** for f.

The converse of Fermat's Theorem (Theorem 4.13) does not hold—that is, a critical value need not be a local extremum: Consider $f(x) = x^3$ at $x = 0$. (The paper [7] amusingly calls such points "duds.") So Fermat's Theorem *only* tells us where to look for local extrema; it *does not* guarantee success. (If someone has caught a fish, then they must have been at a body of water. So suggesting that your friend goes fishing at a body of water is good advice, but this of course does not guarantee success.)

Example 4.14. Let us seek the point(s) on the curve $y = x^2$ closest to the point $(0, 1)$. See see **Fig. 4.3.**

Fig. 4.3 For Example 4.14. Distance d from $(0, 1)$ to $y = x^2$

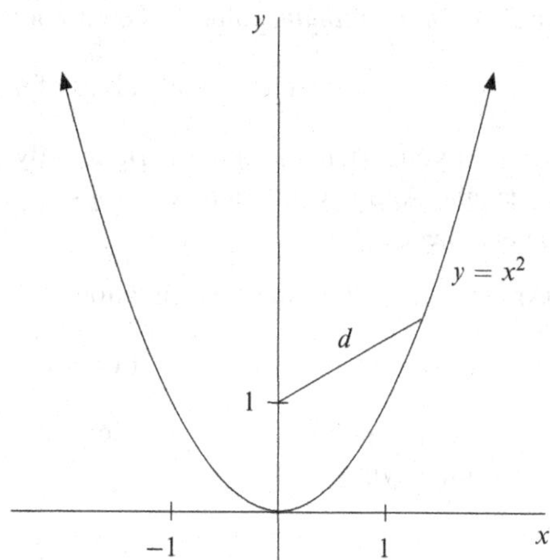

The distance from any point (x, x^2) on the curve to the point $(0, 1)$ is given by

$$d = d(x) = \sqrt{(x-0)^2 + (x^2-1)^2} = \sqrt{x^2 + (x^2-1)^2}\,.$$

Now it is clear from Fig. 4.3, or observe that $d(x) \to +\infty$ as $x \to \pm\infty$, that d indeed has an absolute minimum (and no absolute maximum). An absolute minimum is also a local minimum and by Fermat's Theorem (Theorem 4.13), it must occur at a critical point. By the Chain Rule, and after some simplifying,

$$d'(x) = \frac{x(2x^2-1)}{\sqrt{x^2 + (x^2-1)^2}}\,,$$

and so d has critical points at $x = 0$ and $x = \pm 1/\sqrt{2}$, at which $d' = 0$. Now $d(0) = 1$ and $d(1/\sqrt{2}) = d(-1/\sqrt{2}) = \sqrt{3}/2 < 1$. Therefore the points on $y = x^2$ closest to $(0, 1)$ are $(1/\sqrt{2}, 1/2)$ and $(-1/\sqrt{2}, 1/2)$; each attains the minimum distance $\sqrt{3}/2$. (The value $d(0) = 1$ is a *local* maximum.) ◇

In Example 4.14 we were able to justify, within the context, that Fermat's Theorem (Theorem 4.13) indeed led us to the absolute minimum that we sought. Generally however, deciding which critical points yield absolute extrema can be a delicate matter. We pursue this further in Sect. 5.2.

But if the interval under consideration is $[a, b]$, that is, if it is *closed*, then Fermat's Theorem (Theorem 4.13) and the Extreme Value Theorem (Theorem 3.23) together give a recipe by which the absolute extrema of a continuous function can be found quite easily:

The *absolute maximum* value of a continuous function f on $[a, b]$ is the *largest* of

$$\{\text{the critical values of } f,\ f(a),\ \text{and } f(b)\}.$$

The *absolute minimum* value of a continuous function f on $[a, b]$ is the *smallest* of

$$\{\text{the critical values of } f,\ f(a),\ \text{and } f(b)\}.$$

(Here, if your friend goes to the right body of water then there are definitely fish to be caught. And if your friend employs impeccable fishing techniques, then success is guaranteed!)

Example 4.15. Consider the function

$$f(x) = \mathrm{e}^x \sqrt{5 - 4x}\,,$$

which is continuous on $[-1, 1]$. Here, by the Product and Chain Rules and after some simplifying,

$$f'(x) = \frac{\mathrm{e}^x(3 - 4x)}{\sqrt{5 - 4x}}.$$

The only critical point that f has in $[-1, 1]$ is $x = 3/4$, at which $f' = 0$. Now $f(3/4) \cong 2.99$, $f(-1) \cong 1.1$, and $f(1) = \mathrm{e} \cong 2.718$. Therefore, on $[-1, 1]$, f has a maximum value of about 2.99 and a minimum value of about 1.1 . ◇

Example 4.16. John is on one side of a river 1/2 a mile wide, say at point A. He notices his house burning 2 miles downstream, but it is on the opposite side of the river; naturally, he wants to get to his house as quickly as possible. John can run at 5 miles per hour and he can swim downstream at 3 miles per hour. How should he proceed?

For a solution, consider **Fig. 4.4**, which helps us to obtain an expression for the time T that it takes John to get to his house (point B). We denote by P any point on the opposite bank to which he may swim. The distance from P to the point directly across the river from A (let's call it A') is x.

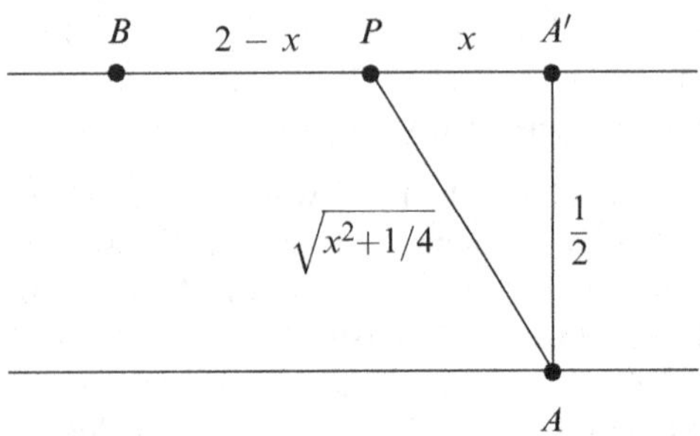

Fig. 4.4 For Example 4.16. John swims from A to P, then runs from P to B

As time is distance divided by speed,

$$T(x) = \frac{\sqrt{x^2 + 1/4}}{3} + \frac{2 - x}{5}.$$

Obviously, we need only consider $T(x)$ for $0 \le x \le 2$. So we want to find a value for x which minimizes $T(x)$ on the closed interval $[0, 2]$. The reader may verify that

$$T'(x) = \frac{5x - 3\sqrt{x^2 + 1/4}}{15\sqrt{x^2 + 1/4}},$$

and so T is differentiable for all x, and $T'(3/8) = 0$. That is, T has a critical point at $x = 3/8$. Now $T(3/8) = 8/15 = 0.5\overline{3}$, $T(0) = 17/30 = 0.5\overline{6}$, and $T(2) = \sqrt{17}/6 \cong 0.687$. Therefore, John should swim to the point exactly 3/8 of a mile downstream, and run the rest of the way. In doing so, it would take him $0.5\overline{3}$ of an hour to get to his house. (If John wanted to allow his house to burn in order to collect insurance money, then try to convince investigators that he did his best in attempting to save his house, he might swim the entire way.) ◇

Remark 4.17. Example 4.16 is a classic [20]. A version of it, in which a man can walk on smooth ground at a certain speed, and walk on plowed ground at a certain (slower) speed, appears in a 1691–1692 manuscript by the Swiss mathematician Johann Bernoulli (1667–1748). The manuscript was published in 1742, just 6 years before Maria Agnesi's book. ○

Exercises

4.1. Prove Lemma 4.6: *If f is differentiable at $x \in I$, then f is continuous at x.*

4.2. Let $\alpha, \beta \in \mathbf{R}$. Use Lemma 1.24 to prove the Linear Combination Rule: *If f and g are each differentiable on I and $\alpha, \beta \in \mathbf{R}$, then $\alpha f + \beta g$ is differentiable on I, with $(\alpha f(x) + \beta g(x))' = \alpha f'(x) + \beta g'(x)$.* So in particular, $(f(x) \pm g(x))' = f'(x) \pm g'(x)$.

4.3. **(a)** Begin by writing

$$f(x)g(x) - f(x_0)g(x_0) = f(x)g(x) - f(x)g(x_0) + f(x)g(x_0) - f(x_0)g(x_0)$$

to prove the Product Rule.

(b) Prove it again, beginning instead with

$$f(x)g(x) - f(x_0)g(x_0) = f(x)g(x) - f(x_0)g(x) + f(x_0)g(x) - f(x_0)g(x_0).$$

4.4. **(a)** Begin by writing

$$\frac{f(x)}{g(x)} - \frac{f(x_0)}{g(x_0)} = \frac{f(x)g(x_0) - f(x_0)g(x_0) + f(x_0)g(x_0) - g(x)f(x_0)}{g(x)g(x_0)}$$

to prove the Quotient Rule.

(b) Prove it again, beginning instead with

$$\frac{f(x)}{g(x)} - \frac{f(x_0)}{g(x_0)} = \frac{f(x)g(x_0) - f(x)g(x) + f(x)g(x) - g(x)f(x_0)}{g(x)g(x_0)}.$$

4.5. **(a)** Prove directly that (wherever $g(x) \neq 0$),

$$\left(\frac{1}{g(x)}\right)' = -\frac{g'(x)}{g(x)^2}.$$

(b) Now prove the Quotient Rule by applying the Product Rule to $\dfrac{f(x)}{g(x)} = f(x)\dfrac{1}{g(x)}$.

4.6. [8]

(a) Show that

$$\frac{(fg)'}{fg} = \frac{f'}{f} + \frac{g'}{g}.$$

(b) Show that

$$\frac{(f/g)'}{f/g} = \frac{f'}{f} - \frac{g'}{g}.$$

4.7. **(a)** Let n be a natural number. Verify that

$$\frac{b^n - a^n}{b - a} = b^{n-1} + ab^{n-2} + a^2b^{n-3} + \cdots + a^{n-3}b^2 + a^{n-2}b + a^{n-1}.$$

(b) Use (a) to prove the Power Rule for Positive Integer Powers.

4.8. [2] Here's a neat direct proof of the Power Rule for Positive Integer Powers.

(a) In the quotient

$$\frac{y^n - x^n}{y - x},$$

make the substitution $q = y/x$ to get

$$\frac{y^n - x^n}{y - x} = x^{n-1}\frac{q^n - 1}{q - 1}.$$

(b) Now write

$$\frac{q^n - 1}{q - 1} = 1 + q + q^2 + \cdots + q^{n-1}, \text{ and let } q \to 1.$$

4.9. [4] Consider the function

$$f(x) = \begin{cases} x^2 & \text{if } x \geq 0 \\ 0 & \text{if } x < 0\,. \end{cases}$$

Show that f' exists at $x = 0$, that f' is continuous at $x = 0$, but that f' is not differentiable at $x = 0$.

4.10. Suppose that f is differentiable on an open interval containing x.

(a) Show that

$$\lim_{h \to 0} \frac{f(x+h) - f(x-h)}{2h} = f'(x).$$

Hint: Consider the average of the right-hand and left-hand derivatives at x.

(b) Show that this limit can exist even when $f'(x)$ does not.

4.11. Consider the function

$$g(x) = \begin{cases} x^2 \sin\left(\frac{1}{x}\right) & \text{if } x \neq 0 \\ \\ 0 & \text{if } x = 0. \end{cases}$$

(a) Show that g is differentiable at $x = 0$.
(b) Show that $g(x_n) = g(0) = 0$ for the sequence $\{x_n\} = \{\frac{1}{n\pi}\}$, which has $x_n \to 0$.
(c) Show that g' is not continuous at $x = 0$.

4.12. [17] Here is a way to obtain the Power Rule for Rational Powers more directly—that is, without the Chain or Product Rules. For $f(x) = x^{p/q}$, write

$$f'(x) = \lim_{h \to 0} \frac{(x+h)^{p/q} - x^{p/q}}{h} = \lim_{h \to 0} \frac{\left((x+h)^{1/q}\right)^p - \left(x^{1/q}\right)^p}{\left((x+h)^{1/q}\right)^q - \left(x^{1/q}\right)^q}$$

then use the formula

$$(a^N - b^N) = (a-b)(a^{N-1} + a^{N-2}b + \cdots + ab^{N-2} + b^{N-1}).$$

4.13. Fill in the details of another proof of the Chain Rule, as follows. Set $g(x_0) = y_0$, and consider the function h defined on I by

$$h(y) = \begin{cases} \dfrac{f(y) - f(y_0)}{y - y_0} & \text{if } y \neq y_0 \\ \\ f'(y_0) & \text{if } y = y_0. \end{cases}$$

(a) Show that h is continuous on I and verify that for $y \in I$,

$$f(y) - f(y_0) = (y - y_0)h(y),$$

so that for $x \in J$,

$$f(g(x)) - f(g(x_0)) = (g(x) - g(x_0))h(g(x)).$$

(b) Now, for $x_n \neq x_0$, consider the quotient

$$\frac{f(g(x_n)) - f(g(x_0))}{x_n - x_0} = \frac{(g(x_n) - g(x_0))h(g(x_n))}{x_n - x_0}.$$

Observe that $h \circ g$ is continuous on J, and let $x_n \to x_0$.

4.14. Use the Chain Rule and the Power Rule for Integer Powers to prove the Power Rule for Rational Powers.

4.15. **(a)** Show that

$$\lim_{h \to 0} \frac{\sin(h)}{h} = 1.$$

(b) Use this to show that

$$\lim_{h \to 0} \frac{1 - \cos(h)}{h} = 0.$$

(c) Now write

$$\sin(x + h) = \sin(x + h) = \sin(x)\cos(h) + \cos(x)\sin(h)$$

to show that

$$\big(\sin(x)\big)' = \cos(x).$$

(d) What if x is in degrees, rather than radians?
(e) In a similar way, show that $(\cos(x))' = -\sin(x)$.
(f) Find the derivatives (where they exist) of the other four trigonometric functions. (See [11,19,25] for neat ways of showing that $(\sin(x))' = \cos(x)$ using some simple geometry then taking a limit.)

4.16. A function f is **strictly increasing** on I if

$$f(x_1) < f(x_2) \quad \text{whenever} \quad x_1, x_2 \in I \quad \text{with} \quad x_1 < x_2.$$

And f is **strictly decreasing** on I if $-f$ is strictly increasing there. A function which is either strictly increasing or strictly decreasing is called **strictly monotonic**. A function $f : I \to J$ is **onto** if for each $y_0 \in J$, there is $x_0 \in I$ such that $f(x_0) = y_0$. If $f : (a,b) \to (p,q)$ is strictly monotonic and onto, then the inverse $f^{-1} : J \to I$ exists. It is defined by $f^{-1}(y) = x \Leftrightarrow y = f(x)$. Suppose that $f : (a,b) \to (p,q)$ is strictly monotonic and onto.

(a) Prove that f is continuous on (a,b).
(b) Prove that f^{-1} is strictly monotonic and onto, and therefore continuous.
(c) Show that if f is also differentiable and $f'(x) \neq 0$ then f^{-1} is differentiable, with

$$(f^{-1})'(y) = \frac{1}{f'(f^{-1}(y))}.$$

(d) [24] Explain what the formula in (c) has to do with **Fig. 4.5**.

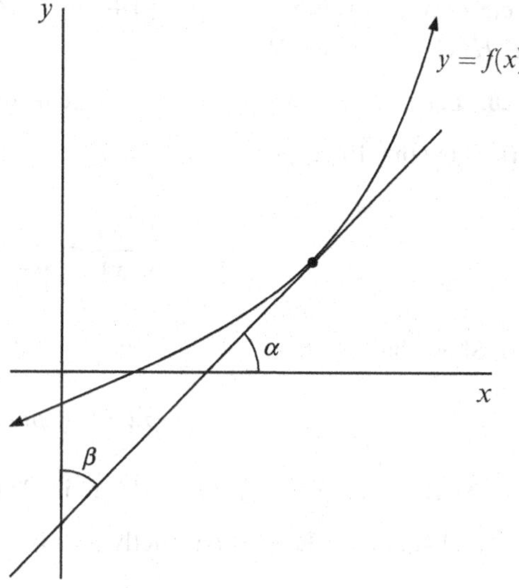

Fig. 4.5 For Exercise 4.16. Observe that $\tan(\beta) = 1/\tan(\alpha)$, since $\alpha + \beta = \pi/2$

4.17. [16] Show that for any n distinct real numbers ($n \geq 4$) there are at least two which satisfy

$$0 < \frac{x-y}{1+xy} < \tan\left(\frac{\pi}{n-1}\right).$$

Hint: Each number can be written as $\tan(u)$, where $-\pi/2 < u < \pi/2$. The trigonometric identity $\tan(A-B) = \frac{\tan(A)-\tan(B)}{1+\tan(A)\tan(B)}$ will be useful.

4.18. **(a)** Verify the identities

$$\arcsin(x) = \arctan\left(\frac{x}{\sqrt{1-x^2}}\right) \quad \text{and} \quad \arccos(x) = \frac{\pi}{2} - \arcsin(x).$$

(b) Use these identities to show that

$$\big(\arcsin(x)\big)' = \frac{1}{\sqrt{1-x^2}}$$

and

$$\big(\arccos(x)\big)' = -\frac{1}{\sqrt{1-x^2}}.$$

4.19. Find derivatives for the following functions:
(a) $f(x) = \arctan\big(\cos(x) + x^3)\big)$, **(b)** $g(x) = e^{\sin^2(x)+\sec(x)}$, **(c)** $h(x) = 3^x + \csc(x) + x^5$, **(d)** $k(x) = \ln\left(\frac{1+x^2}{1-x^2}\right) + \cot(x)$, **(e)** $F(x) = x^{\sqrt{x}}$, **(f)** $G(x) = x^{\ln(x)}$, **(g)** $H(x) = \sin(x)^{\cos(x)}$.

4.20. Let $p(x)$ be a polynomial of degree n, with roots $x_1, x_2, \ldots, x_n$.

(a) Show that for $x \neq x_j$,

$$\frac{p'(x)}{p(x)} = \sum_{j=1}^{n} \frac{1}{x - x_j}.$$

(b) Show that

$$p'(x)^2 \geq p(x)p''(x).$$

Here, $p''(x)$ means $\big(p'(x)\big)'$. (More in Chap. 8 ...)

4.21. [14] Let $f : \mathbf{R} \to \mathbf{R}$ be strictly positive and suppose that

$$\lim_{h\to 0} \left(\frac{f(x+h)}{x}\right)^{1/h} = L \neq 0.$$

(a) Show that f is differentiable at x. **(b)** Find L.

4.22. [12] For $x \in [-1, 1]$, consider the function

$$f(x) = \begin{cases} x^{4/3} \sin\left(\frac{1}{x}\right) & \text{if } x \neq 0 \\ 0 & \text{if } x = 0. \end{cases}$$

Show that f is continuous on $[-1, 1]$, but has unbounded derivative there.

4.23. [3] Let $p(x) = x^n + a_{n-1}x^{n-1} + \cdots + a_1 x + a_0$ be a polynomial with zeros $x_1, x_2, \ldots, x_n$ and denote by $\hat{x}_1, \hat{x}_2, \ldots, \hat{x}_{n-1}$ the zeros of p'. Show that

$$\frac{x_1 + x_2 + \cdots + x_n}{n} = \frac{\hat{x}_1 + \hat{x}_2 + \cdots + \hat{x}_{n-1}}{n-1}.$$

That is, the Arithmetic Mean of the zeros of p equals the Arithmetic Mean of the zeros of p'.

4.24. Prove Fermat's Theorem (Theorem 4.13) for the local minimum case.

4.25. Let $x_1, x_2, \ldots, x_n \in \mathbf{R}$. Find the value of x for which the sum of squares

$$\sum_{j=1}^{n} (x - x_j)^2 \quad \text{is least.}$$

4.26. **(a)** Find the point on the line $y = 2x + 3$ closest to $(2, 1)$ and find the minimum distance thus attained.

(b) Find the point on the line $y = mx + b$ closest to (p, q) and find the minimum distance thus attained. $\left[\text{Answer: } \left(\frac{mq-mb+p}{m^2+1}, m\frac{mq-mb+p}{m^2+1} + b\right); \text{ distance is } \frac{|b-q+mp|}{\sqrt{m^2+1}}.\right]$

4.27. Find the point on the curve $y = \sqrt{x^2 + 2}$ closest to the point $(1, 0)$ and find the minimum distance thus attained.

4.28. Find an equation for the tangent line to the graph of $y = x^3 - 3x^2 + 2x$ which has the least slope.

4.29. [18]

(a) Show that the maximum value of $f(x) = x(b - x)$ on $[0, b]$ is $b^2/4$.

(b) Let $b_1, b_2, \ldots, b_n \in \mathbf{R}$, with $\sum_{j=1}^{n} b_j = b$. Show that $\sum_{j=1}^{n-1} b_j b_{j+1} \le b^2/4$.

4.30. [22] Let f be differentiable on an open interval which contains $[-1, 1]$, with $|f'(x)| \le 1$ for $x \in [-1, 1]$. Show that there is $x_0 \in (-1, 1)$ for which $|f'(x_0)| < 4$. Hint: Consider $g(x) = f(x) + 2x^2$.

4.31. **(a)** Let $f(x) = 2x^2 - x$. Consider a rectangle in the first quadrant with one side on the positive x-axis and inscribed under the graph of $y = f(x)$. Find the rectangle so described which has maximal area.

(b) [15] Let $g(x) = \frac{x}{x^2+1}$. Consider a rectangle in the first quadrant with one side on the positive x-axis and inscribed under the graph of $y = g(x)$. Show that there is no such rectangle which has maximal area.

4.32. A piece of wire L inches long is cut into two pieces—one the shape of a square, and one the shape of a circle. How should the wire be cut so that the total area of the two shapes is as small as possible? How should the wire be cut so that the total area of the two shapes is as large as possible?

4.33. Show, using calculus, that the rectangle with given perimeter which has the greatest area is a square. Show that the rectangle with given area which has the least perimeter is a square. (We've already shown these in Example 2.9, using the AGM Inequality. Using calculus here is rather like killing a mite with a sledgehammer.)

4.34. (a) A rectangular plot of ground is to be enclosed by fencing on three sides, with a long existing wall serving as boundary for the fourth side. Find the dimensions of the plot of greatest area which can be enclosed with 1,000 ft of fencing. Can you do this without using calculus?

(b) Suppose now that we have the same situation as in (a), but that the existing wall is 400 ft long. Find the dimensions of the plot of greatest area which can be enclosed with 1,000 ft of fencing. Can you do this without using calculus? For a thorough study of problems such as these, see [21].

4.35. A box with open top is made from a rectangular piece of cardboard 12 in. by 18 in.; congruent squares are cut from each corner and the edges are folded up. Find the dimensions of such a box which has largest volume.

4.36. (a) Find the dimensions of the rectangle of maximum area that can be inscribed in a circle with radius R.

(b) Find the dimensions of the right circular cylinder of maximum volume that can be inscribed in a sphere with radius R.

4.37. A wooden beam is to be carried horizontally around a corner, from a hallway of width 12 ft into a hallway of width 8 ft Find the length of the longest beam that can be so carried. Can you do this without using calculus?

4.38. A rectangular piece of paper is 6 in. wide and 25 in. long. The paper is folded, creating a crease, so that the lower right corner just touches the left side. Describe the fold which minimizes the length of the crease.

4.39. (e.g., [6]) Let $A > 0$, $a > 0$ and $B > 0$, so that $P = (0, A)$ and $Q = (a, B)$ are points on the positive y-axis and in the first quadrant respectively. Find the point C on the x-axis so that the sum of distances $PC + CQ$ is minimized. The answer is known as *Fermat's Law of Reflection*.

4.40. (e.g., [6]) Let $A > 0$, $a > 0$ and $B < 0$, so that $P = (0, A)$ and $Q = (a, B)$ are points on the positive y-axis and in the fourth quadrant respectively. Suppose that light travels with velocity p above the x-axis and velocity q below the x-axis. Find the path from P to Q which takes the least time. The answer is known as *Snell's Law of Refraction*.

References

1. Alexanderson, G.L.: Maria Gaëtana Agnesi – a divided life. Bull. Am. Math. Soc. **50**, 147–152 (2013)
2. Askey, R.: What do we do about calculus? First, do no harm. Am. Math. Mon. **104**, 738–743 (1997)

3. Bullen, P.S.: Handbook of Means and Their Inequalities. Kluwer Academic, Dordrecht/Boston (2003)
4. Bumcrot, R.: Exactly n-times differentiable functions. Coll. Math. J. **14**, 258–259 (1983)
5. Chapman, S.T.: Editor's endnotes. Am. Math. Mon. **121**, 467–468 (2014)
6. Courant, R.: Differential and Integral Calculus, vol. 1. Wiley Classics Library, Hoboken (1988)
7. Ecker, M.W.: Must a "dud" necessarily be an inflection point? Coll. Math. J. **12**, 332–333 (1981)
8. Eggleton, R., Kustov, V.: The product and quotient rules revisited. Coll. Math. J. **42**, 323–326 (2011)
9. Euler, R.: A note on differentiation. Coll. Math. J. **17**, 166–167 (1986)
10. Gray, S.I.B., Malakyan, T.: The witch of Agnesi; A lasting contribution from the first surviving mathematical work written by a woman. Coll. Math. J. **30**, 258–268 (1999)
11. Hartig, D.: On the differentiation formula for $\sin(\theta)$. Am. Math. Mon. **75**, 787 (1968)
12. Kaptanoglu, H.T.: In praise of $x^\alpha \sin(1/x)$. Am. Math. Mon. **108**, 144–150 (2001)
13. Key, E.: Differentiating the arctangent directly. Coll. Math. J. **40**, 287–289 (2009)
14. Körner, T.W.: A Companion to Analysis. American Mathematical Society, Providence (2004)
15. Latina, M.R: Some maximal rectangles and the realities of applied mathematics. Coll. Math. J. **14**, 248–252 (1983)
16. Long, C.T., Dixon, D.J., Kistner, J.E.: Problem E3121. Am. Math. Mon. **94**, 880–881 (1987)
17. Lindstom, P.A.: A self-contained derivation of the formula $\frac{d}{dx}(rx^r) = rx^{r-1}$ for rational r. Coll. Math. J. **16**, 131–132 (1985)
18. MacDonald, J.E., Tabbe, R.: Problem E1643. Am. Math. Mon. **71**, 914 (1964)
19. McQuillan, D., Poodiack, R.: On the differentiation formulae for sine, tangent, and inverse tangent. Coll. Math. J. **45**, 140–142 (2014)
20. Rickey, V.F.: "Johann Bernoulli's Calculus Texts," a talk given at the A.M.S. & M.A.A. Joint Winter Meetings, San Diego, Jan 10, 1997
21. Schumer, P.: The pen and the barn. Coll. Math. J. **28**, 205–206 (1997)
22. Shapiro, H.S.: Problem E1986. Am. Math. Mon. **75**, 787 (1968)
23. Silverman, R.A.: Introductory Complex Analysis. Dover, Mineola (1972)
24. Snapper, E.: Inverse functions and their derivatives. Am. Math. Mon. **97**, 144–147 (1990)
25. Sridharma, S.: The derivative of $\sin(\theta)$. Coll. Math. J. **30**, 314–315 (1999)
26. Tandra, H.: A yet simpler proof of the chain rule. Am. Math. Mon. **120**, 900 (2013)
27. Thurston, H.: Tangents to graphs. Math. Mag. **61**, 292–294 (1988)
28. Wood, P.G.: The quotient rule. Math. Gaz. **68**, 288 (1984)

Chapter 5
The Mean Value Theorem

Up the airy mountain, Down the rushy glen...

– *The Fairies,* by William Allingham

The main focus of this chapter is the Mean Value Theorem and some of its applications. This is the big theorem in the world of differentiable functions. Many important results in calculus (and well beyond!) follow from the Mean Value Theorem. We also look at an interesting and useful generalization, due to Cauchy.

5.1 The Mean Value Theorem

The following result is named for French mathematician Michel Rolle (1652–1719).

Theorem 5.1. (Rolle's Theorem) *Let f be continuous on $[a,b]$ and differentiable on (a,b), with $f(a) = f(b)$. Then there exists $c \in (a,b)$ such that $f'(c) = 0$.*

Proof. If f is constant on $[a,b]$, then its derivative is zero and so any $c \in (a,b)$ satisfies the conclusion of the theorem. So we assume that f is not constant on $[a,b]$. By the Extreme Value Theorem (Theorem 3.23), f attains an absolute maximum and an absolute minimum on $[a,b]$. Since f is not constant, at least one of these absolute extrema must occur at $c \in (a,b)$. Then since f is differentiable on (a,b), an application of Fermat's Theorem (Theorem 4.13) gives $f'(c) = 0$, as desired.

□

Rolle's Theorem (Theorem 5.1) is fairly obvious, upon drawing a picture: If a differentiable function starts at $f(a)$ then returns to $f(b) = f(a)$, there must be at least one place on its graph at which the tangent line is horizontal. See **Fig. 5.1.**

Here is a another neat proof [2,7,41] of Rolle's Theorem (Theorem 5.1); we leave the details for Exercise 5.3. Since f is continuous and $f(a) = f(b)$, by the Half-Chord Lemma there are $a_1, b_1 \in [a,b]$ with $f(a_1) = f(b_1)$ and $b_1 - a_1 = (b-a)/2$. Again by the Half-Chord Lemma, there are $a_2, b_2 \in [a_1, b_1]$ with $f(a_2) = f(b_2)$ and $b_2 - a_2 = (b-a)/2^2$. Continuing in this way, we obtain a sequence of nested

P.R. Mercer, *More Calculus of a Single Variable*, Undergraduate Texts in Mathematics, DOI 10.1007/978-1-4939-1926-0_5

intervals $[a_n, b_n]$ with $f(a_n) = f(b_n)$ and $b_n - a_n = (b-a)/2^n \to 0$. So by the Nested Interval Property (Theorem 1.41) there is a point $c \in [a,b]$ which belongs to each interval. Since f is differentiable on (a,b), $f'(c) = 0$ (explain!), as desired.

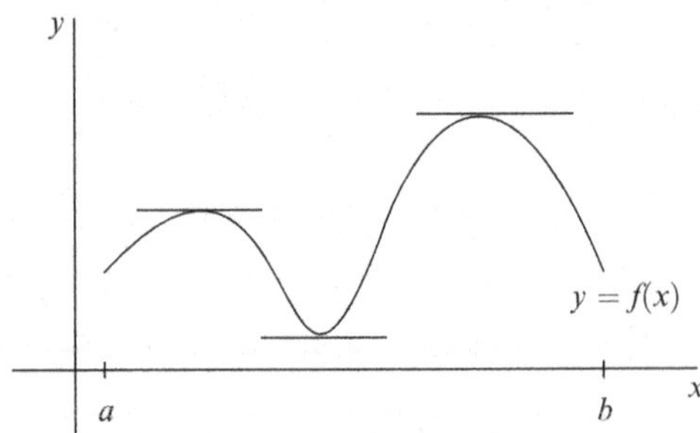

Fig. 5.1 Rolle's Theorem (Theorem 5.1): $f(a) = f(b)$; here there are three horizontal tangent lines

The Mean Value Theorem, which is commonly attributed to French mathematician Joseph-Louis Lagrange (1736–1813), extends Rolle's Theorem to allow $f(a) \neq f(b)$. It is equivalent to Rolle's Theorem but since $f(a) = f(b)$ is generally not the case, it gets used most often in this form. Our proof follows [43]. This proof is a little different from the one found in most textbooks, which we leave for Exercise 5.9.

Theorem 5.2. (Mean Value Theorem) *Let f be continuous on $[a,b]$ and differentiable on (a,b). Then there exists $c \in (a,b)$ such that*

$$f'(c) = \frac{f(b) - f(a)}{b-a}.$$

Proof. The equation of the line L through the origin $(0,0)$ which is parallel to the line through $(a, f(a))$ and $(b, f(b))$ is given by

$$y = \frac{f(b) - f(a)}{b-a}x.$$

See **Fig. 5.2.** Therefore the vertical displacement between $f(x)$ and L is given by the function

$$h(x) = f(x) - \frac{f(b) - f(a)}{b-a}x.$$

It is clear from Fig. 5.2 that $h(a) = h(b)$, or the reader may verify directly that

$$h(a) = h(b) = \frac{bf(a) - af(b)}{b-a}.$$

Now h is continuous on $[a, b]$ and differentiable on (a, b), so by Rolle's Theorem there is $c \in (a, b)$ for which $h'(c) = 0$. Finally,

$$h'(x) = f'(x) - \frac{f(b) - f(a)}{b - a},$$

and so $h'(c) = 0$ gives $f'(c) = \frac{f(b)-f(a)}{b-a}$, as desired. □

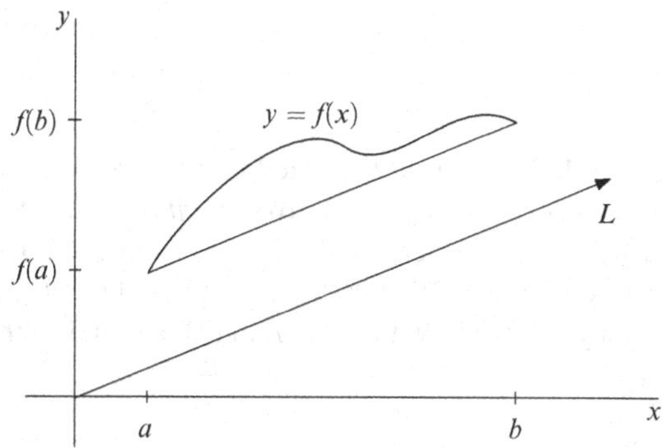

Fig. 5.2 The proof of the Mean Value Theorem (Theorem 5.2): Line L is parallel to the line through $(a, f(a))$ and $(b, f(b))$

Remark 5.3. The quotient $\frac{bf(a)-af(b)}{b-a}$ is the y-intercept of the line through the points $(a, f(a))$ and $(b, f(b))$. We shall meet this quotient again in Sect. 7.3. ∘

Remark 5.4. We outlined just above a proof of Rolle's Theorem (Theorem 5.1) which uses the Half-Chord Lemma (Lemma 3.12) and the Nested Interval Property (Theorem 1.41). In [16], this idea is eloquently modified to prove the Mean Value Theorem (Theorem 5.2). ∘

The Mean Value Theorem (Theorem 5.2) says that between points $(a, f(a))$ and $(b, f(b))$, the graph of a differentiable function must have at least one place $(c, f(c))$ at which the tangent line is parallel to the line through the points $(a, f(a))$ and $(b, f(b))$. See **Fig. 5.3.**

Suppose that over some journey, a car has some particular average speed. Then by the Mean Value Theorem (Theorem 5.2) there must have been an instant during the journey at which the car was travelling at *precisely* that average speed. (There is a rumor that if someone arrives in their car at a toll booth too soon after leaving a previous toll booth, then they could get a ticket for speeding.) But it is not really the full Mean Value Theorem that is required here because a car travels with a continuous position function certainly, but it travels with continuous speed as well. The Mean Value Theorem only requires that the speed function *exists*. It is an interesting fact that there seems to be no simpler proof of the Mean Value Theorem assuming also that f' is continuous—even though this is the context in which it is usually applied [24].

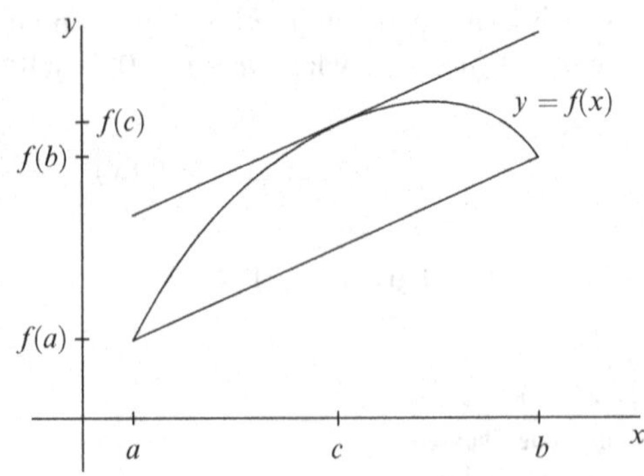

Fig. 5.3 The Mean Value Theorem (Theorem 5.2): There is least one place $(c, f(c))$ at which the tangent line is parallel to the line through $(a, f(a))$ and $(b, f(b))$ (In Fig. 5.2 there are three such places)

Example 5.5. Recall the Fixed Point Lemma (Lemma 3.10): *If* $f : [a,b] \to [a,b]$ *is continuous, then* f *has a fixed point in* $[a,b]$. If we know also that $f'(x) < 1$ for each $x \in [a,b]$, then the fixed point is unique. (This is reasonable upon drawing a picture.) Here's why: If there are two fixed points, say $f(x_0) = x_0$ and $f(y_0) = y_0$, then by the Mean Value Theorem (Theorem 5.2) there is c between x_0 and y_0 such that

$$f'(c) = \frac{f(x_0) - f(y_0)}{x_0 - y_0}.$$

But this reads

$$f'(c) = \frac{x_0 - y_0}{x_0 - y_0} = 1,$$

which contradicts $f' < 1$. So we must have $x_0 = y_0$. ◇

It may seem that we could prove the Mean Value Theorem (Theorem 5.2) by suitably rotating the x- and y-axes, to get $f(a) = f(b)$, and then applying Rolle's Theorem (Theorem 5.1). But the function $f(x) = x^3 - x$, for example, shows that this idea does not work. See **Fig. 5.4**. Here, $f(-2) = -6$ and $f(2) = 6$. If we rotate the axes so that the line through $(-2, -6)$ and $(2, 6)$ is the new x-axis, then the image of the graph of f under this rotation is not a function. Indeed, the new y-axis (which is the old $y = -x/3$ line) intersects the graph of f *three* times. See [17,51].

5.2 Applications

Many of the most basic results in calculus follow from the Mean Value Theorem (Theorem 5.2). A function f defined on (a,b) is **increasing** if $f(x_1) \leq f(x_2)$ whenever $a < x_1 < x_2 < b$, and f is **decreasing** on (a,b) if $f(x_1) \geq f(x_2)$

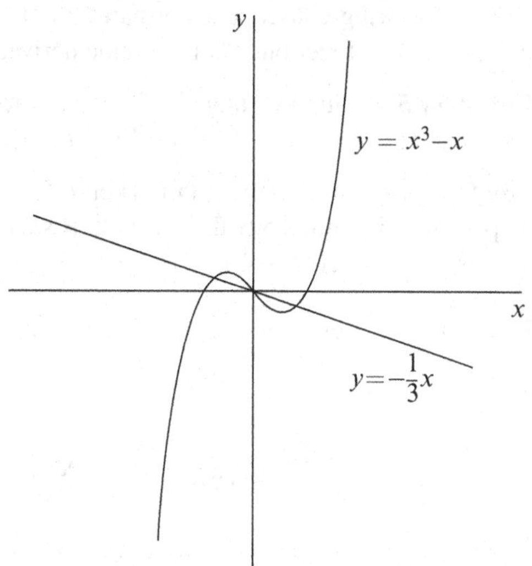

Fig. 5.4 The function $f(x) = x^3 - x$ shows that we cannot simply rotate axes then apply Rolle's Theorem, to prove the Mean Value Theorem (Theorem 5.2)

whenever $a < x_1 < x_2 < b$. (So f is decreasing on $(a, b) \Leftrightarrow -f$ is increasing on (a, b).) The function f being **strictly increasing** means that the $\leq$ above can be replaced with $<$, and the function f being **strictly decreasing** means that the $\geq$ above can be replaced with $>$.

Lemma 5.6. *Suppose that f is differentiable on (a, b).*

(i) *If $f' \geq 0$ on (a, b) then f is increasing on (a, b).*
(ii) *If $f' \leq 0$ on (a, b) then f is decreasing on (a, b).*
(iii) *If $f'(x) = 0$ for every $x \in (a, b)$, then f is constant on (a, b).*

Proof. We let $a < x_1 < x_2 < b$ and apply the Mean Value Theorem (Theorem 5.2) to f on $[x_1, x_2]$. Then there is $c \in (x_1, x_2)$ such that

$$f'(c) = \frac{f(x_2) - f(x_1)}{x_2 - x_1}.$$

That is,

$$f(x_2) - f(x_1) = f'(c)(x_2 - x_1).$$

Now for (i), if $f' \geq 0$ then the right-hand side is ≥ 0, and so the left-hand side is ≥ 0. That is, $f(x_2) \geq f(x_1)$.
For (ii), if $f' \leq 0$ then the right-hand side is ≤ 0, and so the left-hand side is ≤ 0. That is, $f(x_2) \leq f(x_1)$.
For (iii), we must have $f'(c) = 0$ and so the right-hand side is $= 0$. Therefore $f(x_1) = f(x_2)$. This is true for *any* choice of $x_1 < x_2$ in (a, b) and so f must be constant. □

The following consequence of part (iii) of Lemma 5.6 is particularly important. It says that two functions with the same derivative must differ by a constant.

Corollary 5.7. *Suppose that $f'(x) = g'(x)$ for all $x \in (a, b)$. Then there is $C \in \mathbf{R}$ such that $f(x) = g(x) + C$ for all $x \in (a, b)$.*

Proof. Set $h(x) = f(x) - g(x)$. Then $h'(x) = 0$ for every $x \in (a, b)$. Therefore by part (iii) of Lemma 5.6, there is $C \in \mathbf{R}$ such that $h(x) = C$ for every $x \in (a, b)$. That is, $f(x) = g(x) + C$. □

Example 5.8. Let $a_1, a_2, \ldots, a_n$, $b_1, b_2, \ldots, b_n \in \mathbf{R}$, and let $A = \frac{1}{n}\sum_{j=1}^{n} a_j$. For $x \in \mathbf{R}$, the identity

$$\sum_{j=1}^{n}(a_j - A)(b_j - x) = \sum_{j=1}^{n} a_j b_j - \frac{1}{n}\sum_{j=1}^{n} a_j \sum_{j=1}^{n} b_j$$

can be verified by expanding the left-hand side then tidying up. But here is an easier way: The derivative of the left-hand side with respect to x is

$$-\sum_{j=1}^{n}(a_j - A) = -\sum_{j=1}^{n} a_j + \sum_{j=1}^{n} A = 0.$$

The derivative of the right-hand side with respect to x is zero, since x does not appear there. By Corollary 5.7 then, the left-hand and right-hand sides differ by a constant. Setting $x = 0$ reveals that the constant is zero, as we wanted to show. To illustrate one instance in which the identity can be used, we take $b_j = a_j$ for each j. Then

$$0 \le \sum_{j=1}^{n}(a_j - A)^2 = \sum_{j=1}^{n} a_j^2 - \frac{1}{n}\left(\sum_{j=1}^{n} a_j\right)^2.$$

That is,

$$\left(\frac{1}{n}\sum_{j=1}^{n} a_j\right)^2 \le \frac{1}{n}\sum_{j=1}^{n} a_j^2.$$

This can also be obtained from the Cauchy-Schwarz Inequality (Theorem 2.18). We leave this for the reader to verify; see also Exercise 2.40. ◇

With parts (i) and (ii) of Lemma 5.6, we are better equipped to handle many extrema problems. But we continue to rely on Fermat's Theorem (Theorem 4.13) which says that we should look for extrema of a function f defined on (a, b) at the critical points for f. That is, at $c \in (a, b)$ for which either $f'(c) = 0$ or $f'(c)$ does not exist.

Example 5.9. Consider $f(x) = x^{7/3} + x^{4/3} - x^{1/3}$, for $x \in \mathbf{R}$. Then

$$f'(x) = \frac{7}{3}x^{4/3} + \frac{4}{3}x^{1/3} - \frac{1}{3}x^{-2/3} = \frac{7x^2 + 4x - 1}{3x^{2/3}}.$$

So (using the quadratic formula for c_1 and c_3) f has critical points

$$c_1 = \frac{-2 - \sqrt{11}}{7} \cong -0.76, \quad \text{where } f'(c_1) = 0,$$

$$c_2 = 0, \quad \text{where } f'(c_2) \text{ does not exist, and}$$

$$c_3 = \frac{-2 + \sqrt{11}}{7} \cong 0.188, \quad \text{where } f'(c_3) = 0.$$

By Fermat's Theorem, these are the only places where local extrema for f can occur. Now observe that:

On $(-\infty, c_1)$, f' is > 0 (e.g., $f'(-100) > 0$), so f is increasing there.
On (c_1, c_2), f' is < 0 (e.g., $f'(-0.1) < 0$), so f is decreasing there.
On (c_2, c_3), f' is < 0 (e.g., $f'(0.1) < 0$), so f is decreasing there.
On $(c_3, +\infty)$, f' is > 0 (e.g., $f'(100) > 0$), so f is increasing there.

Therefore f has a local maximum value of $f(c_1) \cong 1.079$, and a local minimum value of $f(c_3) \cong -0.445$. Neither of these extrema is absolute; f has no absolute extrema. The graph $y = f(x)$ is shown in **Fig. 5.5**. ◇

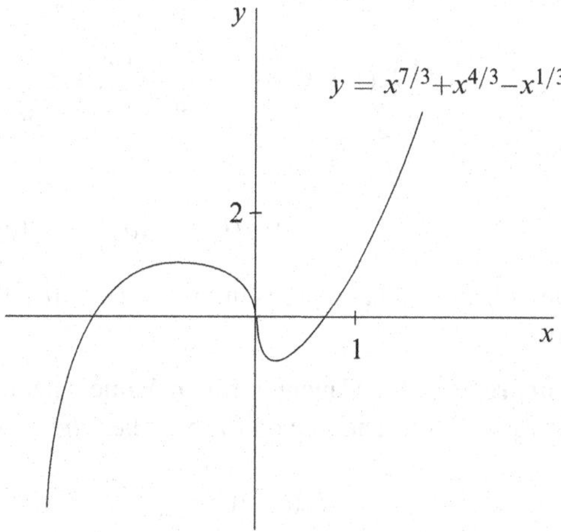

Fig. 5.5 For Example 5.9. The graph of $f(x) = x^{7/3} + x^{4/3} - x^{1/3}$

Example 5.10. Recall Lemma 2.7: *If a and b are nonnegative real numbers then*

$$\sqrt{ab} \le \frac{a+b}{2},$$

and equality occurs if and only if a=b. Here we prove this using calculus (though this is rather like killing a mosquito with a bazooka). Fix $a \ge 0$ and for $x \in [0, \infty)$, consider the function

$$f(x) = \frac{a+x}{2} - \sqrt{ax}.$$

Here,

$$f'(x) = \frac{1}{2} - \frac{\sqrt{a}}{2\sqrt{x}}.$$

So the only critical point for f is $x = a$, at which $f'(a) = 0$. Now observe that:

On $(0, a)$, f' is < 0 (e.g., $f'(a/4) < 0$), so f is decreasing there.
On $(a, +\infty)$, f' is > 0 (e.g., $f'(4a) > 0$), so f is increasing there.

Therefore f has an absolute minimum value of $f(a) = 0$, and so

$$f(x) = \frac{a+x}{2} - \sqrt{ax} > 0 \quad \Leftrightarrow \quad x \ne a. \qquad \diamond$$

5.3 Cauchy's Mean Value Theorem

If we apply the Mean Value Theorem (Theorem 5.2) to each of two functions f and g continuous on $[a, b]$ and differentiable on (a, b), we can conclude that there are $c_1, c_2 \in (a, b)$ such that

$$f'(c_1) = \frac{f(b) - f(a)}{b - a} \quad \text{and} \quad g'(c_2) = \frac{g(b) - g(a)}{b - a}.$$

This gives

$$f'(c_1)\big[g(b) - g(a)\big] = g'(c_2)\big[f(b) - f(a)\big].$$

But it happens that there is in fact *one* $c \in (a, b)$ which works for *both* functions, as follows.

Theorem 5.11. (Cauchy's Mean Value Theorem) *Let f and g be continuous on $[a, b]$ and differentiable on (a, b). Then there exists $c \in (a, b)$ such that*

$$f'(c)[g(b) - g(a)] = g'(c)[f(b) - f(a)].$$

Proof. If $g(a) = g(b)$ then the result holds, by Rolle's Theorem (Theorem 5.1) applied to g. So we assume that $g(a) \neq g(b)$. In our proof of the Mean Value Theorem (Theorem 5.2) we used the auxiliary function

$$h(x) = f(x) - \frac{f(b) - f(a)}{b - a} x \,.$$

The idea here is to replace the identity function x above with $g(x)$. So we consider

$$h(x) = f(x) - \frac{f(b) - f(a)}{g(b) - g(a)} g(x).$$

Then h is continuous on $[a, b]$ and differentiable on (a, b). And one can easily verify that $h(a) = h(b)$. So by Rolle's Theorem (Theorem 5.1) there is $c \in (a, b)$ such that $h'(c) = 0$. Finally, $h'(c) = 0$ gives $f'(c)[g(b) - g(a)] = g'(c)[f(b) - f(a)]$, as desired. □

Evidently, if $g(x) = x$ in Cauchy's Mean Value Theorem (Theorem 5.11) then we get simply the Mean Value Theorem. As long as $g(b) \neq g(a)$, the conclusion of Cauchy's Mean Value Theorem is usually written

$$\frac{f'(c)}{g'(c)} = \frac{f(b) - f(a)}{g(b) - g(a)}.$$

Remark 5.12. We saw the geometrical interpretation of the Mean Value Theorem (Theorem 5.2) in Fig. 5.3. Cauchy's Mean Value Theorem also has a geometrical interpretation, though perhaps not so obvious (e.g., [17,35]). If $P(t) = (g(t), f(t))$ is a point in the xy-plane which depends on t, then $f'(t)/g'(t)$, when it exists, is the slope of the tangent line to the curve that $P(t)$ traces as t varies in $[a, b]$. What Cauchy's Mean Value Theorem says is that as long as $g' \neq 0$, there is at least one place on the curve at which the tangent line is parallel to the line through $(g(a), f(a))$ and $(g(b), f(b))$. See **Fig. 5.6**. If $g(t) = t$ then we recover the Mean Value Theorem (Theorem 5.2) and its geometric interpretation. ∘

Probably the best known application of Cauchy's Mean Value Theorem (Theorem 5.11) is L'Hospital's Rule, as follows. (But we shall see others.)

Theorem 5.13. (L'Hospital's Rule) *Let f and g be continuous on $[a, b]$ and differentiable on (a, b) and let $x_0 \in (a, b)$. Let $f(x_0) = g(x_0) = 0$, but suppose that $g \neq 0$ at all other points of (a, b). Suppose also that $g' \neq 0$ on (a, b). Then*

$$\lim_{x \to x_0} \frac{f'(x)}{g'(x)} = L \quad \Rightarrow \quad \lim_{x \to x_0} \frac{f(x)}{g(x)} = L \,.$$

Proof. Since $f(x_0) = g(x_0) = 0$, for $x \neq x_0$ we may write

$$\frac{f(x)}{g(x)} = \frac{f(x) - f(x_0)}{g(x) - g(x_0)}.$$

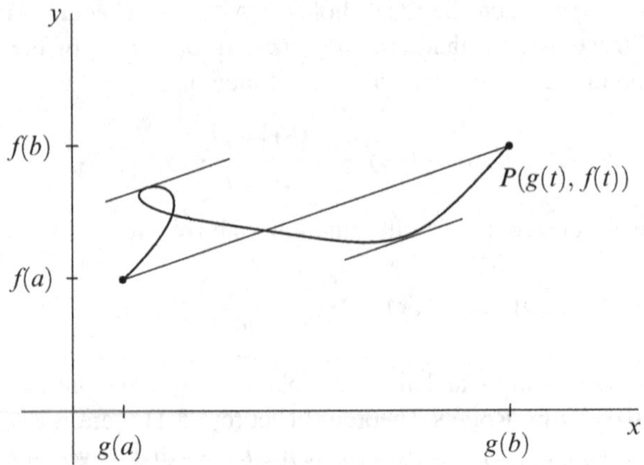

Fig. 5.6 Cauchy's Mean Value Theorem (Theorem 5.11): There is at least one place on the curve at which the tangent line is parallel to the line through $(g(a), f(a))$ and $(g(b), f(b))$. In this picture, there are two such places

Then by Cauchy's Mean Value Theorem (Theorem 5.11) there is c between x and x_0 such that

$$\frac{f(x)}{g(x)} = \frac{f(x) - f(x_0)}{g(x) - g(x_0)} = \frac{f'(c)}{g'(c)}.$$

Now as $x \to x_0$, we must also have $c \to x_0$ because c is between x and x_0. So if

$$\lim_{x \to x_0} \frac{f'(x)}{g'(x)} = L,$$

then

$$\lim_{x \to x_0} \frac{f'(c)}{g'(c)} = L.$$

Therefore

$$\lim_{x \to x_0} \frac{f(x)}{g(x)} \to L,$$

as desired. □

Example 5.14. By L'Hospital's Rule applied twice,

$$\lim_{x \to 0} \left(\frac{1}{x} - \frac{1}{\sin(x)} \right) = \lim_{x \to 0} \frac{\sin(x) - x}{x \sin(x)} = \lim_{x \to 0} \frac{\cos(x) - 1}{\sin(x) + x\cos(x)}$$
$$= \lim_{x \to 0} \frac{-\sin(x)}{\cos(x) + \cos(x) - x\sin(x)} = 0.$$

As regards the statement of L'Hospital's Rule, we point out that the second and third equals signs above are not really justified until the fourth one has been reached. ◇

In Exercises 5.43 and 5.44 we consider the roles that some of the various hypotheses in L'Hospital's Rule play. There are also many variants of L'Hospital's Rule, a few of which we explore in Exercises 5.45 and 5.46.

Remark 5.15. L'Hospital's Rule is in fact due to Swiss mathematician Johann Bernoulli (1667–1748). Guillaume de L'Hospital (1661–1704) was a French student of Bernoulli's who, with permission, published notes from his teacher's lectures in 1696. This was the first-ever calculus textbook. ∘

Remark 5.16. [9] There is some disagreement among historians of mathematics on the spelling of L'Hospital's name. He himself spelled it, at times, Lhospital. That is, without the apostrophe and with a lower case h. R.P. Boas Jr. used this spelling on occasion (e.g., [5]). On the cover of the 1696 calculus book, it is spelled l'Hospital. The official French national bibliographic entry is L'Hospital, which is what most historians choose. ∘

Exercises

5.1. **(a)** Show that a polynomial of degree n cannot have more than n real zeros.
(b) Show that if a polynomial p has n distinct real zeros then p' has $n-1$ distinct real zeros.

5.2. [30] Let p be a cubic polynomial with real zeros $a_1 < a_2 < a_3$.

(a) Show that p has a critical point c, with $a_1 < c < a_2$.
(b) Show that c is closer to a_1 than to a_2.

5.3. **(a)** Fill in the details of the proof of Rolle's Theorem (Theorem 5.1) outlined in Sect. 5.1, which uses the Half-Chord Lemma (Lemma 3.12) and the Nested Interval Property of **R** (Theorem 1.41).
(b) Is the "$c \in (a,b)$" from our proof of Rolle's Theorem (Theorem 5.1) necessarily the same as the "$c \in (a,b)$" from the proof in (a) ? Explain.

5.4. [54]

(a) Let f be continuous on $[a,b]$ and differentiable on (a,b) such that $f(a) = f(b) = 0$. Show that there is $c \in (a,b)$ such that $f'(c) = f(c)$.
Hint: Consider $g(x) = \mathrm{e}^{-x} f(x)$.
(b) Interpret the result in (a) geometrically.

5.5. [25]

(a) Let f be continuous on $[a,b]$ and differentiable on (a,b) such that $f(a) = f(b) = 0$, but f is not the zero function. Show that for any real number $r \neq 0$ there is $c \in (a,b)$ such that $rf'(c) + f(c) = 0$. Hint: Consider $g(x) = \mathrm{e}^{x/r} f(x)$.
(b) Interpret the result in (a) geometrically.

5.6. [50]

(a) Let f be continuous on $[a, b]$ and differentiable on (a, b). Set

$$F(x) = f(x)(x - a)(x - b)$$

to show that there is $c \in (a, b)$ such that

$$f'(c) + \frac{f(c)}{c - a} + \frac{f(c)}{c - b} = 0\,.$$

(b) Interpret the result in (a) geometrically.

5.7. [48] Suppose that f is a quadratic, say $f(x) = a_2x^2 + a_1x + a_0$. Show that the point c as given by the Mean Value Theorem (Theorem 5.2) applied to f on $[a, b]$ is the midpoint $(a + b)/2$. (This property in fact characterizes parabolas [13,40].)

5.8. Let f be continuous on $[a, b]$ and differentiable on (a, b). Is it always possible, for each $c \in (a, b)$, to find $x_1, x_2 \in [a, b]$ with $x_1 < x_2$ such that

$$\frac{f(x_2) - f(x_1)}{x_2 - x_1} = f'(c)\,?$$

5.9. The proof of the Mean Value Theorem (Theorem 5.2) given in the text is slightly different from the one given in most textbooks. Typically, one uses the auxiliary function

$$h(x) = f(x) - f(a) - \frac{f(b) - f(a)}{b - a}(x - a).$$

(a) Prove the Mean Value Theorem using this h.
(b) What is the significance of this h, geometrically?
(c) Is the "$c \in (a, b)$" from the proof supplied in the text necessarily the same as the "$c \in (a, b)$" from the proof in (a)?

5.10. [53] Apply Rolle's Theorem (Theorem 5.1) to

$$g(x) = (f(x) - f(a))(x - b) - (f(x) - f(b))(x - a)$$

to obtain another proof of the Mean Value Theorem (Theorem 5.2). The function $g(x)$ is $\pm$ twice the area of the triangle determined by the points $(a, f(a))$, $(x, f(x))$, and $(b, f(b))$. See Exercise 1.12.

5.11. Show that the conclusion of the Mean Value Theorem (Theorem 5.2) can be written

$$f'(a + t(b - a)) = \frac{f(b) - f(a)}{b - a},$$

for some $t \in (0, 1)$.

5.12. [11,33] Let f be differentiable on $[0,1]$ with $f(0) = f(1) = 0$.

(a) For a given positive integer n, show that there are distinct $x_1, x_2, \dots, x_n$ such that

$$\sum_{j=1}^{n} \frac{1}{f'(x_j)} = n.$$

(b) For a given positive integer n and positive numbers $k_1, k_2, \dots, k_n$ show that there are distinct $x_1, x_2, \dots, x_n$ such that

$$\sum_{j=1}^{n} \frac{k_j}{f'(x_j)} = \sum_{j=1}^{n} k_j.$$

5.13. [52] Let $b_1, b_2, \dots, b_n$ be nonnegative integers such that $\prod_{j=1}^{n} (1 + b_j) > 2^t$. Show that $\sum_{j=1}^{n} b_j > t$. Hint: Apply the Mean Value Theorem to $\ln(1 + x)$ on $[1, b_j]$.

5.14. [27] Prove the following converse to the Mean Value Theorem (Theorem 5.2). Let F and f be defined on (a,b) and let f be continuous there. Suppose that for every $x, y \in (a,b)$ there is c between x and y such that

$$F(x) - F(y) = (x - y) f(c).$$

Show that F is differentiable on (a,b), and that f is its derivative.

5.15. [29] Let $a, b > 0$ and $x > 0$. Prove, as follows, that

$$\left(\frac{a+x}{b+x}\right)^{b+x} > \left(\frac{a}{b}\right)^{b}.$$

(a) But first, set $x = 1$, $a = n+1$, and $b = n$ here to show (again) that $\left\{ \left(1 + \frac{1}{n}\right)^n \right\}$ is an increasing sequence.
Now set $f(x) = \left(\frac{a+x}{b+x}\right)^{b+x}$.
(b) Show that $f'(x) = f(x) \left(\frac{b-a}{a+x} + \ln\left(\frac{a+x}{b+x}\right)\right)$.
(c) Show that $g'(x) < 0$, where $g(x) = \frac{b-a}{a+x} + \ln\left(\frac{a+x}{b+x}\right)$.
(d) Conclude that f is strictly increasing and so $\left(\frac{a+x}{b+x}\right)^{b+x} > \left(\frac{a}{b}\right)^{b}$.

5.16. Suppose that f and g are continuous on $[a,b]$ and differentiable on (a,b), with $f(a) = g(a)$ and $f(b) = g(b)$.

(a) Show that there exists $c \in (a,b)$ such that $f'(c) = g'(c)$.
(b) Explain how this is a generalization of the Mean Value Theorem (Theorem 5.2).

5.17. Prove part (iii) of Lemma 5.6 by going at it in the contrapositive direction. That is, show that if f is not constant then there is a $c \in (a,b)$ for which $f'(c) \neq 0$. (For another interesting proof, which uses a bisection algorithm and not the Mean Value Theorem, see [39].)

5.18. Suppose that f is differentiable on (a,b) with $f'(x) \neq 0$ for every $x \in (a,b)$. Prove that f is one-to-one on (a,b).

5.19. [36] Have a look again at Example 5.8. Under what conditions do the real numbers $a_1, a_2, \ldots, a_n$ satisfy

$$\sum_{j=1}^{n} a_j^2 = \frac{1}{n}\left(\sum_{j=1}^{n} a_j\right)^2 ?$$

5.20. [1] Here's a slick way of verifying the trigonometric identities

$$\sin(x+y) = \sin(x)\cos(y) + \sin(y)\cos(x) \quad \text{and}$$
$$\cos(x+y) = \cos(x)\cos(y) - \sin(x)\sin(y).$$

(a) Fix y and set

$$u(x) = \sin(x+y) - \big[\sin(x)\cos(y) + \sin(y)\cos(x)\big],$$
$$v(x) = \cos(x+y) - \big[\cos(x)\cos(y) + \sin(x)\sin(y)\big],$$

then verify that $u'(x) = v(x)$ and $v'(x) = -u(x)$.

(b) For $f(x) = [u(x)]^2 + [v(x)]^2$, show that $f' \equiv 0$.

(c) Therefore f is constant. Find the constant, then conclude that $u \equiv v \equiv 0$.

5.21. [21,28]

(a) Show that $2\big(\sin^6(x) + \cos^6(x)\big) - 3\big(\sin^4(x) + \cos^4(x)\big)$ is constant.

(b) Show that $\big(\tan(x)+1\big)^2 + \big(\cot(x)+1\big)^2 - \big(\sec(x)+\csc(x)\big)^2$ is constant.

(c) Show that $\cos(x)\cos(x+2) - \big(\cos(x+1)\big)^2$ is constant.

(d) Find the constant in each of (a), (b), and (c).

5.22. Let $a_1, a_2, \ldots, a_n$ be real numbers, $A = \frac{1}{n}\sum_{j=1}^{n} a_j$, and let x be any real number.

(a) Show that

$$\sum_{j=1}^{n}\left(a_j^2 - A^2\right) = \sum_{j=1}^{n}\left((a_j - x)^2 - (A-x)^2\right).$$

(b) Conclude that $\sum_{j=1}^{n}(a_j^2 - A^2) = \sum_{j=1}^{n}(a_j - A)^2$.

(c) Conclude that

$$\sum_{j=1}^{n}(a_j - x)^2 \geq \sum_{j=1}^{n}(a_j - A)^2.$$

5.23. Let f be differentiable on (a, b) and suppose that there are $m, M \in \mathbf{R}$ such that $m \leq f'(x) \leq M$ for every $x \in (a, b)$. Show that for all $x, y \in (a, b)$,

$$m\,|x - y| \leq |f(x) - f(y)| \leq M\,|x - y|.$$

5.24. Find any absolute extrema for the following functions.
(a) $f(x) = x^{4/3} + x^{1/3}$, **(b)** $g(x) = x^{1/3}(x+1)^{-2/3}$,
(c) $h(x) = 2x/(x^2+1)$, **(d)** $k(x) = x/(x^2+8)^2$.

5.25. [38] Let f be differentiable on $(0, \infty)$. Show that if $f'(x) < f(x)/x$ on $(0, \infty)$ then $f(x)/x$ is decreasing.

5.26. [45] Let $0 < a < b$ and for $x \in [\frac{-4a}{b-a}, \frac{4b}{b-a}]$, let

$$M(x) = \frac{1}{2}\sqrt{x(b-a)^2 + 4ab}.$$

(a) Show that $f(x) = M(x)^2$ is increasing and conclude that $M(x)$ is increasing.
(b) Evaluate

$$M\left(\tfrac{-4ab}{(a+b)^2}\right),\quad M(0),\quad M(1),\quad \text{and} \quad M(2).$$

(c) Conclude that $H < G < A < R$, where H is the Harmonic Mean, G is the Geometric Mean, A is the Arithmetic Mean, and R is the Root Mean Square, of a, b.

5.27. [47] Let $0 < a < b$. Show that $h(x) = \dfrac{\ln(ax+1)}{\ln(bx+1)}$ is increasing for $x > 0$.

5.28. [44] Show that for $x > 0$,

$$\arctan(x) > \frac{3x}{1 + 2\sqrt{1+x^2}}.$$

5.29. [12] Let $x \in (0, 1)$.

(a) Show that $\tan^2(x) - 4\tan(x) + 4x > 0$.
(b) Conclude that

$$\frac{1 + \sqrt{1-x}}{2x} < \cot(x).$$

5.30. **(a)** Show that the equation $x^7 + x^3 + x + 2 = 0$ has exactly one solution.
(b) Show that the equation $x^3 - x - \cos(x) = 0$ has exactly one solution in $[0, \pi]$.

5.31. [31] Let $x, y > 0$. Prove, as follows, that

$$x^y y^x \leq \left(\frac{x+y}{2}\right)^{x+y} \leq x^x y^y,$$

with equality if and only if $x = y$. For $0 < t < 1$, set

$$f(t) = (tx + (1-t)y)^x (ty + (1-t)x)^y,$$

and consider f'. This in fact proves a bit more - explain.

5.32. [42] Let $0 < a < b$. Apply the Mean Value Theorem (Theorem 5.2), on $[a, b]$, to the function

$$f(x) = x\ln(x) - x$$

to show that

$$\left(\frac{a}{b}\right)^b < \frac{\mathrm{e}^a}{\mathrm{e}^b} < \left(\frac{a}{b}\right)^a.$$

5.33. [46] Show that $x = 2$ is the only $x > -1$ for which

$$(1+x)^x + (2+x)^x = (3+x)^x.$$

5.34. [14,20] Let n be a positive integer.

(a) Find $a > 1$ such that $a^x = x^n$ has exactly one positive solution.
(b) Find the solution in (a). [Answer: $a = \mathrm{e}^{n/\mathrm{e}}$; solution is $x = \mathrm{e}$.]

5.35. [19,37] Denote by ψ the positive root of $x^2+x-1 = 0$. This is the reciprocal of the golden mean φ (the positive root of $x^2-x-1 = 0$), which we met in Sect. 1.5 and in Exercises 1.40 and 1.42. Let $0 \leq a \leq b$. Prove **Dalzell's Theorem**:

$$\left|\frac{b}{a+b} - \psi\right| \leq \left|\frac{a}{b} - \psi\right|.$$

This result implies the well-known fact that for any Fibonacci-like sequence $f_1 = a$, $f_2 = b$, $f_{n+2} = f_n + f_{n+1}$, it is the case that $f_n/f_{n+1} \to \psi$. (And so $f_{n+1}/f_n \to \varphi$.) The classic Fibonacci sequence has $a = b = 1$.

5.36. Let f be continuous on $\mathbf{R}$. Let $x_0 \in \mathbf{R}$ and consider the **fixed point iteration scheme**:

$$x_n = f(x_{n-1}) \quad \text{for } n = 1, 2, 3, \ldots.$$

(a) Show that *if* it is the case that the sequence $\{x_n\}$ converges to some number p, then p is necessarily a fixed point for f.

(b) Now suppose that $f : [a,b] \to [a,b]$ has a continuous derivative. Let $0 < k < 1$ and suppose that $|f'(x)| \le k$ for all $x \in (a,b)$. Show that for any $x_0 \in [a,b]$, the iteration scheme defined in (a) converges.

(c) Verify the hypotheses in (b), for $f(x) = \frac{1}{1+x}$ on $[0,1]$. To what number does the fixed point iteration scheme converge in this case? (Take $x_0 = 0$, say.)

5.37. [22] Let $x, y, z \ge 0$. **Schur's Inequality** is: *For any* $\lambda \in \mathbf{R}$,

$$(x-y)(x-z)x^\lambda + (y-x)(y-z)y^\lambda + (z-x)(z-y)z^\lambda \ge 0,$$

with equality if and only if $x = y = z = 0$. Prove Schur's Inequality, as follows.

(a) Show that we may assume, with no loss of generality, that $x < y < z$ and $\lambda > 0$.

(b) For such x, y, z and λ, apply the Mean Value Theorem (Theorem 5.2) to the function $f(t) = (t-x)t^\lambda$ on $[y,z]$.

5.38. Try proving L'Hospital's Rule using only the Mean Value Theorem (Theorem 5.2), i.e., not using Cauchy's Mean Value Theorem (Theorem 5.11). What goes wrong?

5.39. [15]

(a) Verify that for $x \ne 1$,

$$\sum_{k=0}^{n} x^k = \frac{x^{n+1}-1}{x-1}.$$

(b) Differentiate both sides of the equation in (a) with respect to x, then let $x \to 1$ to obtain the familiar formula

$$\sum_{k=1}^{n} k = \frac{n(n+1)}{2}.$$

(c) For the intrepid reader: Use similar reasoning to show that

$$\sum_{k=1}^{n} k^2 = \frac{n(n+1)(2n+1)}{6} \quad \text{and} \quad \sum_{k=1}^{n} k^3 = \frac{n^2(n+1)^2}{4}.$$

5.40. [32] Suppose that the function f is such that f' is defined at c, and that $\lim_{x\to c} f'(x)$ exists. Show that f is continuous at c.

5.41. Evaluate the following limits.

(a) $\lim_{x\to 0} \frac{\sin(5x)}{x}$, **(b)** $\lim_{x\to 0} \frac{x^2-\sin^2(x)}{x^2\sin^2(x)}$, **(c)** $\lim_{x\to 0} \frac{x-\tan x}{x-\sin x}$, **(d)** $\lim_{x\to\pi} \frac{1+\cos(x)}{(x-\pi)^2}$.

5.42. [18] Evaluate $\lim_{x\to 0+} \dfrac{x}{\sqrt{1-e^{-x^2}}}$. Hint: Consider $\lim_{x\to 0+} \dfrac{x^2}{1-e^{-x^2}}$.

5.43. [10]

(a) Let $f(x) = x^2 \sin(1/x)$ [with $f(0) = 0$] and $g(x) = \sin x$. Show that $\lim_{x\to 0} \dfrac{f'(x)}{g'(x)}$ does not exist, but $\lim_{x\to 0} \dfrac{f(x)}{g(x)} = 0$.

(b) Let $f(x) = x\sin(1/x)$ [with $f(0) = 0$] and $g(x) = \sin x$. Show that neither $\lim_{x\to 0} \dfrac{f'(x)}{g'(x)}$ nor $\lim_{x\to 0} \dfrac{f(x)}{g(x)}$ exists.

(c) Explain how the examples in (a) and (b) fit in with L'Hospital's Rule.

5.44. [6,8] Let $g(x) = 2/x + \sin(2/x)$ and $g(x) = 1/x \sin(1/x) + \cos(1/x)$.

(a) Show that $\lim_{x\to 0} \dfrac{f'(x)}{g'(x)} = 0$, but that $\lim_{x\to 0} \dfrac{f(x)}{g(x)}$ does not exist.

(b) Explain how this example fits in with L'Hospital's Rule.

5.45. [3]

(a) Define what we would mean by $\lim_{x\to x_0} f(x) = +\infty$ and $\lim_{x\to x_0} g(x) = -\infty$.

(b) Prove the following variant of L'Hospital's Rule. *Let f and g be differentiable on (a,b), except possibly at $x_0 \in (a,b)$. Suppose that $g'(x) \neq 0$ on (a,b), except possibly at x_0. Suppose that $\lim_{x\to x_0} f(x) = \lim_{x\to x_0} g(x) = \pm\infty$. Then*

$$\lim_{x\to x_0} \frac{f'(x)}{g'(x)} = L \quad \Rightarrow \quad \lim_{x\to x_0} \frac{f(x)}{g(x)} = L\,.$$

(c) Evaluate $\lim_{x\to 0+} \dfrac{\ln(\sin(x))}{1/x}$ and $\lim_{x\to 0+} \dfrac{1}{x(\ln(x))^2}$.

5.46. [3] For a function f defined on an interval $[a,\infty)$, $\lim_{x\to +\infty} f(x)$ means $\lim_{x\to 0+} f(1/x)$.

(a) State and prove a variant of L'Hospital's Rule which allows $x_0 = +\infty$.

(b) Evaluate the limit referred to in Remark 2.14 in the context of isoperimetric inequalities, namely

$$\lim_{n\to\infty} \left(n \tan(\tfrac{\pi}{n})\right).$$

(c) Evaluate $\lim_{x\to +\infty} \dfrac{\tan(1/x)}{1/x}$ and $\lim_{x\to +\infty} \dfrac{\arctan(4/x)}{\sin(3/x)}$.

(d) State and prove variants of L'Hospital's Rule as in (a) which also allow $L = \pm\infty$.

(e) Evaluate $\lim_{x\to +\infty} \dfrac{x}{\ln(x)}$ and $\lim_{x\to +\infty} \dfrac{e^x}{x^n}$.

5.47. [3] Evaluate the following limits.

(a) $\lim_{x\to 0+} x^x$, **(b)** $\lim_{x\to 0+} \left(1+\frac{1}{x}\right)^x$, **(c)** $\lim_{x\to +\infty} \left(1+\frac{1}{x}\right)^x$, **(d)** $\lim_{x\to +\infty} x^{1/x}$.

5.48. [26] Show that if

$$\lim_{x\to\infty} \left(f(x)+f'(x)\right) = a$$

then

$$\lim_{x\to\infty} f(x) = a \qquad \text{and} \qquad \lim_{x\to\infty} f'(x) = 0.$$

Hint: Write $f(x) = \mathrm{e}^x f(x)/\mathrm{e}^x$.

5.49. [23] Suppose that $x \in (0,1)$ and that m and n are integers with $m > n \geq 1$. Prove, as follows, that

$$(m+n)(1+x^m) \geq 2n\frac{1-x^{m+n}}{1-x^n}.$$

(a) Let

$$F(x) = \frac{(1-x^n)(1+x^m)}{1-x^{m+n}}.$$

Use Cauchy's Mean Value Theorem (Theorem 5.11) to show that there is $t \in (x,1)$ such that

$$F(x) = \frac{nt^{-m} + m + n - mt^{-n}}{m+n}.$$

(b) Now show that $nt^{-m} + m + n - mt^{-n} > 2n$. Hint: Show that

$$g(t) = (m-n)t^m - mt^{m-n} + n$$

is strictly decreasing on $(0,1)$ and observe that $g(1) = 0$ to conclude that $g(t) > 0$.

5.50. [34]

(a) Prove the following generalization of Cauchy's Mean Value Theorem (Theorem 5.11). *Let f, g, p, q be continuous on $[a,b]$ and differentiable on (a,b). Then there exist $\xi_1, \xi_2 \in (a,b)$ such that*

$$\frac{f(a)g(b)-f(b)g(a)}{p(a)q(b)-p(b)q(a)} = \frac{f(\xi_1)g'(\xi_2)-f'(\xi_2)g(\xi_1)}{p(\xi_1)q'(\xi_2)-p'(\xi_2)q(\xi_1)}.$$

(b) Verify that if $g = q = 1$ we get Cauchy's Mean Value Theorem, and if further $p(x) = x$, we get the Mean Value Theorem (Theorem 5.2).

5.51. [4,49]

(a) Let $\alpha, \beta \in \mathbf{R}$, with $\alpha + \beta = 1$. Apply Rolle's Theorem to

$$\begin{aligned} F(x) &= \alpha\big[f(b) - f(a)\big]\big[h(b) - h(a)\big]\big[g(x) - g(a)\big] \\ &\quad + \beta\big[f(b) - f(a)\big]\big[g(b) - g(a)\big]\big[h(x) - h(a)\big] \\ &\quad - \big[g(b) - g(a)\big]\big[h(b) - h(a)\big]\big[f(x) - f(a))\big] \end{aligned}$$

to prove the following generalization of Cauchy's Mean Value Theorem (Theorem 5.11). *Let $f, g,$ and h be continuous on $[a,b]$ and differentiable on (a,b). Then there is $c \in (a,b)$ such that*

$$\begin{aligned} f'(c)\big[g(b) - g(a)\big]\big[h(b) - h(a)\big] &= \alpha g'(c)\big[f(b) - f(a)\big]\big[h(b) - h(a)\big] \\ &\quad + \beta h'(c)\big[f(b) - f(a)\big]\big[g(b) - g(a)\big]. \end{aligned}$$

(b) Verify that if $\beta = 0$, we get Cauchy's Mean Value Theorem.

References

1. Apostol, T.M.: Calculus, vol. 1. Blaisdell Publishing Co., New York (1961)
2. Aziz, A.K., Diaz, J.B.: On Pompeiu's proof of the mean value theorem. Contrib. Differ. Equ. **1**, 467–471 (1963)
3. Bartle, R.G., Sherbert, D.R.: Introduction to Real Analysis, 2nd edn. Wiley, New York (1992)
4. Beck, R.: Letter to the editor. Coll. Math. J. **35**, 384 (2004)
5. Boas, R.P., Jr.: Lhospital's rule without mean-value theorems. Am. Math. Mon. **76**, 1051–1053 (1969)
6. Boas, R.P., Jr.: Indeterminate forms revisited. Math. Mag. **63**, 155–159 (1980)
7. Boas, R.P., Jr.: A Primer of Real Functions. Mathematical Association of America, Washington, DC (1981)
8. Boas, R.P., Jr.: Counterexamples to L'Hospital's rule. Am. Math. Mon. **93**, 644–645 (1986)
9. Bressoud, D.: The math forum at Drexel (2014). http://mathforum.org/
10. Bumcrot, R.J.: Some subtleties in L'Hospital's rule. Coll. Math. J. **15**, 51–52 (1984)
11. Chang, G.Z., Vowe, M.: Problem 1125. Math. Mag. **55**, 241 (1982)
12. Chao, W.W., Doster, D.: Problem 1477. Math. Mag. **69**, 227–228 (1996)
13. Chorlton, F.: A fixed feature of the mean value theorem. Math. Gazette **67**, 49–50 (1983)
14. Couch, E.: An overlooked calculus question. Coll. Math. J. **33**, 399–400 (2002)
15. Dunham, W.: When Euler met l'Hôpital. Math. Mag. **82**, 16–25 (2009)
16. Easton, R.J.: The sliding interval technique. Am. Math. Mon. **75**, 886–888 (1968)
17. Evans, J.P.: The extended law of the mean by a translation-rotation of axes. Am. Math. Mon. **67**, 580–581 (1960)
18. Feuerman, M., Miller, A.R.: Problem Q756. Math. Mag. **62**, 344, 349 (1989)
19. Fischler, R.: How to find the "golden number" without really trying. Fibonacci Q. **19**, 406–410 (1981)

20. Fisk, R.S.: Problem 168. Coll. Math. J. **12**, 343–344 (1981)
21. Hoyt, J.P.: Problem Q721. Math. Mag. **60**, 115, 120 (1987)
22. Ivády, P.: An application of the mean value theorem. Math. Gazette **67**, 126–127 (1983)
23. Klamkin, M.S., Zwier, P.: Problem E2483. Am. Math. Mon. **82**, 758–759 (1975)
24. Körner, T.W.: A Companion to Analysis. American Mathematical Society, Providence (2004)
25. Kung, S., Boas, R.P., Jr.: Problem 1274. Math. Mag. **61**, 263–264 (1988)
26. Landau, M.D., Jones, W.R.: A Hardy old problem. Math. Mag. **56**, 230–232 (1983)
27. Levi, H.: Integration, anti-differentiation and a converse to the mean value theorem. Am. Math. Mon. **74**, 585–586 (1967)
28. Lord, N.: David and Goliath proofs. Math. Gazette **75**, 194–196 (1991)
29. Mitrinovic, D.S.: Elementary Inequalities. P. Noordhoff Ltd., Groningen (1964)
30. Moran, D.A., Williams, R., Waterhouse, W.C.: Problem 459. Coll. Math. J. **23**, 345–346 (1992)
31. Murty, V.N., Deland, D.: Problem 123. Coll. Math. J. **11**, 164–165 (1980)
32. Noland, H.: Problem Q782. Math. Mag. **74**, 275, 281 (1991)
33. Orno, P., Clark, R.: Problem 1053. Math. Mag. **53**, 51 (1980)
34. Polya, G., Szegő, G.: Problems and Theorems in Analysis II. Springer, New York (2004)
35. Powderly, M.: A geometric proof of Cauchy's generalized law of the mean. Coll. Math. J. **11**, 329–330 (1980)
36. Purdy, G., Perry, G.M.: Problem E1775. Am. Math. Mon. **73**, 546 (1966)
37. Putz, J.F.: The golden section and the piano sonatas of Mozart. Math. Mag. **68**, 275–282 (1995)
38. Rathore, S.P.S.: On subadditive and superadditive functions. Am. Math. Mon. **72**, 653–654 (1965)
39. Richmond, D.E.: An elementary proof of a theorem in calculus. Am. Math. Mon. **92**, 589–590 (1985)
40. Richmond, M.B., Richmond, T.A.: How to recognize a parabola. Am. Math. Mon. **116**, 910–922 (2009)
41. Samelson, H.: On Rolle's theorem. Am. Math. Mon. **86**, 486 (1979)
42. Schaumberger, N.: Problem Q917. Math. Mag. **75**, 64,69 (2002)
43. Silverman, H.: A simple auxiliary function for the mean value theorem. Coll. Math. J. **20**, 323 (1989)
44. Shafer, R.E., Grinstein, L.S.: Problem E1867. Am. Math. Mon. **74**, 726–727 (1967)
45. Slay, J.C., Solomon, J.L.: A mean generating function. Coll. Math. J. **12**, 27–29 (1981)
46. Smarandache, F., Melissen, J.B.M.: Problem E3097. Am. Math. Mon. **94**, 1002–1003 (1987)
47. Smith, W.E., Berry, P.M.: Problem E1774. Am. Math. Mon. **73**, 545 (1966)
48. Talman, L.A.: Simpson's rule is exact for cubics. Am. Math. Mon. **113**, 144–155 (2006)
49. Tong, J.: Cauchy's mean value theorem involving n functions. Coll. Math. J. **35**, 50–51 (2004)
50. Tong, J.: A property possessed by every differentiable function. Coll. Math. J. **35**, 216–217 (2004)
51. Wang, C.L.: Proof of the mean value theorem. Am. Math. Mon. **65**, 362–364 (1958)
52. Wang, E.T.H., Hurd, C.: Problem E2447. Am. Math. Mon. **82**, 78–80 (1975)
53. Yates, R.C.: The law of the mean. Am. Math. Mon. **66**, 579–580 (1959)
54. Zhang, G.Q., Roberts, B.: Problem E3214. Am. Math. Mon. **96**, 739–740 (1980)

Chapter 6
The Exponential Function

The greatest shortcoming of the human race is our inability to understand the exponential function.

—Albert A. Bartlett

By now we know Euler's number $e = e^1$ quite well. In this chapter we define the exponential function e^x for *any* $x \in \mathbf{R}$, and its inverse the natural logarithmic function $\ln(x)$, for $x > 0$. (In the first section of the chapter we take a concise approach to the exponential function; in the second section we do things carefully.) These functions enable us to extend many of our previous results to allow for real exponents. For example, we obtain the Power Rule for real exponents, we extend Bernoulli's Inequality, and we obtain a more strapping version of the AGM Inequality. We also meet the Logarithmic Mean, the Harmonic series and its close relatives the Alternating Harmonic series and p-series, and Euler's constant γ.

6.1 The Exponential Function, Quickly

In this section we take a concise approach (e.g., [52]) to the exponential function, while omitting some details of rigor. In the next section we offer an entirely rigorous and self-contained approach. The reader may choose to concentrate on this section or on the next before proceeding to Sect. 6.3, but understanding both would be best.

We begin with the basic *assumption* that there exists a function $\phi(x)$ defined for all $x \in \mathbf{R}$ such that

$$\phi(0) = 1 \quad \text{and} \quad \phi(x) = \phi'(x) \quad \text{for all } x \in \mathbf{R}.$$

Then we verify below $\phi(x)$ that has the following five properties:

(i) $\phi(x)\phi(-x) = 1$ for all $x \in \mathbf{R}$,
(ii) $\phi(x) > 0$ for all $x \in \mathbf{R}$,
(iii) $\phi(x)\phi(y) = \phi(x + y)$ for all $x, y \in \mathbf{R}$,
(iv) $\phi(x)$ is unique,
(v) $1 + x < \phi(x)$ for $x \neq 0$.

P.R. Mercer, *More Calculus of a Single Variable*, Undergraduate Texts in Mathematics, DOI 10.1007/978-1-4939-1926-0_6

For (i), set $f(x) = \phi(x)\phi(-x)$. Since $\phi' = \phi$, the Product Rule and the Chain Rule give

$$f'(x) = \phi'(x)\phi(-x) - \phi(x)\phi'(-x) = 0.$$

Therefore f is a constant function, by Corollary 5.7. Now $f(0) = \phi(0)\phi(0) = 1$, and so we must have $f(x) = 1$ for all $x \in \mathbf{R}$.
For (ii), we first notice that $\phi(x)$ is never zero, by property (i). Now since ϕ is differentiable it is continuous, by Lemma 4.6. So by the Intermediate Value Theorem (Theorem 3.17) ϕ is either positive or negative. Since $\phi(0) = 1$, $\phi(x)$ must be positive.
For (iii), fix x and for $t \in \mathbf{R}$, consider the function

$$g(t) = \frac{\phi(x+t)}{\phi(t)}.$$

Again since $\phi' = \phi$, the Quotient Rule and the Chain Rule give

$$g'(t) = \frac{\phi(t)\phi'(x+t) - \phi'(t)\phi(x+t)}{\phi(t)^2} = 0.$$

Therefore g is constant (Corollary 5.7) and so it equals $g(0) = \phi(x)$ for every t, and in particular for $t = y$. That is, $\frac{\phi(x+y)}{\phi(y)} = \phi(x)$.
For (iv), suppose ψ is a function which also satisfies $\psi(0) = 1$ and $\psi(x) = \psi'(x)$ for all $x \in \mathbf{R}$. Consider then the function

$$h(x) = \frac{\phi(x)}{\psi(x)}.$$

Then by the Quotient Rule,

$$h'(x) = \frac{\psi(x)\phi'(x) - \psi'(x)\phi(x)}{\psi(x)^2} = 0.$$

Therefore h is constant (Corollary 5.7) and so it equals $h(0) = 1$ everywhere. That is, $\phi(x) = \psi(x)$ for all $x \in \mathbf{R}$.
For (v), we first let $x > 0$. Then the Mean Value Theorem (Theorem 5.2) applied to $\phi(t)$ on $[0, x]$ gives a $c \in (0, x)$ such that

$$\frac{\phi(x) - \phi(0)}{x - 0} = \phi'(c) = \phi(c).$$

Now $\phi(0) = 1$ and by item (ii), ϕ is increasing. Therefore

$$\frac{\phi(x) - 1}{x} = \phi(c) > \phi(0) = 1,$$

which yields the desired result. (The case $x < 0$ is handled similarly.) □

The number e is then *defined* by

$$e = \phi(1).$$

(This number is unique because ϕ is strictly increasing: $\phi'(x) = \phi(x) > 0$.)

To obtain an approximation for $e = \phi(1)$, we divide the interval $[0, 1]$ into n equal subintervals:

$$\left[0, \tfrac{1}{n}\right], \left[\tfrac{1}{n}, \tfrac{2}{n}\right], \left[\tfrac{2}{n}, \tfrac{3}{n}\right], \ldots, \left[\tfrac{(n-1)}{n}, 1\right].$$

By the definition of the derivative we have (for n large):

$$\phi\left(\tfrac{1}{n}\right) - \phi(0) \cong \phi'(0)\left(\tfrac{1}{n} - 0\right).$$

Then since $\phi(0) = 1$ and $\phi(x) = \phi'(x)$,

$$\phi\left(\tfrac{1}{n}\right) \cong \phi(0) + \phi'(0)\left(\tfrac{1}{n} - 0\right) = 1 + \tfrac{1}{n}.$$

Likewise

$$\phi\left(\tfrac{2}{n}\right) \cong \phi\left(\tfrac{1}{n}\right) + \phi'\left(\tfrac{1}{n}\right)\left(\tfrac{2}{n} - \tfrac{1}{n}\right) \cong 1 + \tfrac{1}{n} + \left(1 + \tfrac{1}{n}\right)\tfrac{1}{n} = \left(1 + \tfrac{1}{n}\right)^2.$$

And continuing in this way, we get

$$\phi\left(\tfrac{n}{n}\right) = \phi(1) \cong \left(1 + \tfrac{1}{n}\right)^n.$$

Taking n as large as we please we can obtain $e \cong 2.71828$, say.

The symbol e is used in honor of the Swiss mathematician Leonhard Euler (1701–1783). It is often called **Euler's number**.

Remark 6.1. The scheme used above for approximating e is a special case of *Euler's method of tangent lines*. This is a method for obtaining approximate solutions to differential equations, like $\phi'(x) = \phi(x)$. See also Exercise 6.2. ∘

Finally, because $\phi(1) = e$, and because of item (iii) which is evocative of the "same base add the exponents" rule, it is customary to write

$$\phi(x) = e^x.$$

The most important properties of this, the **exponential function** e^x, are (arguably):

$$e^x e^y = e^{x+y}, \qquad (e^x)' = e^x,$$

and

$$1 + x \le e^x \quad \text{for } x \in \mathbf{R} \quad (\text{with strict inequality for } x \neq 0).$$

This last inequality is tremendously useful, as we shall see many times. A graph of $1 + x$ and e^x over $[-1, 2]$ is shown in **Fig. 6.1**.

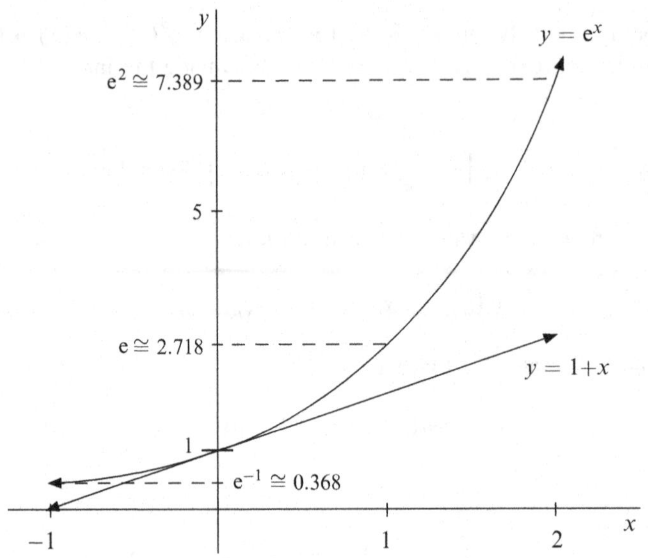

Fig. 6.1 The tremendously useful inequality $1 + x \leq e^x$

6.2 The Exponential Function, Carefully

We saw in Examples 1.32 and 1.43, then again after Example 2.6, that

$$\left(1 + \frac{1}{n}\right)^n < e < \left(1 + \frac{1}{n}\right)^{n+1} \quad \text{for } n = 1, 2, 3, \ldots. \tag{6.1}$$

And we saw that $\{(1 + \frac{1}{n})^n\}$ increases, and $\{(1 + \frac{1}{n})^{n+1}\}$ decreases, to the common limit $e \cong 2.71828$ (Euler's number).

Obviously $e^1 = e$ and $e^0 = 1$. In this section we define what e^x means, for *any* real number x. The approach we take has been influenced mainly by [23, 24, 54] but see also [32, 53, 78, 81].

Looking at the right-hand side of (6.1), since

$$\left(1 + \frac{1}{n}\right)^{n+1} = \left(1 - \frac{1}{n+1}\right)^{-(n+1)},$$

we might just as well have considered

$$\left(1+\frac{1}{n}\right)^n < \mathrm{e} < \left(1-\frac{1}{n}\right)^{-n} \quad \text{for } n = 2,3,4,\ldots.$$

Indeed, this latter form will be more suitable for the present purposes. Among other things, its obvious symmetry will be useful.

For a given $x \in \mathbf{R}$, we consider now the sequences

$$\left\{\left(1+\frac{x}{n}\right)^n\right\} \quad \text{and} \quad \left\{\left(1-\frac{x}{n}\right)^{-n}\right\}, \quad \text{for natural numbers } n > |x|.$$

Now because $\left(1+\frac{x}{n}\right)^n\left(1-\frac{x}{n}\right)^n = \left(1-\frac{x^2}{n^2}\right)^n < 1$, we have immediately that

$$\left(1+\frac{x}{n}\right)^n < \left(1-\frac{x}{n}\right)^{-n}.$$

But we can say much more, as follows in the next two results.

Lemma 6.2. *Let $x \in \mathbf{R}$. Then $\{(1+\frac{x}{n})^n\}$ is increasing and $\{(1-\frac{x}{n})^{-n}\}$ is decreasing, for natural numbers $n > |x|$.*

Proof. Since $n > |x|$, each member of the following two lists of $n+1$ numbers is positive:

$$1, \left(1+\frac{x}{n}\right), \left(1+\frac{x}{n}\right), \ldots, \left(1+\frac{x}{n}\right) \quad \text{and} \quad 1, \left(1-\frac{x}{n}\right), \left(1-\frac{x}{n}\right), \ldots, \left(1-\frac{x}{n}\right).$$

So we may apply the AGM Inequality (Theorem 2.10) to each list (using $\pm$ for brevity) to get:

$$\begin{aligned}\left((1)\left(1\pm\frac{x}{n}\right)^n\right)^{\frac{1}{n+1}} &\le \frac{1}{n+1}\left(1+\left(1\pm\frac{x}{n}\right)+\left(1\pm\frac{x}{n}\right)+\cdots+\left(1\pm\frac{x}{n}\right)\right)\\ &= \frac{1}{n+1}\left(1+n\pm n\frac{x}{n}\right) = 1\pm\frac{x}{n+1}.\end{aligned}$$

Therefore

$$\left(1\pm\frac{x}{n}\right)^n \le \left(1\pm\frac{x}{n+1}\right)^{n+1}.$$

Taking the +'s, this says that $\left\{\left(1+\frac{x}{n}\right)^n\right\}$ is increasing. And taking the −'s, we get $\left(1-\frac{x}{n+1}\right)^{-(n+1)} \le \left(1-\frac{x}{n}\right)^{-n}$, which says that $\left\{\left(1-\frac{x}{n}\right)^{-n}\right\}$ is decreasing. □

Remark 6.3. The paper [73] contains the interesting fact that $\{(1+\frac{x}{n})^n\}$ being increasing *implies* the AGM Inequality (Theorem 2.10). ∘

Lemma 6.4. *Let* $x \in \mathbf{R}$. *Then* $\left(1 - \frac{x}{n}\right)^{-n} - \left(1 + \frac{x}{n}\right)^{n} \to 0$ *as* $n \to \infty$.

Proof. Observe that

$$0 < \left(1 - \frac{x}{n}\right)^{-n} - \left(1 + \frac{x}{n}\right)^{n} = \left(1 - \frac{x}{n}\right)^{-n} \left(1 - \left(1 - \frac{x^2}{n^2}\right)^{n}\right).$$

Taking n large enough, we shall have $n > |x|$. Then we can apply Bernoulli's Inequality (Lemma 2.1) to get:

$$1 - \left(1 - \frac{x^2}{n^2}\right)^{n} \le \frac{x^2}{n}.$$

Therefore

$$0 < \left(1 - \frac{x}{n}\right)^{-n} - \left(1 + \frac{x}{n}\right)^{n} \le \left(1 - \frac{x}{n}\right)^{-n} \frac{x^2}{n}.$$

And since $\{\left(1 - \frac{x}{n}\right)^{-n}\}$ is decreasing (Lemma 6.2), the right hand side here $\to 0$ as $n \to \infty$, and the proof is complete. □

Lemmas 6.2 and 6.4 together with the Nested Interval Property of $\mathbf{R}$ (Theorem 1.41), allow us to denote by $\phi(x)$ the common limit to which $\{\left(1 + \frac{x}{n}\right)^{n}\}$ increases and $\{\left(1 - \frac{x}{n}\right)^{-n}\}$ decreases. So $\phi(0) = 1$, $\phi(1) = \mathrm{e}$, and

$$\left(1 + \frac{x}{n}\right)^{n} \le \phi(x) \le \left(1 - \frac{x}{n}\right)^{-n} \quad \text{for natural numbers } n > |x|. \tag{6.2}$$

In particular, for $|x| < 1$, we have the estimates

$$1 + x \le \left(1 + \frac{x}{n}\right)^{n} \le \phi(x) \le \left(1 - \frac{x}{n}\right)^{-n} \le \frac{1}{1 - x} \quad \text{for } n = 1, 2, 3, \ldots. \tag{6.3}$$

But also, for any $x \ge -1$, we have $1 + x \le \left(1 + \frac{x}{n}\right)^{n}$ by Bernoulli's Inequality (Lemma 2.1). So taking n as large as we please in (6.2), we get

$$1 + x \le \phi(x) \quad \text{for } x \ge -1.$$

Two very important properties of ϕ are contained in the next two results.

Lemma 6.5. *The function* ϕ *satisfies the functional equation*

$$\phi(x)\phi(y) = \phi(x + y) \quad \textit{for all } x, y \in \mathbf{R}. \tag{6.4}$$

Proof. We show that

$$\frac{\left(1 + \frac{x}{n}\right)^{n} \left(1 + \frac{y}{n}\right)^{n}}{\left(1 + \frac{x+y}{n}\right)^{n}} \to 1 \quad \text{as } n \to \infty,$$

from which the result follows. One can verify by cross multiplying that

$$\frac{\left(1+\frac{x}{n}\right)\left(1+\frac{y}{n}\right)}{\left(1+\frac{x+y}{n}\right)} = \left(1+\frac{xy}{n(n+x+y)}\right).$$

And setting $h = \frac{xy}{n+x+y}$, we get

$$\left(1+\frac{xy}{n(n+x+y)}\right)^n = \left(1+\frac{h}{n}\right)^n.$$

Taking n large we can ensure that $|h| < 1$, and so by (6.3),

$$1+h \le \left(1+\frac{h}{n}\right)^n \le \frac{1}{1-h}.$$

In fact, by taking n large enough, we can make $|h|$ as small as we please. Therefore $\left(1+\frac{h}{n}\right)^n = \frac{\left(1+\frac{x}{n}\right)^n\left(1+\frac{y}{n}\right)^n}{\left(1+\frac{x+y}{n}\right)^n} \to 1$, as we wanted to show. □

Lemma 6.6. *$\phi'(x) = \phi(x)$ for all $x \in \mathbf{R}$.*

Proof. Using the functional equation (6.4),

$$\frac{\phi(x+h)-\phi(x)}{h} = \frac{\phi(x)\phi(h)-\phi(x)}{h} = \phi(x)\frac{\phi(h)-1}{h}.$$

And for $|h| < 1$ the estimates (6.3) give

$$1+h \le \phi(h) \le \frac{1}{1-h} = 1+\frac{h}{1-h}.$$

Therefore

$$1 \le \frac{\phi(h)-1}{h} \le \frac{1}{1-h}.$$

The result now follows upon letting $h \to 0$. □

Now if we take $y = -x$ in the functional equation (6.4), we get

$$\phi(x)\phi(-x) = 1 \quad \text{for all } x \in \mathbf{R}.$$

Therefore ϕ is never zero. Also, since ϕ is differentiable (Lemma 6.6), it is continuous (Lemma 4.6). So by the Intermediate Value Theorem (Theorem 3.17) ϕ is either positive or negative. Since $\phi(0) = 1$, ϕ must be positive.

We have already seen that $1 + x \leq \phi(x)$ for $x \geq -1$. Since ϕ is positive, we must therefore have

$$1 + x \leq \phi(x) \quad \text{for all } x \in \mathbf{R}.$$

Now because ϕ satisfies the functional equation $\phi(x)\phi(y) = \phi(x + y)$ which is evocative of the "same base add the exponents" rule, and because $\phi(1) = \mathrm{e}$, it is customary to write

$$\phi(x) = \mathrm{e}^x.$$

The most important properties of this, the **exponential function** e^x, are (arguably) the contents of Lemmas 6.5 and 6.6:

$$\mathrm{e}^x\mathrm{e}^y = \mathrm{e}^{x+y}, \qquad (\mathrm{e}^x)' = \mathrm{e}^x,$$

and

$$1 + x \leq \mathrm{e}^x \quad \text{for } x \in \mathbf{R} \text{ (with strict inequality for } x \neq 0). \tag{6.5}$$

The inequality (6.5) is tremendously useful, as we shall see many times. A graph of $1 + x$ and e^x over $[-1, 2]$ is shown in **Fig. 6.1** in Sect. 6.1.

6.3 The Natural Logarithmic Function

In this section we show that the exponential function has an inverse. To do so, we establish a few more of its properties.

Lemma 6.7. *The exponential function* e^x *has the following properties:*

(i) e^x *is strictly increasing on* $(-\infty, +\infty)$,
(ii) $\mathrm{e}^x \to +\infty$ *as* $x \to +\infty$,
(iii) $\mathrm{e}^x \to 0$ *as* $x \to -\infty$.

Proof. For (i), we have seen that $\mathrm{e}^x > 0$. Then since $(\mathrm{e}^x)' = \mathrm{e}^x$, we must have $(\mathrm{e}^x)' > 0$. Therefore e^x is strictly increasing, by Lemma 5.6.
For (ii), we saw in (6.5) that $1 + x \leq \mathrm{e}^x$. As such, $\mathrm{e}^x \to +\infty$ as $x \to +\infty$.
For (iii), by (ii) we have $\mathrm{e}^x \to +\infty$ as $x \to +\infty$. Then since $\mathrm{e}^{-x} = 1/\mathrm{e}^x$ by the functional equation (6.4), we have $\mathrm{e}^{-x} \to 0$ as $x \to +\infty$. That is, $\mathrm{e}^x \to 0$ as $x \to -\infty$. □

Example 6.8. As regards item (ii) of Lemma 6.7, much more can be said:

$$\text{For any } n \in \mathbf{N}, \qquad \lim_{x \to +\infty} \frac{\mathrm{e}^x}{x^n} = +\infty.$$

This follows from applying L'Hospital's Rule (Theorem 5.13) n times. It says that as $x \to +\infty$, $e^x \to +\infty$ faster than *any* polynomial. ◇

Lemma 6.7 shows that $\phi(x) = e^x$ has an inverse, defined on $(0, \infty)$. This inverse is denoted by $\phi^{-1}(x) = \ln(x)$, and its range is $(-\infty, +\infty)$. This is the **natural logarithmic function**. Being the inverse of e^x, $\ln(x)$ satisfies:

$$e^{\ln(x)} = x \quad \text{for } x > 0 \qquad \text{and} \qquad \ln(e^x) = x \quad \text{for } x \in \mathbf{R}.$$

Graphs of e^x and $\ln(x)$ are shown in **Fig. 6.2**.

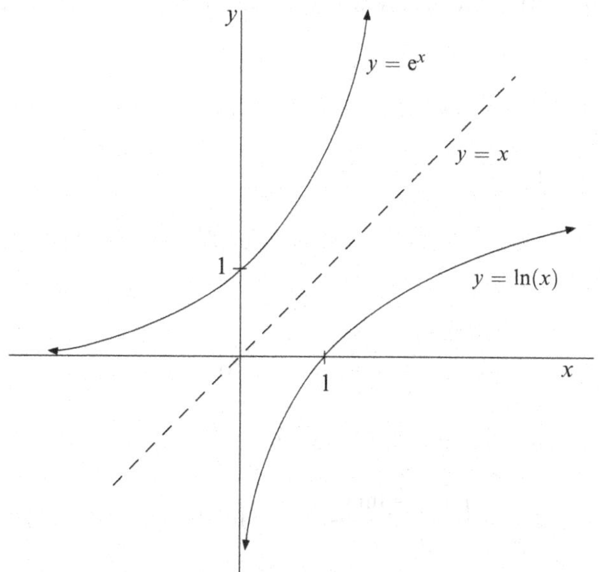

Fig. 6.2 The graphs of $y = e^x$ and $y = \ln(x)$. Each is the graph of the other, reflected the line $y = x$

Any property of the exponential function gives rise to a property of the natural logarithmic function, since the latter is the inverse of the former. We list some of these properties below and leave their proofs as an exercise. We shall use them freely without explicit mention.

Lemma 6.9. *The natural logarithmic function* $\ln(x)$ *has the following properties:*

(i) $\ln(ab) = \ln(a) + \ln(b)$ *for* $a, b > 0$,
(ii) $\ln(a/b) = \ln(a) - \ln(b)$ *for* $a, b > 0$,
(iii) $\ln(a^r) = r\ln(a)$ *for* $a > 0$ *and* $r \in \mathbf{R}$,
(iv) $\ln(x)$ *is a strictly increasing function,*
(v) $\ln(x) \to +\infty$ *as* $x \to +\infty$,
(vi) $\ln(x) \to -\infty$ *as* $x \to 0^+$.

Proof. This is Exercise 6.15. □

We saw in Lemma 6.6 that $(e^x)' = e^x$ for all x and so by the Chain Rule,

$$\left(e^{\alpha(x)}\right)' = \alpha'(x)e^{\alpha(x)}, \quad \text{for differentiable functions } \alpha(x).$$

Then the relationship $e^{\ln(x)} = x$ appears to imply that $\left(e^{\ln(x)}\right)' = e^{\ln(x)}(\ln(x))' = 1$, and so $(\ln(x))' = 1/x$ for $x > 0$. But this only shows that *if* $(\ln(x))'$ exists, then it equals $1/x$ (for $x > 0$). One really must show that $(\ln(x))'$ indeed exists. This can be done using Exercise 4.16, but here we do so more directly—similarly in spirit to how we obtained the derivative of $\arctan(x)$ in Sect. 4.3.

In inequality (6.5), we replace x with $u - v$ and then with $v - u$ to get

$$\begin{aligned} 1 + (u - v) &< e^{u-v} && \text{and} \\ 1 + (v - u) &< e^{v-u}, && \text{both for } u \neq v. \end{aligned}$$

Together these read

$$1 - e^{u-v} < v - u < e^{v-u} - 1,$$

or

$$e^{-v} < \frac{v-u}{e^v - e^u} < e^{-u} \quad \text{for } u < v.$$

Now setting $u = \ln(x)$ and $v = \ln(y)$ we get

$$\frac{1}{y} < \frac{\ln(y) - \ln(x)}{y - x} < \frac{1}{x} \quad \text{for } 0 < x < y.$$

Therefore, letting $y \to x$ (or $x \to y$), we get

$$\left(\ln(x)\right)' = \frac{1}{x} \quad \text{for } x > 0.$$

For $x < 0$ we have $\ln(|x|) = \ln(-x)$, and the so Chain Rule gives

$$\left(\ln(|x|)\right)' = \frac{1}{x} \qquad \text{for } x \neq 0.$$

Applying the Chain Rule further, we get

$$\left(\ln|\beta(x)|\right)' = \frac{\beta'(x)}{\beta(x)}, \quad \text{for differentiable functions } \beta(x) \text{ which are never zero.}$$

We close this section with two more examples (the first is very simple), in which a property of the exponential function gives rise to a corresponding property of the natural logarithmic function.

Example 6.10. The reader may verify by taking logarithms then dividing by n and $n+1$ in turn, that the estimates (6.1), i.e., $\left(1+\frac{1}{n}\right)^n < \mathrm{e} < \left(1+\frac{1}{n}\right)^{n+1}$ are equivalent to the equally useful estimates

$$\frac{1}{n+1} < \ln\left(1+\frac{1}{n}\right) < \frac{1}{n}. \tag{6.6}$$

◇

Example 6.11. Again, the inequality (6.5) reads $\mathrm{e}^x \geq 1+x$ for all real x. Since $\ln(t)$ is increasing, $x \geq \ln(1+x)$ for $x > -1$. Or, replacing x with $x-1$ here, we get

$$\ln(x) \leq x-1 \quad \text{for} \quad x > 0. \tag{6.7}$$

And replacing x with $1/x$ in (6.7) we get

$$1-\frac{1}{x} \leq \ln(x) \quad \text{for} \quad x > 0,$$

or

$$\frac{\mathrm{e}}{x} \leq \mathrm{e}^{\frac{1}{x}} \quad \text{for} \quad x > 0.$$

The inequalities $1-\frac{1}{x} \leq \ln(x) \leq x-1$ are shown in **Fig. 6.3**. ◇

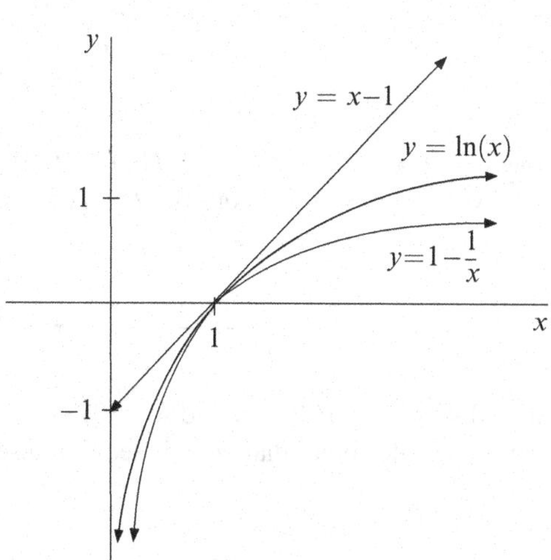

Fig. 6.3 Example 6.11: The inequalities $1-\frac{1}{x} \leq \ln(x) \leq x-1$, for $x > 0$

6.4 Real Exponents

The exponential function (along with its inverse, the natural logarithmic function) enables us to define x^r for *any* exponent $r \in \mathbf{R}$, not only for $r \in \mathbf{Q}$. By item (iii) of Lemma 6.9,

$$x^r = \mathrm{e}^{\ln(x^r)} = \mathrm{e}^{r\ln(x)} \quad \text{for } x > 0.$$

And now we can establish the following important fact, which we only *stated* in Sect. 4.2.

Power Rule for Real Exponents: *Let $r \in \mathbf{R}$ and for $x > 0$, let $f(x) = x^r$. Then f is differentiable for $x > 0$, and $f'(x) = rx^{r-1}$.*

Proof. By the Chain Rule,

$$(x^r)' = (\mathrm{e}^{r\ln(x)})' = \mathrm{e}^{r\ln(x)}\frac{r}{x} = x^r\frac{r}{x} = rx^{r-1}$$

as desired. □

As the next two examples show, we can extend other results which contain integer or rational exponents by allowing them to contain real exponents.

Example 6.12. We improve Bernoulli's Inequality (Lemma 2.1) to allow it to contain real exponents ≥ 1, not just integers ≥ 1: *Let $\alpha \geq 1$ be any real number. Then for $x > -1$,*

$$(1 + x)^\alpha \geq 1 + \alpha x.$$

Here's a proof. We take $x > 0$ and consider $f(t) = (1 + t)^\alpha$ for $t \in [0, x]$. By the Mean Value Theorem (Theorem 5.2) there is $c \in (0, x)$ such that

$$\frac{f(x) - f(0)}{x} = \alpha(1 + c)^{\alpha-1}.$$

Now $\alpha(1 + c)^{\alpha-1} > \alpha$ and so $f(x) - f(0) = (1 + x)^\alpha - 1 > \alpha x$, as desired. Now we take $x < 0$ and consider $t \in [x, 0]$. By the Mean Value Theorem there is $c \in (x, 0)$ such that

$$\frac{f(0) - f(x)}{-x} = \alpha(1 + c)^{\alpha-1}.$$

Here, $0 < \alpha(1+c)^{\alpha-1} < \alpha$ and so $f(0) - f(x) = 1 - (1 + x)^\alpha < -\alpha x$, as desired. (For values of α other than $\alpha \geq 1$, see Exercise 6.5.) ◇

Example 6.13. We saw in (6.1) that

$$\left(1+\frac{1}{n}\right)^n < \mathrm{e} < \left(1+\frac{1}{n}\right)^{n+1} \quad \text{for } n = 1, 2, 3, \ldots,$$

and that each of the sequences $\{(1+\frac{1}{n})^n\}$ and $\{(1+\frac{1}{n})^{n+1}\}$ converges to e. The former sequence is increasing and the latter sequence is decreasing. But rather more is true: the natural number n can be replaced with $x \in \mathbf{R}$, as follows. For $x > 0$, the Mean Value Theorem (Theorem 5.2) applied to $f(t) = \ln(t)$ on $[x, x+1]$ gives a $c \in (x, x+1)$ such that

$$f'(c) = \frac{1}{c} = \frac{\ln(x+1) - \ln(x)}{(x+1) - x} = \ln\left(\frac{x+1}{x}\right).$$

Since $x < c < x+1$, we have $\frac{1}{x+1} < \frac{1}{c} < \frac{1}{x}$, and so

$$\frac{1}{x+1} < \ln\left(\frac{x+1}{x}\right) < \frac{1}{x} \quad \text{for } x > 0. \tag{6.8}$$

These inequalities extend those of (6.6), and imply

$$\left(1+\frac{1}{x}\right)^x < \mathrm{e} < \left(1+\frac{1}{x}\right)^{x+1} \quad \text{for all } x \in \mathbf{R}, \tag{6.9}$$

which extend (6.1). One can also verify that

$$\left(x \ln\left(\frac{x+1}{x}\right)\right)' = \ln\left(\frac{x+1}{x}\right) - \frac{1}{x+1} > 0$$

by the left-hand side of (6.8), and that

$$\left((x+1) \ln\left(\frac{x+1}{x}\right)\right)' = \ln\left(\frac{x+1}{x}\right) - \frac{1}{x} < 0$$

by the right-hand side of (6.8) (cf. [30], sec 4.2). Therefore, $\left(1+\frac{1}{x}\right)^x$ is increasing and $\left(1+\frac{1}{x}\right)^{x+1}$ is decreasing. Finally,

$$\left(1+\frac{1}{x}\right)^{x+1} - \left(1+\frac{1}{x}\right)^x = \left(1+\frac{1}{x}\right)^x \left(1 - \left(1+\frac{1}{x}\right)\right) = \frac{1}{x}\left(1+\frac{1}{x}\right)^x < \frac{\mathrm{e}}{x}$$

and so each of $\left(1+\frac{1}{x}\right)^x$ and $\left(1+\frac{1}{x}\right)^{x+1}$ converges to e as $x \to +\infty$. ◇

6.5 The AGM Inequality Again

In this section use the exponential function to extend the AGM Inequality (Theorem 2.10) as well.

Let $p_1, p_2, \ldots, p_n$ be positive numbers. For $a_1, a_2, \ldots, a_n \in \mathbf{R}$, we saw in Sect. 3.4 that their associated weighted Arithmetic Mean is the expression

$$\frac{1}{\sum_{k=1}^{n} p_k} \sum_{j=1}^{n} p_j a_j \,.$$

By setting

$$w_j = \frac{p_j}{\sum_{k=1}^{n} p_k},$$

this reads

$$\sum_{j=1}^{n} w_j a_j, \quad \text{with} \quad \sum_{j=1}^{n} w_j = 1.$$

In this context, the w_j*'s* are called **weights**. For the special case in which $p_1 = p_2 = \cdots = p_n = 1$, we get $w_j = \frac{1}{n}$ for each j and so indeed $\sum_{j=1}^{n} w_j = 1$, and $\sum_{j=1}^{n} w_j a_j$ is the ordinary Arithmetic Mean $A = \frac{a_1 + a_2 + \cdots + a_n}{n}$.

Example 6.14. In the weighted Arithmetic Mean

$$\frac{2a_1 + 7a_2 + a_3 + 5a_4}{15},$$

we set $w_1 = 2/15$, $w_2 = 7/15$, $w_3 = 1/15$, and $w_4 = 5/15$. Then

$$\frac{2a_1 + 7a_2 + a_3 + 5a_4}{15} = \tfrac{2}{15}a_1 + \tfrac{7}{15}a_2 + \tfrac{1}{15}a_3 + \tfrac{1}{3}a_4 = \sum_{j=1}^{4} w_j a_j \,. \qquad \diamond$$

The AGM Inequality (Theorem 2.10) can be extended without too much difficulty to allow for positive *rational* weights. This is Exercise 6.25.

But there is an even more general version of the AGM Inequality, as follows, which allows all positive real numbers as weights, not only positive rationals. We provide the beautiful proof [30, 81] by American (Hungarian born) mathematician George Polya (1887–1985) which uses (6.5) (or see (6.7)) in the form:

$$x \le e^{x-1} \quad \text{for } x \in \mathbf{R} \quad (\text{with strict inequality for } x \ne 1)\,.$$

Theorem 6.15. (Weighted AGM Inequality) *Let $a_1, a_2, \ldots, a_n$ be positive real numbers and let $w_1, w_2, \ldots, w_n$ be positive real numbers satisfying $\sum_{j=1}^{n} w_j = 1$. Then*

$$\prod_{j=1}^{n} a_j^{w_j} \leq \sum_{j=1}^{n} w_j a_j ,$$

and equality occurs here $\Leftrightarrow$ $a_1 = a_2 = \cdots = a_n$.

Proof. Set $A = \sum_{j=1}^{n} w_j a_j$. Since $x \leq \mathrm{e}^{x-1}$, we have

$$\frac{a_j}{A} \leq \mathrm{e}^{\frac{a_j}{A}-1} \quad \text{for } j = 1, 2, \ldots, n.$$

Therefore

$$\begin{aligned}
\prod_{j=1}^{n} \left(\frac{a_j}{A}\right)^{w_j} &\leq \prod_{j=1}^{n} \left(\mathrm{e}^{\frac{a_j}{A}-1}\right)^{w_j} \\
&= \prod_{j=1}^{n} \left(\mathrm{e}^{\frac{w_j a_j}{A}-w_j}\right) \\
&= \mathrm{e}^{\sum_{j=1}^{n}\left(\frac{w_j a_j}{A}-w_j\right)} = \mathrm{e}^{1-1} = \mathrm{e}^{0} = 1.
\end{aligned}$$

That is, $\prod_{j=1}^{n} a_j^{w_j} \leq A$, as desired. Moreover, $x = \mathrm{e}^{x-1}$ if and only if $x = 1$, so equality occurs here if and only if $a_1 = a_2 = \cdots = a_n \ (= A)$. □

Some other proofs of the weighted AGM Inequality (Theorem 6.15) are explored in Exercises 6.26–6.28 and 6.30. We isolate below an important special case, which extends Lemma 2.7. (Lemma 2.7 is the result below, with $t = 1/2$.)

Corollary 6.16. (Weighted AGM Inequality with $n = 2$) *Let $a, b \geq 0$ and* $0 \leq t \leq 1$. Then

$$a^t b^{1-t} \leq ta + (1-t)b,$$

and equality occurs here $\Leftrightarrow$ $a = b$.

Proof. This is Theorem 6.15, with $n = 2$, $a_1 = a$, $a_2 = b$, $w_1 = t$, and $w_2 = 1-t$. □

Example 6.17. [16] Here we use the weighted AGM Inequality with $n = 2$ (Corollary 6.16) to obtain the neat inequality

$$\left(\sin x\right)^{\sin x} < \left(\cos x\right)^{\cos x} \quad \text{for } 0 < x < \pi/4.$$

For $0 < x < \pi/4$, we have $0 < \tan x < 1$. So we may apply the weighted AGM inequality for $n = 2$ to $\tan^2 x$ and $1 + \tan^2 x$, and with weights $\tan x$ and $1 - \tan x$ to get

$$\left(\tan^2 x\right)^{\tan x}\left(1 + \tan^2 x\right)^{1-\tan x} \leq \left(\tan x\right)\left(\tan^2 x\right) + \left(1 - \tan x\right)\left(1 + \tan^2 x\right).$$

The left-hand side here is $\left(\sin^2 x\right)^{\tan x} / \cos^2 x$, and the right-hand side is

$$1 + \tan^2 x - \tan x < 1.$$

Therefore $\left(\sin^2 x\right)^{\tan x} < \cos^2 x$. The desired inequality now follows by raising each side to the power $(\cos x)/2$. ◇

In Sect. 2.3 we saw how Lemma 2.7 can be used to obtain the Cauchy–Schwarz Inequality (Theorem 2.18). In an entirely similar way, the weighted AGM Inequality with $n = 2$ (Corollary 6.16) yields the following extension of the Cauchy–Schwarz Inequality. (In it, taking $p = q = 2$ gives the Cauchy–Schwarz Inequality.)

Lemma 6.18. (Hölder's Inequality) *Let $a_1, a_2, \ldots, a_n$, and $b_1, b_2, \ldots, b_n$ be positive real numbers and let $p, q > 1$ satisfy $\frac{1}{p} + \frac{1}{q} = 1$ (p and q are called* ***conjugate exponents****). Then*

$$\sum_{j=1}^{n} a_j b_j \leq \left(\sum_{j=1}^{n} a_j^p\right)^{1/p} \left(\sum_{j=1}^{n} b_j^q\right)^{1/q}.$$

Proof. If $\sum\limits_{j=1}^{n} a_j^p = 0$ or $\sum\limits_{j=1}^{n} b_j^q = 0$ then the equality holds. Otherwise, set $a = \dfrac{a_k^p}{\sum\limits_{j=1}^{n} a_j^p}$ and $b = \dfrac{b_k^q}{\sum\limits_{j=1}^{n} b_j^q}$ in the weighted AGM Inequality with $n = 2$ (Corollary 6.16), taking $t = \frac{1}{p}$ there, to obtain

$$\frac{a_k}{\left(\sum\limits_{j=1}^{n} a_j^p\right)^{1/p}} \frac{b_k}{\left(\sum\limits_{j=1}^{n} b_j^q\right)^{1/q}} \leq \frac{1}{p} \frac{a_k^p}{\sum\limits_{j=1}^{n} a_j^p} + \frac{1}{q} \frac{b_k^q}{\sum\limits_{j=1}^{n} b_j^q}.$$

Then summing from $k = 1$ to n we get

$$\frac{\sum_{k=1}^{n} a_k b_k}{\left(\sum_{j=1}^{n} a_j^p\right)^{1/p}\left(\sum_{j=1}^{n} b_j^q\right)^{1/q}} \leq \frac{1}{p}\frac{\sum_{k=1}^{n} a_k^p}{\sum_{j=1}^{n} a_j^p} + \frac{1}{q}\frac{\sum_{k=1}^{n} b_k^q}{\sum_{j=1}^{n} b_j^q} = \frac{1}{p} + \frac{1}{q} = 1,$$

as desired. □

Sometimes the weighted AGM Inequality (Corollary 6.16) is more convenient to apply in the following (equivalent) form.

Corollary 6.19. (Young's Inequality) *Let* $a, b \geq 0$ *and* $p, q > 1$ *with* $\frac{1}{p} + \frac{1}{q} = 1$. *Then*

$$ab \leq \frac{a^p}{p} + \frac{b^q}{q},$$

and equality occurs here $\Leftrightarrow a^p = b^q$.

Proof. in the weighted AGM Inequality with $n = 2$ (Corollary 6.16), replace a with $a^{1/t}$, b with $b^{1/(1-t)}$, and take $t = 1/p$, $1 - t = 1/q$. □

Naturally then, Young's Inequality can also be used to obtain Hölder's Inequality (Lemma 6.18). We leave the details for Exercise 6.34.

In Exercise 6.37 we see how the weighted AGM Inequality (Theorem 6.15) can be used to extend Hölder's Inequality. In Exercise 6.38 we see how the weighted AGM Inequality (Theorem 6.15) can be used to extend Young's Inequality.

6.6 The Logarithmic Mean

The **Logarithmic Mean** of the positive numbers a and b is given by

$$L = L(a,b) = \begin{cases} \dfrac{b-a}{\ln(b)-\ln(a)} & \text{if } a \neq b \\ a & \text{if } a = b. \end{cases}$$

As well as having intrinsic interest, the Logarithmic Mean arises in problems dealing with heat transfer and fluid mechanics. Since $L(a,b) = L(b,a)$, we may suppose that $a \leq b$. Applying Cauchy's Mean Value Theorem (Theorem 5.11) to

$$f(x) = x - a \quad \text{and} \quad g(x) = \ln(x) - \ln(a)$$

on $[a, b]$, there is $c \in (a, b)$ such that

$$\frac{b-a}{\ln(b)-\ln(a)} = \frac{f(b)-f(a)}{g(b)-g(a)} = \frac{f'(c)}{g'(c)} = \frac{1}{1/c} = c.$$

Therefore

$$\min\{a,b\} \leq L \leq \max\{a,b\},$$

and so L is indeed a *mean*. And this justifies the choice $L(a,a) = a$, making L continuous. (The reader might also verify that $L(ta,tb) = tL(a,b)$, for $t > 0$.)

Of course we didn't really use *Cauchy's* Mean Value Theorem here, just the Mean Value Theorem (Theorem 5.2) upside down. But the idea of using Cauchy's Mean Value Theorem can give us more, as follows. First, recall that for positive numbers a and b, their Arithmetic Mean is

$$A = \frac{a+b}{2}$$

and their Geometric Mean is

$$G = \sqrt{ab}.$$

Lemma 6.20. *Let a and b be positive real numbers. Then*

$$G \leq L \leq A.$$

Proof. For the right-hand inequality, for $x \in [a,b]$ we set

$$f(x) = \frac{x-a}{x+a} \quad \text{and} \quad g(x) = \ln(x) - \ln(a).$$

Then by Cauchy's Mean Value Theorem (Theorem 5.11), there is $c \in (a,b)$ such that

$$\frac{\frac{b-a}{b+a}}{\ln(b) - \ln(a)} = \frac{f(b) - f(a)}{g(b) - g(a)} = \frac{f'(c)}{g'(c)}.$$

It is easily verified that

$$\frac{f'(c)}{g'(c)} = \frac{2ac}{(c+a)^2},$$

and this is $\leq \frac{1}{2}$, by Lemma 2.7. Therefore $L \leq A$.

For the left-hand inequality, for $x \in [a,b]$ we set

$$f(x) = \frac{x-a}{\sqrt{xa}} \quad \text{and} \quad g(x) = \ln(x) - \ln(a).$$

Then by Cauchy's Mean Value Theorem, there is $c \in (a,b)$ such that

$$\frac{\frac{b-a}{\sqrt{ba}}}{\ln(b)-\ln(a)} = \frac{f(b)-f(a)}{g(b)-g(a)} = \frac{f'(c)}{g'(c)}.$$

Here it is easily verified that

$$\frac{f'(c)}{g'(c)} = \frac{\frac{c+a}{2}}{\sqrt{ac}},$$

and this is ≥ 1, by Lemma 2.7. Therefore $G \leq L$. □

Remark 6.21. The inequalities in Lemma 6.20 were first obtained (by more complicated methods) in [60]; see also [11,15]. In Exercises 6.43, 6.44, and 6.46 we obtain refinements of these inequalities, also via Cauchy's Mean Value Theorem. ∘

Setting $a = x$ and $b = x + 1$ in Lemma 6.20 , we get

$$\sqrt{x(x+1)} \leq \frac{1}{\ln(x+1)-\ln(x)} \leq \frac{2x+1}{2}.$$

After a little manipulation, this yields

$$\left(1+\frac{1}{x}\right)^{\sqrt{x(x+1)}} \leq \mathrm{e} \leq \left(1+\frac{1}{x}\right)^{x+1/2}. \tag{6.10}$$

These estimates improve (6.9) considerably. See also Exercises 6.9 and 6.45.

Since $\sqrt{x(x+1)}$ is the Geometric Mean of $x+1$ and x, and $x+1/2$ is their Arithmetic Mean, we point out the rather satisfying fact that (6.10) reads:

$$\left(\frac{x+1}{x}\right)^{G(x+1,x)} \leq \mathrm{e} \leq \left(\frac{x+1}{x}\right)^{A(x+1,x)}.$$

6.7 The Harmonic Series and Some Relatives

(i) The **Harmonic series** is the infinite series

$$\sum_{n=1}^{\infty} \frac{1}{n} = 1 + \frac{1}{2} + \frac{1}{3} + \frac{1}{4} + \frac{1}{5} + \cdots.$$

Each term (after the first) is the Harmonic Mean of the term just before it and the term just after it. As with any infinite series, this expression denotes the limit (if it exists) of the *sequence of partial sums* $\{S_N\}$. In this case, $\{S_N\} = \left\{\sum_{n=1}^{N} \frac{1}{n}\right\}$.

We saw in the course of obtaining (6.7) that $\ln(1 + x) \leq x$ for $x > -1$. Therefore, as in [7, 48] for example,

$$\ln(N+1) = \sum_{n=1}^{N} \left[\ln(n+1) - \ln(n) \right]$$
$$= \sum_{n=1}^{N} \ln\left(1 + \frac{1}{n}\right) \leq \sum_{n=1}^{N} \frac{1}{n}.$$

Now since $\ln(N + 1) \to +\infty$ as $N \to +\infty$, we must also have $S_N \to +\infty$. So we write

$$\sum_{n=1}^{\infty} \frac{1}{n} = +\infty.$$

We might say that the Harmonic series *diverges to* $+\infty$. Exercises 6.47–6.50 contain several other proofs of this important fact.

Remark 6.22. The partial sums of the Harmonic series grow without bound, but they do so very slowly. For example, $S_{10{,}000} \cong 9.8$, the smallest N for which $S_N > 20$ is 272400600, and the smallest N for which $S_N > 1{,}000$ is greater than 10^{434}. ∘

(ii) Euler's constant. In (6.6) we saw that

$$\frac{1}{n+1} < \ln\left(\frac{n+1}{n}\right) < \frac{1}{n} \qquad \text{for } n = 1, 2, 3, \ldots.$$

For such n, set

$$\gamma_n = \sum_{k=1}^{n} \frac{1}{k} - \ln(n).$$

Then

$$\gamma_{n+1} - \gamma_n = \left(\sum_{k=1}^{n+1} \frac{1}{k} - \ln(n+1) \right) - \left(\sum_{k=1}^{n} \frac{1}{k} - \ln(n) \right) = \frac{1}{n+1} - \ln\left(\frac{n+1}{n}\right).$$

And $\frac{1}{n+1} - \ln\left(\frac{n+1}{n}\right) < 0$, by the left-hand side of (6.6). Therefore $\{\gamma_n\}$ is a decreasing sequence. Moreover,

$$\gamma_n = \frac{1}{n} + \sum_{k=1}^{n-1} \frac{1}{k} - \ln(n) = \frac{1}{n} + \sum_{k=1}^{n-1} \left(\frac{1}{k} - \ln\left(\frac{k+1}{k}\right) \right),$$

and this is $> \frac{1}{n}$, by the right-hand side of (6.6). Therefore $\{\gamma_n\}$ is bounded below. So by the Increasing Bounded Sequence Property (Theorem 1.34), $\{\gamma_n\}$ converges to some real number $\gamma \geq 0$. The number γ is called **Euler's constant**. Since γ_n is decreasing, $\gamma < \gamma_1 = 1$, $\gamma < \gamma_2 \cong 0.807$, $\gamma < \gamma_3 \cong 0.735$ etc. In fact,

$$\gamma \cong 0.577216.$$

Remark 6.23. Euler's constant arises often, in apparently disparate mathematical contexts. Mathematicians typically rate γ just below π and e in its overall importance in mathematical analysis. Still, γ remains elusive. It is not even known whether γ is irrational. Another approach to γ can be found in [12]. Also, the book [31] is highly recommended. ◦

(iii) The **Alternating Harmonic series** is the infinite series

$$\sum_{n=1}^{\infty}(-1)^{n+1}\frac{1}{n} = 1 - \frac{1}{2} + \frac{1}{3} - \frac{1}{4} + \cdots.$$

Here $\{S_N\} = \left\{\sum_{n=1}^{N}(-1)^{n+1}\frac{1}{n}\right\}$, and so

$$\begin{aligned} S_{2N} &= 1 - \frac{1}{2} + \frac{1}{3} - \frac{1}{4} + \frac{1}{5} - \frac{1}{6} + \cdots + \frac{1}{2N-1} - \frac{1}{2N} \\ &= \left(1 - \frac{1}{2}\right) + \left(\frac{1}{3} - \frac{1}{4}\right) + \left(\frac{1}{5} - \frac{1}{6}\right) + \cdots + \left(\frac{1}{2N-1} - \frac{1}{2N}\right). \end{aligned}$$

Therefore $\{S_{2N}\}$ is increasing. But also,

$$S_{2N} = 1 - \left(\frac{1}{2} - \frac{1}{3}\right) - \left(\frac{1}{4} - \frac{1}{5}\right) - \cdots - \left(\frac{1}{2N-2} - \frac{1}{2N-1}\right) - \frac{1}{2N},$$

and therefore $\{S_{2N}\}$ is bounded above, by 1. So by the Increasing Bounded Sequence Property (Theorem 1.34), $\{S_{2N}\}$ converges to some real number $S \leq 1$. Now

$$S_{2N+1} = S_{2N} + \frac{1}{2N+1},$$

so that $\{S_{2N+1}\}$ must converge to S also. Therefore, $\{S_N\}$ converges to S.

Using Euler's constant γ, we can find S as follows. Observe that

$$1 - \frac{1}{2} + \frac{1}{3} - \frac{1}{4} + \frac{1}{5} - \frac{1}{6} + \cdots - \frac{1}{2n}$$

$$= 1 + \left(\frac{1}{2} - \frac{2}{2}\right) + \frac{1}{3} + \left(\frac{1}{4} - \frac{2}{4}\right) + \frac{1}{5} + \left(\frac{1}{6} - \frac{2}{6}\right) + \cdots + \left(\frac{1}{2n} - \frac{2}{2n}\right)$$

$$= 1 + \frac{1}{2} + \frac{1}{3} + \frac{1}{4} + \frac{1}{5} + \frac{1}{6} + \cdots + \frac{1}{2n} - \left(1 + \frac{1}{2} + \frac{1}{3} + \frac{1}{4} + \frac{1}{5} + \frac{1}{6} + \cdots + \frac{1}{n}\right)$$

$$= \left(\ln(2n) + \gamma_{2n}\right) - \left(\ln(n) + \gamma_n\right).$$

Notice that $\ln(2n) - \ln(n) = \ln(2)$ and $\gamma_n \to$ Euler's constant γ. So we have at hand the sum of the Alternating Harmonic series:

$$\sum_{n=1}^{\infty} (-1)^{n+1} \frac{1}{n} = \ln(2) \cong 0.693147.$$

We shall show in Corollary 12.6 that $\ln(2)$ is irrational.

Remark 6.24. For the Alternating Harmonic series we have (for example)

$$S = 1 - \frac{1}{2} + \frac{1}{3} - \frac{1}{4} + \frac{1}{5} - \frac{1}{6} + \cdots$$

$$= \left(1 - \frac{1}{2}\right) - \frac{1}{4} + \left(\frac{1}{3} - \frac{1}{6}\right) - \frac{1}{8} + \left(\frac{1}{5} - \frac{1}{10}\right) - \frac{1}{12} + \cdots$$

$$= \frac{1}{2} - \frac{1}{4} + \frac{1}{6} - \frac{1}{8} + \frac{1}{10} - \frac{1}{12} + \cdots$$

$$= \frac{1}{2}\left(1 - \frac{1}{2} + \frac{1}{3} - \frac{1}{4} + \frac{1}{5} - \frac{1}{6} + \cdots\right)$$

$$= \frac{1}{2} S.$$

This means that the sum S is dependent on how the terms are arranged! There is a theorem (see [42] or [61]) due to the German mathematician Bernhard Riemann (1826–1866) which implies that for any real number S, there is a *rearrangement* of the Alternating Harmonic series which sums to S. And there are rearrangements which diverge to each of $\pm\infty$ as well. We address this phenomenon a little more in Exercise 6.57 and in Sect. 10.1. But for more about rearrangements of the Alternating Harmonic series, see for example, [5, 8, 20, 46]. ∘

(iv) For any real number p, the associated **p-series** is the infinite series

$$\sum_{n=1}^{\infty} \frac{1}{n^p} = 1 + \frac{1}{2^p} + \frac{1}{3^p} + \frac{1}{4^p} + \frac{1}{5^p} + \cdots .$$

For $p = 1$ this is simply the Harmonic series and so it diverges. It is easy to see, and we leave this for Exercise 6.58, that a p-series diverges also for $p < 1$.

Here we show, as in [19], that a p-series converges for $p > 1$. For the Nth partial sum

$$S_N = \sum_{n=1}^{N} \frac{1}{n^p},$$

we have

$$\begin{aligned} S_{2N+1} &= 1 + \left(\frac{1}{2^p} + \frac{1}{4^p} + \cdots + \frac{1}{(2N)^p}\right) + \left(\frac{1}{3^p} + \frac{1}{5^p} + \cdots + \frac{1}{(2N+1)^p}\right) \\ &< 1 + \left(\frac{1}{2^p} + \frac{1}{4^p} + \cdots + \frac{1}{(2N)^p}\right) + \left(\frac{1}{2^p} + \frac{1}{4^p} + \cdots + \frac{1}{(2N)^p}\right) \\ &= 1 + \frac{1}{2^p} S_N + \frac{1}{2^p} S_N = 1 + \frac{1}{2^{p-1}} S_N \ < \ 1 + \frac{1}{2^{p-1}} S_{2N+1} . \end{aligned}$$

That is,

$$\left(1 - \frac{1}{2^{p-1}}\right) S_{2N+1} < 1.$$

Therefore, the increasing sequence $\{S_N\}$ is bounded above by $\left(1 - \frac{1}{2^{p-1}}\right)^{-1}$ and so it converges, by the Increasing Bounded Sequence Property (Theorem 1.34). We have then:

$$\text{the } p\text{-series} \quad \sum_{n=1}^{\infty} \frac{1}{n^p} \quad \text{converges} \Leftrightarrow p > 1.$$

Taking $p = 2$ in the analysis above, we get

$$\sum_{n=1}^{\infty} \frac{1}{n^2} \ < \ 2.$$

We shall see in Theorem 12.7 that in fact,

$$\sum_{n=1}^{\infty} \frac{1}{n^2} = \frac{\pi^2}{6} \cong 1.645\,.$$

Being the first to find the sum of this series was one of Euler's many great triumphs.

Remark 6.25. Taking $p = 3$, it is the case that

$$\sum_{n=1}^{\infty} \frac{1}{n^3} < \left(1 - \frac{1}{4}\right)^{-1} = \frac{4}{3},$$

but the precise value of this sum is not known. It was proved in only 1979, by the French mathematician Roger Apéry (1916–1994), that $\sum\limits_{n=1}^{\infty} \frac{1}{n^3} \cong 1.202$ is irrational. The values of the sums $\sum\limits_{n=1}^{\infty} \frac{1}{n^p}$ are known if p is a positive even integer. ○

Remark 6.26. We have seen (essentially) that

$$\lim_{p\to 1+} \sum_{n=1}^{\infty} \frac{1}{n^p} = +\infty.$$

We finish by pointing out that, however, it is the case that

$$\lim_{p\to 1+} \left(\sum_{n=1}^{\infty} \frac{1}{n^p} - \frac{1}{p-1} \right) = \gamma.$$

Pretty cool—see [31], for example. ○

Exercises

6.1. Show that $C e^{kx}$ is the is the only function which satisfies

$$f(0) = C \quad \text{and} \quad f'(x) = k f(x) \quad \text{for all } x \in \mathbf{R}.$$

6.2. Consider the differential equation $y' = y$ with initial condition $y(0) = 1$. Apply Euler's method of tangent lines on $[0, x]$, for $x > 0$, with n equal subintervals to obtain the approximation $y(x) \cong \left(1 + \frac{x}{n}\right)^n$. What happens as $n \to \infty$?

6.3. [1] Show that $\phi \equiv 0$ and $\phi \equiv 1$ are the only functions defined on $(0, \infty)$ which satisfy $\phi(x)\phi(y) = \phi(x - y)$.

6.4. Show that $(\mathrm{e}^x)^y = \mathrm{e}^{xy}$ for all $x, y \in \mathbf{R}$.

6.5. Consider again the improved Bernoulli's Inequality from Example 6.12:

$$(1+x)^\alpha \geq 1 + \alpha x\,, \quad \text{for } \alpha > 1 \text{ and } x > -1.$$

Show that this inequality persists for $\alpha < 0$, and is reversed for $0 < \alpha < 1$.

6.6. **(a)** Show that for $x \geq 0$,

$$\mathrm{e}^x \geq 1 + x + \frac{1}{2}x^2, \quad \text{with equality only for } x = 0.$$

(b) Show that for $x \geq 0$,

$$\mathrm{e}^x \geq 1 + x + \frac{1}{2}x^2 + \frac{1}{3!}x^3 + \cdots + \frac{1}{n!}x^n, \quad \text{with equality only for } x = 0.$$

(c) Conclude that $\left(\frac{n}{\mathrm{e}}\right)^n < n!$

(d) For which n is $\mathrm{e}^x \geq 1 + x + \frac{1}{2}x^2 + \frac{1}{3!}x^3 + \cdots + \frac{1}{n!}x^n$ true for all $x \in \mathbf{R}$?

6.7. [65]

(a) Show that

$$\mathrm{e}^x > \left(1 + \frac{x}{y}\right)^y \quad \text{for } x, y > 0.$$

(b) Set $x = \pi - \mathrm{e}$ and $y = \mathrm{e}$ to conclude that $\mathrm{e}^\pi > \pi^\mathrm{e}$.

6.8. [67]

(a) Show that $f(x) = x^{1/x}$ $(x > 0)$ has an absolute maximum at $x = \mathrm{e}$.
(b) Conclude that $\mathrm{e}^x \geq x^\mathrm{e}$ for all x. (In particular, $\mathrm{e}^\pi > \pi^\mathrm{e}$.)
(c) Use $\mathrm{e}^x \geq x^\mathrm{e}$ to prove the AGM Inequality (Theorem 2.10) as follows: Set $x = \frac{a_j \mathrm{e}}{G}$ for each of $j = 1, 2, \ldots n$, then multiply all these together.

6.9. [35, 36, 77] We saw in (6.1) that for $n = 1, 2, 3, \ldots$,

$$\left(1 + \frac{1}{n}\right)^n < \mathrm{e} < \left(1 + \frac{1}{n}\right)^{n+1}.$$

(a) Show that $\alpha = 1/2$ is the least α for which

$$\mathrm{e} \leq \left(1 + \frac{1}{n}\right)^{n+\alpha} \quad \text{for } n = 1, 2, 3, \ldots.$$

(b) Find the largest β for which

$$\left(1+\frac{1}{n}\right)^{n+\beta} \leq \mathrm{e} \quad \text{for } n = 1, 2, 3, \ldots.$$

6.10. [22] Evaluate

$$\lim_{n\to\infty}\left(n^2\left[\left(1+\frac{1}{n+1}\right)^{n+1} - \left(1+\frac{1}{n}\right)^n\right]\right).$$

6.11. [62]

(a) Apply Cauchy's Mean Value Theorem (Theorem 5.11) to the functions $f(x) = (x+1)^\alpha$ and $g(x) = x^\alpha$ on $[n, n+1]$ to show that

$$\left(\frac{n+2}{n+1}\right)^\alpha \leq \frac{(n+2)^{\alpha+1} - (n+1)^{\alpha+1}}{(n+1)^{\alpha+1} - n^{\alpha+1}} \leq \left(\frac{n+1}{n}\right)^\alpha.$$

(b) Conclude, in particular, that

$$\mathrm{e} < \left(1+\frac{1}{n+1}\right)^{n+1} \leq \frac{(n+2)^{n+2} - (n+1)^{n+2}}{(n+1)^{n+1} - n^{n+1}} \leq \left(1+\frac{1}{n}\right)^{n+1} < \mathrm{e}\left(1+\frac{1}{n}\right).$$

6.12. [63]

(a) Show that

$$\lim_{n\to\infty}\frac{(n!)^{1/n}}{n} = \frac{1}{\mathrm{e}}.$$

Hint: In $\left(1+\frac{1}{k}\right)^k < \mathrm{e} < \left(1+\frac{1}{k}\right)^{k+1}$, take the product for $k = 1, 2, \ldots n$.

(b) Denote by A_n the Arithmetic Mean and by G_n the Geometric Mean, of the first n natural numbers. Show that the result in (a) is the same as

$$\lim_{n\to\infty}\frac{G_n}{A_n} = \frac{2}{\mathrm{e}}.$$

Other approaches to this problem can be found in [13, 44, 82]. It is generalized in various directions in [43, 68, 71, 83].

6.13. Suppose that a certain population at year $t \geq 0$ is given (approximately) by $P(t) = C\mathrm{e}^{kt}$, and that the population's growth is r % per year. Show that the population doubles in size every $\ln(2)/\ln(1 + r/100)$ years.

6.14. Let $x_1 > 1$ and for $n = 1, 2, \ldots$, let $x_{n+1} = \dfrac{x_n}{\ln(x_n)}$. Show that $\{x_n\}$ converges and find the limit.

6.15. Prove Lemma 6.9.

6.16. [41] Here's a way to show that $(\ln(x))' = \frac{1}{x}$, assuming we already know that $\left(1+\frac{1}{u}\right)^u \to \mathrm{e}$ as $u \to \infty$. Verify that

$$\frac{\ln(x+h)-\ln(x)}{h} = \frac{1}{x}\ln\left(1+\frac{h}{x}\right)^{x/h}, \quad \text{then let } h \to 0.$$

6.17. Show that for $a > 0$,

$$(a^x)' = a^x \ln(a)$$

and for differentiable functions $\alpha(x)$,

$$\left(a^{\alpha(x)}\right)' = a^{\alpha(x)} \ln(a)\alpha'(x).$$

6.18. **(a)** Use the Chain Rule to find $(\ln(ax))'$.
(b) Conclude that $\ln(ab) = \ln(a) + \ln(b)$ for $a, b > 0$.

6.19. The logarithmic function with base $a > 0$ but $a \neq 1$ is defined by

$$y = \log_a(x) \quad \Leftrightarrow \quad a^y = x.$$

For example, $\log_{\mathrm{e}}(x) = \ln(x)$.

(a) Prove the change of base formula (for $b > 0$)

$$\log_a(x) = \frac{\log_b(x)}{\log_b(a)}.$$

So in principle, by taking $b = \mathrm{e}$, one only needs to know *natural* logarithms. For example, $\log_{10}(x) = \ln(x)/\ln(10)$.

(b) Show that for $x > 0$ we have $(\log_a(x))' = \frac{\log_a(\mathrm{e})}{x}$ and that (for functions $\alpha > 0$ differentiable),

$$(\log_a \alpha(x))' = \frac{\log_a(e)}{\alpha(x)}\alpha'(x).$$

(c) Show that for $x > 1$ and $L > 0$,

$$\frac{d}{dx}\log_x(L) = -\frac{\log_x L}{x\ln(x)}.$$

6.20. [28] Fill in the details of the following proof that $\ln(x)$ is not a rational function. If it were, we could write $\ln(x) = \frac{p(x)}{q(x)}$, where p and q are polynomials *with no common factors*. Now differentiate both sides of this expression to obtain a contradiction.

6.21. **(a)** Use L'Hospital's Rule (Theorem 5.13) to show that $\lim_{x\to+\infty} \ln\left(1+\frac{1}{x}\right)^x = 1$.
(b) Conclude that $\lim_{x\to+\infty} \left(1+\frac{1}{x}\right)^x = \mathrm{e}$.

6.22. Evaluate the following limits (n is a positive integer).

(a) $\lim_{x\to+\infty} x(1-\mathrm{e}^{1/x})$, **(b)** $\lim_{x\to+\infty} \dfrac{\ln(1+\mathrm{e}^x)}{2x}$,
(c) $\lim_{x\to+\infty} \dfrac{x^n}{\mathrm{e}^x}$, **(d)** $\lim_{x\to+\infty} \dfrac{x^n}{\ln(x)}$.

6.23. [49]

(a) Show how we might (to some extent) improve inequality (6.7), namely $\ln(x) \le x-1$ for $x > 0$, by writing $\ln(x) = 2\ln\sqrt{x}$.
(b) On the way to inequality (6.7) we saw that $\ln(x+1) \le x$ for $x > -1$. Show that

$$\frac{x}{1+x} \le \ln(1+x) \quad \text{for} \quad x > -1.$$

(c) Show that

$$\frac{x}{1+\frac{1}{2}x} \le \ln(1+x) \quad \text{for} \quad x \ge 0.$$

6.24. [56,72]

(a) In Example 6.11 we saw that $x\ln(x) \ge x-1$ for $x > 0$. In this inequality, set $x = p_j/q_j$ then sum to obtain the following inequality, which is basic in Information Theory: *Let* $p_j, q_j > 0$ *for* $j = 1, 2, \ldots, n$, *with* $\sum_{j=1}^n p_j = \sum_{j=1}^n q_j$. *Then*

$$\sum_{j=1}^n p_j \ln(p_j) \ge \sum_{j=1}^n p_j \ln(q_j).$$

(b) Show that for $a, b, c > 0$,

$$a^a b^b c^c \ge \left(\frac{a+b}{2}\right)^a \left(\frac{b+c}{2}\right)^b \left(\frac{a+c}{2}\right)^c.$$

Can you extend this to more than three numbers?

6.25. [75] Use the AGM Inequality (Theorem 2.10) to prove the AGM Inequality with rational weights: *Let* $a_1, a_2, \ldots, a_n$ *be distinct nonnegative real numbers and let* $w_1, w_2, \ldots, w_n$ *be positive rational numbers satisfying* $\sum_{j=1}^n w_j = 1$. *Then*

$$\prod_{j=1}^n a_j^{w_j} < \sum_{j=1}^n w_j a_j.$$

Hint: Let M denote a common denominator for the fractions $w_1, w_2, \ldots, w_n$. Now apply the AGM Inequality to a suitable collection of M numbers, which contains (perhaps lots of) repetition.

6.26. [30] Fill in the details of another proof of the weighted AGM Inequality (Theorem 6.15), as follows. Set $A = \sum_{j=1}^{n} w_j a_j$ and $x = a_j/A$ in (6.7), i.e., in

$$\ln(x) \leq x - 1 \quad \text{for } x > 0.$$

Now multiply by w_j, and sum. This is a 1930 proof by Hungarian mathematician Frigyes Riesz (1880–1956). It is the logarithmic companion of the proof we gave (i.e., G. Polya's) of Theorem 6.15.

6.27. [50,64,66] Fill in the details of another proof of the weighted AGM Inequality (Theorem 6.15), as follows. Set $G = \prod_{j=1}^{n} a_j^{w_j}$ and $x = a_j/G$ in (6.7): $\ln(x) \leq x-1$ for $x > 0$. Now multiply by w_j, and sum. This is the Geometric Mean companion to Riesz's proof from Exercise 6.26.

6.28. [69] Fill in the details of another proof of the weighted AGM Inequality (Theorem 6.15), as follows.

(a) Verify that

$$\frac{nA - nG}{G} = \sum_{j=1}^{n} \left(\frac{a_j}{G} - 1\right).$$

(b) Now use this in (6.7), i.e., $\ln(x) \leq x - 1$ for $x > 0$.

6.29. [38,39] Let $a, b \geq 0$, $0 \leq t \leq 1$, and $r = \min\{t, 1-t\}$. Prove the following refinements of the weighted AGM Inequality with $n = 2$ (Corollary 6.16).

(a) $a^t b^{1-t} + r\left(\sqrt{a} - \sqrt{b}\right)^2 \leq ta + (1-t)b$.

(b) $\left(a^t b^{1-t}\right)^2 + r\left(a - b\right)^2 \leq \left(ta + (1-t)b\right)^2$.
Hint: In each case, treat $t \leq 1/2$ and $t > 1/2$ separately.

(c) Is the inequality in (a) better than the one in (b)? Or vise-versa? Or neither?

6.30. [4] In Exercise 6.5 we showed that the improved Bernoulli's Inequality ($\alpha > 1$)

$$(1+x)^\alpha \geq 1 + \alpha x, \quad \text{for } x > -1$$

persists for $\alpha < 0$, and is reversed for $0 < \alpha < 1$. Use the $0 < \alpha < 1$ case to prove the weighted AGM Inequality (Theorem 6.15), as follows.

(a) For $a_1, a_2 > 0$, substitute $x = \frac{a_1}{a_2} - 1$ to obtain the $n = 2$ case.

(b) Proceed by induction: Assume the result holds for $a_1, a_2, \ldots, a_n > 0$ then replace a_n with $a_n^{w_n/(w_n+w_{n+1})} a_{n+1}^{w_{n+1}/(w_n+w_{n+1})}$.

6.31. [74] Show that if a, b and c are the side lengths of a triangle, then

$$(a+b-c)^a(b+c-a)^b(a+c-b)^c \le a^a b^b c^c.$$

6.32. [70] Show that if $a, b, c \in (0,1)$ and $a+b+c=2$, then

$$a^{1-a}b^{1-b}c^{1-c} + a^{1-b}b^{1-c}c^{1-a} + a^{1-c}b^{1-a}c^{1-2} \le 2.$$

6.33. [55] Let $x_1, x_2, \ldots, x_n > 0$. Use the weighted AGM Inequality (Theorem 6.15) two different ways to show that

$$\left(\prod_{j=1}^{n} x_j\right)^{\frac{1}{n}\sum_{j=1}^{n} x_j} \le \prod_{j=1}^{n} x_j^{x_j} \le \left(\frac{\sum_{j=1}^{n} x_j^2}{\sum_{j=1}^{n} x_j}\right)^{\sum_{j=1}^{n} x_j}.$$

6.34. [76] Here is an ostensibly different proof of Hölder's Inequality (Lemma 6.18). **(a)** Dispense with the cases $a_1 = a_2 = \cdots = a_n = 0$ or $b_1 = b_2 = \cdots = b_n = 0$. **(b)** In Young's Inequality (Corollary 6.19), set

$$a = a_j\left(\sum_{j=1}^{n} a_j^p\right)^{-1/p} \quad \text{and} \quad b = b_j\left(\sum_{j=1}^{n} b_j^q\right)^{-1/q},$$

then sum from 1 to n. (Compare with Exercise 2.42.)

6.35. [40] Use Hölder's Inequality (Lemma 6.18) to show that

$$\frac{xyz}{(x+y+a)(y+z+a)(x+z+a)} \le \frac{1}{81a} \quad \text{for } a > 0 \text{ and } x, y, z > 0.$$

6.36. Find necessary and sufficient conditions for equality to hold in Hölder's Inequality (Lemma 6.18).

6.37. [30] We used the weighted AGM Inequality with $n = 2$ (Corollary 6.16) to obtain Hölder's Inequality (Lemma 6.18). Use the full weighted AGM Inequality (Theorem 6.15) to obtain the following **extension of Hölder's Inequality**. *Let* $a_{11}, a_{21}, a_{31}, \ldots, a_{n1}, a_{12}, a_{22}, a_{32} \ldots, a_{n2}, \ldots, a_{1m}, a_{2m}, a_{3m} \ldots, a_{nm}$, *be* nm *nonnegative real numbers and let* $p_1, p_2, \ldots, p_m > 1$ *satisfy* $\frac{1}{p_1} + \frac{1}{p_2} + \cdots + \frac{1}{p_m} = 1$. *Then*

$$\sum_{j=1}^{n}\left(\prod_{k=1}^{m} a_{jk}\right) \le \prod_{k=1}^{m}\left(\sum_{j=1}^{n} a_{jk}^{p_k}\right)^{1/p_k}.$$

6.38. We saw that Young's Inequality (Corollary 6.19)) follows directly from the weighted AGM Inequality with $n = 2$ (Corollary 6.16). Use the full weighted AGM Inequality (Theorem 6.15) to obtain the following **extension of Young's Inequality**. (See also [2].) *Let* $a_1, a_2, \ldots, a_n$ *be nonnegative and let* $p_1, p_2, \ldots, p_n > 1$ *satisfy* $\frac{1}{p_1} + \frac{1}{p_2} + \cdots + \frac{1}{p_n} = 1$. *Then*

$$\prod_{j=1}^{n} a_j \le \sum_{j=1}^{n} \frac{a_j^{p_j}}{p_j}.$$

6.39. In Exercise 2.53 we used the Cauchy–Schwarz Inequality (Theorem 2.18) to prove Minkowski's Inequality. Use Hölder's Inequality (Lemma 6.18) to prove the following **extension of Minkowski's Inequality**: *Let* $a_1, a_2, \ldots, a_n$, *and* $b_1, b_2, \ldots, b_n$ *be real numbers and let* $p > 0$. *Then*

$$\left(\sum_{j=1}^{n} (a_j + b_j)^p\right)^{1/p} \le \left(\sum_{j=1}^{n} a_j^p\right)^{1/p} + \left(\sum_{j=1}^{n} b_j^p\right)^{1/p}.$$

Hint: Write $(a_j + b_j)^p = a_j (a_j + b_j)^{p-1} + b_j (a_j + b_j)^{p-1}$, then sum, then apply Hölder's Inequality to each piece.

6.40. In Exercise 2.54 we proved **Chebyshev's Inequality**: *Let* $\{a_1, a_2, \ldots, a_n\}$, $\{b_1, b_2, \ldots, b_n\}$ *be two sequences of real numbers, with either both increasing or both decreasing. Then*

$$\frac{1}{n}\sum_{j=1}^{n} a_j \cdot \frac{1}{n}\sum_{j=1}^{n} b_j \le \frac{1}{n}\sum_{j=1}^{n} a_j b_j.$$

(a) Prove the **weighted Chebyshev's Inequality**:
Let $\{\alpha_1, \alpha_2, \ldots, \alpha_n\}$ *and* $\{\beta_1, \beta_2, \ldots, \beta_n\}$ *be sequences of real numbers, with either both increasing or both decreasing. Let* $w_1, w_2, \cdots, w_n$ *be any* n *nonnegative numbers. Then*

$$\sum_{j=1}^{n} w_j \alpha_j \cdot \sum_{j=1}^{n} w_j \beta_j \le \sum_{j=1}^{n} w_j \cdot \sum_{j=1}^{n} w_j \alpha_j \beta_j.$$

(b) Use this to prove the Cauchy–Schwarz Inequality (Theorem 2.18). Hint: First dispense with any $b_j^{\prime}s$ which equal zero (and explain). Then set $w_j = \beta_j^2$, and $\alpha_j = \beta_j = a_j / b_j$.

6.41. [9, 10] In Lemma 6.20 we saw that for $a, b > 0$,

$$\sqrt{ab} \le \frac{a - b}{\ln(a) - \ln(b)} \le \frac{a + b}{2} \quad \text{i.e} \quad G \le L \le A.$$

(a) Set $a = x$ and $b = 1/x$ to show that

$$\frac{1}{x^2+1} \le \frac{\ln(x)}{x^2-1} \le \frac{1}{2x} \quad \text{for } x > 0.$$

(b) How do the inequalities in (a) compare with

$$1 - \frac{1}{x} \le \ln(x) \le x - 1 \quad \text{for } x > 0\,?$$

(c) Suppose now that $0 < a \le b$ and let $b/a = 1 + x$ to show that

$$\frac{x}{1+\frac{1}{2}x} \le \ln(1+x) \le \frac{x}{\sqrt{1+x}}.$$

6.42. [80] Denote by $L(a,b)$ the Logarithmic Mean of $a, b > 0$.

(a) Show that for $n \ge 7$,

$$L\big(\ln(n+1), \ln(n)\big) > 2.$$

(b) Conclude that for $n \ge 7$,

$$n^{\sqrt{n+1}} > (n+1)^{\sqrt{n}}.$$

(c) How about $n < 7$?

6.43. Let $a, b > 0$ and denote by L and A the Logarithmic and Arithmetic Means of a and b respectively. Apply Cauchy's Mean Value Theorem (Theorem 5.11) to

$$f(x) = \frac{x-a}{\frac{2}{3}\sqrt{xa} + \frac{1}{3}\frac{a+x}{2}} = \frac{6(x-a)}{4\sqrt{xa} + x + a} \quad \text{and} \quad g(x) = \ln(x) - \ln(a)$$

on $[a,b]$ to show that $L \le \frac{2}{3}G + \frac{1}{3}A \le A$. This refines the inequality $L \le A$ from Lemma 6.20. (This was first obtained by other methods in [60].)

6.44. Let $a, b > 0$ and denote by G and L the Geometric and Logarithmic Means of a and b respectively. Apply Cauchy's Mean Value Theorem (Theorem 5.11) to

$$f(x) = \frac{x-a}{(\sqrt{xa})^{2/3}\left(\frac{x+a}{2}\right)^{1/3}} = \frac{2^{1/3}(x-a)}{(x^2a + a^2x)^{1/3}} \quad \text{and} \quad g(x) = \ln(x) - \ln(a)$$

on $[a,b]$ to show that

$$G \le G^{2/3}A^{1/3} \le L.$$

This refines the inequality $G \le L$ from Lemma 6.20. (This was first obtained, by rather sophisticated methods, in [45].) Hint: Near the end, you will use the weighted AGM Inequality with $n = 2$ (Corollary 6.16) applied to A^2 and G^2: $\left(A^2\right)^{1/3}\left(G^2\right)^{2/3} \le \frac{1}{3}A^2 + \frac{2}{3}G^2$.

6.45. [27] Taken together, Exercises 6.43 and 6.44 give

$$G^{2/3}A^{1/3} \le L \le \frac{2}{3}G + \frac{1}{3}A.$$

(a) Apply these to x and $x+1$ to obtain the further improvement of (6.9):

$$\left(1+\frac{1}{x}\right)^{\frac{2}{3}\sqrt{x(x+1)}+\frac{1}{3}\frac{2x+1}{2}} \le \mathrm{e} \le \left(1+\frac{1}{x}\right)^{(\sqrt{x(x+1)})^{2/3}\left(\frac{2x+1}{2}\right)^{1/3}}.$$

(b) Show that $1 \le \left(\cosh(t)\right)^{1/3} \le \frac{\sinh(t)}{t} \le \frac{2}{3} + \frac{1}{3}\sinh(t) \le \cosh(t)$.

6.46. [47] For $x, y > 0$, consider the **Lorentz Mean**

$$M_{1/3} = \left(\frac{x^{1/3}+y^{1/3}}{2}\right)^3,$$

which comes up in the theory of equations of state for gases. Show that this is indeed a *mean* then show, as follows, that

$$L < M_{1/3} \quad \text{for } x \neq y,$$

where L is the Logarithmic Mean of x and y.

(a) For $t \ge 1$, set

$$f(t) = \frac{3}{8}\ln(t) - \frac{t^3-1}{(t+1)^3}.$$

Show that $f'(t) > 0$ for $t > 1$, so that f is increasing.

(b) Conclude that $f(t) > 0$ for $t > 1$.

(c) Assume that $0 < y < x$ and substitute $t = x^{1/3}/y^{1/3}$.

(This argument is due to American mathematician Harley Flanders (1925–2013.))

6.47. [18, 19, 25, 26]

(a) Fill in the details of the following proof, due to American mathematician Leonard Gillman (1917–2009), that the Harmonic series diverges:

$$\begin{aligned} S &= \left(1+\frac{1}{2}\right)+\left(\frac{1}{3}+\frac{1}{4}\right)+\left(\frac{1}{5}+\frac{1}{6}\right)+\cdots \\ &> \left(\frac{1}{2}+\frac{1}{2}\right)+\left(\frac{1}{4}+\frac{1}{4}\right)+\left(\frac{1}{6}+\frac{1}{6}\right)+\cdots \\ &= S. \end{aligned}$$

(b) And here's a similar argument, though slightly more complicated. If $S = \sum_{n=1}^{\infty} \frac{1}{n}$, then $\frac{1}{2}S = \frac{1}{2}\sum_{n=1}^{\infty}\frac{1}{n} = \sum_{n=1}^{\infty}\frac{1}{2n}$ and so we must have $\sum_{n=1}^{\infty}\frac{1}{2n-1} = \frac{1}{2}S$ also. Show that this leads to a contradiction.

6.48. [79] Fill in the details of another proof that the Harmonic series diverges: If $S = \sum_{n=1}^{\infty}\frac{1}{n}$ exists then $S_{2N} - S_N \to 0$, where $S_N = \sum_{n=1}^{N}\frac{1}{n}$. Show that $S_{2N} - S_N > \frac{1}{2}$ to obtain a contradiction.

6.49. [21] Fill in the details of another proof that the Harmonic series diverges: Obtain a contradiction by observing that

$$\begin{aligned} \sum_{n=1}^{\infty}\frac{1}{n} &= \sum_{n=0}^{\infty}\left(\frac{1}{2n+1}+\frac{1}{2n+2}\right) \\ &= \sum_{n=0}^{\infty}\left(\frac{1}{n+1}+\frac{1}{(2n+1)(2n+2)}\right). \end{aligned}$$

6.50. [17] Fill in the details of another proof that the Harmonic series diverges:

(a) Prove (or at least recall) the well-known fact (e.g., Exercise 5.35) that for the Fibonacci sequence $f_1 = 1$, $f_2 = 1$, $f_{n+2} = f_n + f_{n+1}$, it is the case that $f_{n+1}/f_n \to \varphi$, where φ is the golden mean, i.e., the positive root of $x^2 - x - 1 = 0$.

(b) By collecting successive blocks of the Harmonic series whose lengths are the Fibonacci numbers, show that

$$\sum_{n=1}^{\infty}\frac{1}{n} \geq 1 + \sum_{n=1}^{\infty}\frac{f_{n-1}}{f_{n+1}}.$$

6.51. [34]

(a) Show that $\sum_{n=1}^{\infty}\left(n^{1/n}-1\right)$ diverges to $+\infty$.

(b) Show that $\sum_{n=1}^{\infty}\left(\mathrm{e}-\left(1+\frac{1}{n}\right)^n\right)$ diverges to $+\infty$.

6.52. [3] Here's another approach to Euler's constant. Define a_k by

$$\mathrm{e}=a_k\left(1+\frac{1}{k}\right)^k,$$

(a) Verify that $1=\ln(a_k)+k\big(\ln(k+1)-\ln(k)\big)$.

(b) Sum these from $k=1$ to n to get

$$\sum_{k=1}^{n}\frac{1}{k}-\ln(n+1)=\ln\left(\prod_{k=1}^{n}a_k^{1/k}\right).$$

(c) Show that $\left\{\ln\left(\prod_{k=1}^{n}a_k^{1/k}\right)\right\}$ is increasing.

(d) On the way to inequality (6.7) we saw that $\ln(x+1)\le x$ for $x>-1$. Use this to show that

$$\ln\left(\prod_{k=1}^{n}a_k^{1/k}\right)<\sum_{k=1}^{n}\frac{1}{k^2},$$

so that $\left\{\ln\left(\prod_{k=1}^{n}a_k^{1/k}\right)\right\}$ is bounded above (by $\pi^2/6$).

6.53. (a) Show that we may define Euler's constant by way of

$$\gamma_n=\sum_{k=1}^{n}\frac{1}{k}-\ln(n+\alpha),$$

for any $\alpha>-n$. Therefore

$$\gamma=\lim_{n\to\infty}\left(\sum_{k=1}^{n}\frac{1}{k}-\ln(n+\alpha)\right).$$

(b) Take $\alpha=1$ in (a) then show that

$$\gamma=\lim_{n\to\infty}\left(\sum_{k=1}^{n}\frac{1}{k}-\ln\left(1+\frac{1}{k}\right)\right).$$

6.54. [33] Let K_j be the least integer for which $\sum_{n=1}^{K_j} \frac{1}{n} \geq j$. Show that

$$\lim_{j\to\infty} \frac{K_{j+1}}{K_j} = \mathrm{e}.$$

6.55. [58]

(a) Show that $x - \frac{x^3}{6} \leq \sin(x) \leq x$, for $0 < x < \pi/2$.

(b) Set $x = 1/k$ and use $\sum_{j=1}^{n} j = \frac{n(n+1)}{2}$ to show that

$$\lim_{n\to\infty} \left(\frac{1}{n^2} \sum_{j=1}^{n} \csc\left(\frac{1}{k}\right) \right) = \frac{1}{2}.$$

6.56. [14,59] The Alternating Harmonic series has Nth partial sum

$$S_N = \sum_{n=1}^{N} \frac{(-1)^{n+1}}{n}.$$

Show that

$$|S - S_N| \leq \frac{1}{2N}.$$

6.57. In Sect. 6.7 we considered $\gamma_{2n} - \gamma_n$ to show that

$$\ln(2) = \sum_{n=1}^{\infty} \frac{(-1)^{n+1}}{n} = 1 - \frac{1}{2} + \frac{1}{3} - \frac{1}{4} + \cdots.$$

(a) Consider $\gamma_{2n} - \frac{1}{2}\gamma_n - \frac{1}{2}\gamma_{2n}$ to sum a certain rearrangement of this series.
(b) Can you sum other rearrangements in a similar way?

6.58. **(a)** Show that $\sum_{n=1}^{\infty} \frac{1}{n^p}$ diverges for $p \leq 1$.

(b) Use

$$S_N = \sum_{n=1}^{N} \frac{1}{n^2} = 1 + \sum_{n=2}^{N} \frac{1}{n^2} < 1 + \sum_{n=2}^{N} \frac{1}{(n-1)n}$$

to show that $\sum_{n=1}^{\infty} \frac{1}{n^2} \leq 2$. (The series on the right-hand side is a *telescoping* series).

(c) Use (b) to show that $\sum_{n=1}^{\infty} \frac{1}{n^p}$ converges for $p \geq 2$. (We showed in Sect. 6.7 that it also converges for $1 < p < 2$.)

6.59. [57] Here is an extension of Example 1.32, which shows that $\sum_{n=1}^{\infty} \frac{1}{n^p}$ converges for $p > 1$.

(a) Verify that for natural numbers $r > 1$,

$$\frac{b^r - a^r}{b - a} = b^{r-1} + ab^{r-2} + a^2b^{r-3} + \ldots + a^{r-3}b^3 + a^{r-2}b^2 + a^{r-1}.$$

(b) Conclude that for $a < b$, $b^r - a^r \leq rb^{r-1}(b - a)$.
(c) Set $a = n^{1/r}$ and $b = (n + 1)^r$ to get

$$1 = \sum_{n=1}^{\infty} \left(\frac{1}{n^{1/r}} - \frac{1}{(n+1)^{1/r}} \right) \geq \frac{1}{r} \sum_{n=1}^{\infty} \frac{1}{(n+1)^{1+1/r}}.$$

(d) Conclude that $\sum_{n=1}^{\infty} \frac{1}{(n+1)^{1+1/r}}$ is convergent.
(e) Finally, for $p > 1$ observe that there is a natural number $r > 1$ such that $1 + \frac{1}{r} < p$.

6.60. [51] Evaluate $\lim_{n\to\infty} \left(1 + \frac{1}{2} + \frac{1}{3} + \cdots + \frac{1}{n} - \left(\frac{1}{n+1} + \frac{1}{n+2} + \cdots + \frac{1}{n^2}\right)\right)$.

6.61. [37]

(a) Show that the sequence $\{\delta_n\}$ is increasing for $n \geq 2$, where

$$\delta_n = \ln(n) - 2\left[\frac{1}{3} + \frac{1}{5} + \frac{1}{7} + \cdots + \frac{1}{2n-1}\right].$$

(b) Show that $\delta_n \to -\gamma + 2\big(1 - \ln(2)\big)$, as $n \to \infty$ (where γ is Euler's constant).

6.62. [29] Show that if S is the sum of a particular p-series (of course $p > 1$) then

$$\frac{2^p - 1}{2^p - 2} < S < \frac{2^p}{2^p - 2}.$$

6.63. [6] Fill in the details of the following neat visual description of why

$$\sum_{n=1}^{\infty} \frac{1}{n} = +\infty, \quad \sum_{n=1}^{\infty} \frac{1}{n^2} < 2, \quad \text{and} \quad \sum_{n=1}^{\infty} \frac{1}{n^3} < \frac{3}{2}.$$

Suppose that we stack cubes with side lengths $1, 1/2, 1/3, 1/4, 1/5\dots$ together, as shown in **Fig. 6.4**.

(a) Looking at the side view, the height of each vertical stack is $> 1/2$ and so we have $\sum_{n=1}^{\infty} \frac{1}{n} = +\infty$.

(b) Looking again at the side view, the total area obtained by taking one face of each cube gives $\sum_{n=2}^{\infty} \frac{1}{n^2} < 1$, and so $\sum_{n=1}^{\infty} \frac{1}{n^2} < 2$.

(c) Looking at the full view, all of the cubes are inside a $1 \times 1 \times \frac{3}{2}$ box. Therefore their total volume gives $\sum_{n=1}^{\infty} \frac{1}{n^3} < \frac{3}{2}$.

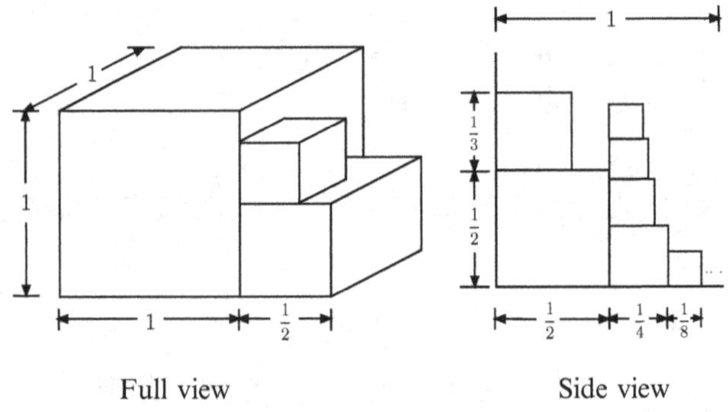

Fig. 6.4 For Exercise 6.63: $\sum_{n=1}^{\infty} \frac{1}{n} = +\infty$, $\sum_{n=1}^{\infty} \frac{1}{n^2} < 2$, and $\sum_{n=1}^{\infty} \frac{1}{n^3} < \frac{3}{2}$

References

1. Ahuja, M., Palmer, L., Mayne, D.C.: Problem 102. Coll. Math. J. **10**, 129–130 (1979)
2. Akhlaghi, M.R.: Another proof of Young's inequality for products. Am. Math. Mon. **120**, 518 (2013)
3. Barnes, C.W.: Euler's constant and e. Am. Math. Mon. **91**, 428–430 (1984)
4. Beckenbach, E.F., Bellman, R.: Inequalities. Springer, Berlin (1965)
5. Beigel, R.: Rearranging terms in alternating series. Math. Mag. **54**, 244–246 (1981)
6. Berkove, E.: Another look at some p-series. Coll. Math. J. **37**, 385–386 (2006)
7. Bradley, D.M.: The harmonic series. Am. Math. Mon. **107**, 615 (2000)
8. Brown, F., Cannon, L.O., Elich, F., Wright, D.G.: On rearrangements of the alternating harmonic series. Coll. Math. J. **16**, 135–138 (1985)
9. Bruce, I.: The logarithmic mean. Math. Gaz. **81**, 89–92 (1997)
10. Bullen, P.S.: Handbook of Means and Their Inequalities. Kluwer, Dordrecht/Boston (2003)
11. Burk, F.: By all means. Am. Math. Mon. **107**, 50 (1985)
12. Burk, F.: Euler's constant. Coll. Math. J. **16**, 79 (1985)

13. Burk, F., Goel, S.K., Rodriguez, D.M.: Using Riemann sums in evaluating a familiar limit. Coll. Math. J. **17**, 170–171 (1986)
14. Calabrese, P.: A note on alternating series. Am. Math. Mon. **69**, 215–217 (1962)
15. Carlson, B.C.: The logarithmic mean. Am. Math. Mon. **79**, 615–618 (1972)
16. Chao, W.W., Seiffert, H.J.: Problem 10261. Am. Math. Mon. **101**, 690 (1994)
17. Chen, H., Kennedy, C.: Harmonic series meets Fibonacci series. Coll. Math. J. **43**, 237–243 (2012)
18. Chowdhury, M.R.: The harmonic series again. Math. Gaz. **59**, 186 (1975)
19. Cohen, T., Knight, W.J.: Convergence and divergence of $\sum 1/n^p$. Math. Mag. **52**, 178 (1979)
20. Cowen, C., Davidson, K., Kaufman, R.P.: Rearranging the alternating harmonic series. Am. Math. Mon. **87**, 817–819 (1980)
21. Cusumano, A.: The harmonic series diverges. Am. Math. Mon. **105**, 608 (1998)
22. Cusumano, A., The Western Maryland College Problems Group: Problem 1331. Math. Mag. **64**, 61–63 (1991)
23. Darst, R.B.: Simple proofs of two estimates for e. Am. Math. Mon. **80**, 194 (1973)
24. Dunkel, O.: Relating to the exponential function. Am. Math. Mon. **24**, 244–246 (1917)
25. Ecker, M.W.: Divergence of the harmonic series by rearrangement. Coll. Math. J. **28**, 209–210 (1997)
26. Editor's note. Coll. Math. J. **28**, 411 (1997)
27. Halliwell, G.T., Mercer, P.R.: A refinement of an inequality from information theory. J. Inequal. Pure Appl. Math. **5**, 29–31 (2004)
28. Hamming, R.W.: An elementary discussion of the transcendental nature of the elementary transcendental functions. Am. Math. Mon. **77**, 294–297 (1970)
29. Hansheng, Y.: Another proof of the p-series test. Coll. Math. J. **36**, 235–237 (2005)
30. Hardy, G.H., Littlewood, J.E., Polya, G.: Inequalities. Cambridge University Press, Cambridge/New York (1967).
31. Havil, J.: Gamma; Exploring Euler's Constant. Princeton University Press, Princeton/Oxford (2003)
32. Ho, W.K., Ho, F.H., Lee, T.Y.: Exponential function and its derivative revisited. Int. J. Math. Ed. Sci. Tech. **44**, 423–428 (2013)
33. Howard, J., Kane, S., Searcy, C.: Problem 280. Coll. Math J. **17**, 188 (1986)
34. Johnsonbaugh, R., Chouteau, C.: Problem E2361. Am. Math. Mon. **80**, 693–694 (1973)
35. Just, E., Schaumberger, N., Klamkin, M.S.: Problem E2406. Am. Math. Mon. **81**, 291–292 (1974)
36. Kang, C.X.: The convergence behavior of $f_\alpha(x) = (1 + 1/x)^{x+\alpha}$. Coll. Math. J. **38**, 385–387 (2007)
37. Khajeh-Khalili, P., Krishnapriyan, H.K.: Problem 505. Coll. Math. J. **25**, 245–246 (1994)
38. Kittaneh, F., Manasrah, Y.: Matrix Young inequalities for the Hilbert-Schmidt norm. Linear Algebr. Appl. **308**, 77–84 (2000)
39. Kittaneh, F., Manasrah, Y.: Improved Young and Heinz inequalities for matrices. J. Math. Anal. Appl. **361**, 262–269 (2010)
40. Klamkin, M.S.: Problem Q942. Math. Mag. **77**, 234 (2004)
41. Knuth, D.: Algorithmic thinking and mathematical thinking. Am. Math. Mon. **92**, 170–181 (1985)
42. Körner, T.W.: A Companion to Analysis. AMS, Providence (2004)
43. Kubelka, R.: Means to an end. Math. Mag. **74**, 141–142 (2001)
44. Lang, S.: A First Course in Calculus. Addison Wesley, Reading (1964)
45. Leach, E.B., Sholander, M.C.: Extended mean values II. J. Math. Anal. Appl. **92**, 207–223 (1983)
46. Lesko, J.: A seies for $\ln(K)$. Coll. Math. J. **32**, 119–122 (2001)
47. Lin, T.P.: The power mean and the logarithmic mean. Am. Math. Mon. **81**, 879–883 (1974)
48. Lord, N.: Seeing the divergence of the Harmonic series. Math. Gaz. **87**, 125–126 (2003)
49. Love, E.R.: Some logarithm inequalities. Math. Gaz. **64**, 55–57 (1980)

50. Lucht, L.G.: On the arithmetic-geometric mean inequality. Am. Math. Mon. **102**, 739–740 (1995)
51. Macys, J.J.: A new problem. Am. Math. Mon. **119**, 82 (2012)
52. McGehee, O.C.: Properties of the exponential function derived from its differential equation (2014). www.math.lsu.edu/~mcgehee
53. Melzak, Z.A.: On the exponential function. Am. Math. Mon. **82**, 842–844 (1975)
54. Mendelsohn, N.S.: An application of a famous inequality. Am. Math. Mon. **58**, 563 (1951)
55. Milalković, Ž.M., Keller, J.B., Lyons, R.: Problem E2691. Am. Math. Mon. **86**, 224–225 (1979)
56. Mitrinovic, D.S.: Analytic Inequalities. Springer, New York (1970)
57. Nillsen, R.: A proof that the series $\sum_{n=1}^{\infty} \frac{1}{n^p}$ is convergent for $p > 1$. Math. Gaz. **97**, 273–274 (2013)
58. Pfaff, T.J., Hartman, J.: Problem 726. Coll. Math. J. **34**, 240–241 (2003)
59. Pinsky, M.: Averaging an alternating series. Math. Mag. **51**, 235–237 (1978)
60. Polya, G., Szegő, G.: Isoperimetric Inequalities in Mathematical Physics. Princeton Univercity Press, Princeton (1951)
61. Rudin, W.R.: Principles of Mathematical Analysis, 3rd edn. McGraw-Hill, New York (1976)
62. Sandor, J.: On an inequality of Alzer. J. Math. Anal. Appl. **192**, 1034–1035 (1995)
63. Schaumberger, N.: Alternate approaches to two familiar results. Coll. Math. J. **15**, 422–423 (1984)
64. Schaumberger, N.: More applications of the mean value theorem. Coll. Math. J. **16**, 397–398 (1985)
65. Schaumberger, N.: An instant proof of $e^{\pi} > \pi^{e}$. Coll. Math. J. **16**, 280 (1985)
66. Schaumberger, N.: A general form of the arithmetic-geometric mean inequality via the mean value theorem. Coll. Math. J. **19**, 172–173 (1988)
67. Schaumberger, N.: The AM-GM inequality via $x^{1/x}$. Coll. Math. J. **20**, 320 (1989)
68. Schaumberger, N.: A generalization of $\lim \sqrt[n]{n!}/n = e^{-1}$. Coll. Math. J. **20**, 416–418 (1989)
69. Schaumberger, N.: Mathematics without words. Coll. Math. J. **31**, 68 (2000)
70. Schaumberger, N.: Problem Q930. Math. Mag. **76**, 152, 155 (2003)
71. Schaumberger, N., Boivin, S.: Problem 275. Coll. Math. J. **17**, 97 (1986)
72. Schaumberger, N., Howard, J.: Problem 524. Coll. Math. J. **26**, 161–162 (1995)
73. Scott, J.A.: The arithmetic-geometric mean from an e^x sequence. Math. Gaz. **71**, 40–41 (1987)
74. Selby, M., Yip, L.W.: Problem 233. Coll. Math. J. **15**, 272–273 (1984)
75. Steele, J.M.: The Cauchy-Schwarz Master Class. Mathematical Association of America. Cambridge University Press, Washington DC - Cambridge/New York (2004).
76. Tolsted, E.: An elementary derivation of the Cauchy, Hölder, and Minkowski inequalities from Young's inequality. Math. Mag. **37**, 2–12 (1964)
77. Truitt, S.A., Easley, D.: Problem 442. Coll. Math. J. **23**, 71–72 (1992)
78. van Yzeren, J.: A rehabilitation of $(1 + z/n)^n$. Am. Math. Mon. **77**, 995–999 (1970)
79. Ward, A.J.B.: Divergence of the harmonic series. Math. Gaz. **54**, 277 (1970)
80. Webb, J.H., Levine, E., Smith, J.T.: Problem 360. Coll. Math. J. **20**, 345–346 (1989)
81. Wetzel, J.E.: On the functional inequality $f(x + y) \geq f(x)f(y)$. Am. Math. Mon. **74**, 1065–1068 (1967)
82. Wilker, J.B.: Stirling ideas for freshman calculus. Math. Mag. **57**, 209–214 (1984)
83. Xu, C.: A GM-AM ratio. Math. Mag. **83**, 49–50 (2010)

Chapter 7
Other Mean Value Theorems

One cannot fix one's eyes on the commonest natural production without finding food for a rambling fancy.

—*Mansfield Park,* by Jane Austen

In this chapter, which is independent of all subsequent chapters, we allow ourselves a brief diversion. We have met and used Rolle's Theorem (Theorem 5.1), its extension the Mean Value Theorem (Theorem 5.2), and *its* extension Cauchy's Mean Value Theorem (Theorem 5.11). Here we consider other Mean Value – type theorems. Each of these, as with their namesake, has an appealing geometric interpretation. For convenience we recall below the Mean Value Theorem.

Theorem 5.2. (Mean Value Theorem): *Let f be continuous on $[a,b]$ and differentiable on (a,b). Then there exists $c \in (a,b)$ such that*

$$f'(c) = \frac{f(b) - f(a)}{b - a}. \qquad \square$$

7.1 Darboux's Theorem

We begin with a preliminary result, which extends Rolle's Theorem (Theorem 5.1), to cases in which $f'(b)$ exists. It was obtained by D.H. Trahan [18] in 1966.

Lemma 7.1. *Let f be continuous on $[a,b]$ and differentiable on $(a,b]$, with*

$$\big[f(b) - f(a)\big]f'(b) \leq 0.$$

Then there exists $c \in (a,b]$ such that $f'(c) = 0$.

Proof. If $f(b) - f(a) = 0$ then the result holds, by Rolle's Theorem (Theorem 5.1). If $f'(b) = 0$ then the result holds, with $c = b$. Otherwise, we consider the function g defined on $[a,b]$ via

P.R. Mercer, *More Calculus of a Single Variable*, Undergraduate Texts in Mathematics, DOI 10.1007/978-1-4939-1926-0_7

$$g(x) = \begin{cases} \frac{f(b)-f(x)}{b-x} & \text{if } x \neq b \\ f'(b) & \text{if } x = b. \end{cases}$$

Then g is continuous on $[a, b]$, and by hypothesis it satisfies

$$g(a)g(b) = \frac{f(b) - f(a)}{b - a} f'(b) < 0.$$

So by Bolzano's Theorem (Theorem 3.7) there is $\xi \in (a, b)$ such that $g(\xi) = 0$. That is, $f(\xi) = f(b)$. Then by Rolle's Theorem (Theorem 5.1), there is $c \in (\xi, b)$ such that $f'(c) = 0$, as desired. □

Figure 7.1 contains a generic picture for Lemma 7.1.

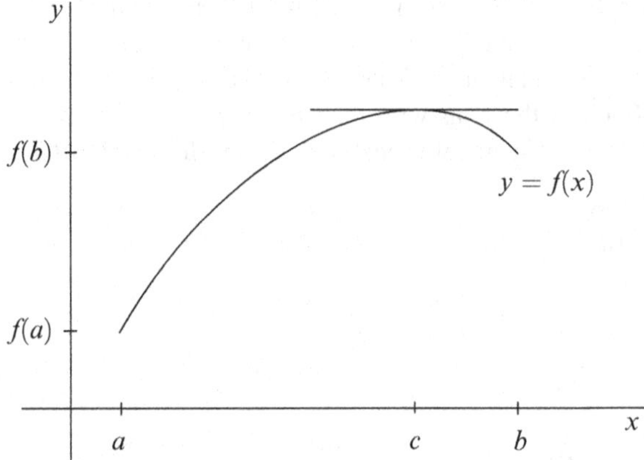

Fig. 7.1 Lemma 7.1: Here $f(b) - f(a) > 0$, $f'(b) < 0$, and $f'(c) = 0$

The Intermediate Value Theorem (Theorem 3.17) says, in short, that a continuous function satisfies the Intermediate Value Property. It is a rather surprising fact that derivatives (which need not be continuous) also satisfy the Intermediate Value Property. This discovery was made in 1875 by French mathematician J.G. Darboux (1842–1917).

We provide a 2004 proof due to L. Olsen [10], which is a natural extension of the proof of Lemma 7.1: the g in the proof of Lemma 7.1 is the g_1 in the proof below. This is different from the proof found in most textbooks, which we leave for Exercise 7.3. (See [4, 8] for two other clever proofs.)

Theorem 7.2. (Darboux's Theorem) *Let f be differentiable on $[a, b]$. Let y be between $f'(a)$ and $f'(b)$. Then there is $c \in (a, b)$ such that $f'(c) = y$.*

Proof. We may suppose that y is strictly between $f'(a)$ and $f'(b)$. The functions

$$g_1(x) = \begin{cases} \frac{f(b)-f(x)}{b-x} & \text{if } x \neq b \\ f'(b) & \text{if } x = b \end{cases} \quad \text{and} \quad g_2(x) = \begin{cases} f'(a) & \text{if } x = a \\ \frac{f(x)-f(a)}{x-a} & \text{if } x \neq a \end{cases}$$

are each continuous on $[a,b]$ and satisfy $g_1(a) = g_2(b)$. So since y lies between $f'(a) = g_2(a)$ and $f'(b) = g_1(b)$, y must either lie between $g_1(a)$ and $g_1(b)$ or between $g_2(a)$ and $g_2(b)$. If y lies between $g_1(a)$ and $g_1(b)$ then by the Intermediate Value Theorem (Theorem 3.17) there is $p \in [a,b)$ such that

$$y = g_1(p) = \frac{f(b) - f(p)}{b - p}.$$

Then by the Mean Value Theorem (Theorem 5.2) there is $c \in (p,b)$ such that

$$\frac{f(b) - f(p)}{b - p} = f'(c).$$

Therefore $f'(c) = y$, as we wanted to show. The case of y lying between $g_2(a)$ and $g_2(b)$, which is left to the reader, is handled in a similar fashion. □

Example 7.3. Consider the function

$$f(x) = \begin{cases} x^2 \sin(\frac{1}{x}) & \text{if } x \neq 0 \\ 0 & \text{if } x = 0. \end{cases}$$

By definition,

$$f'(0) = \lim_{x \to 0} \frac{f(x) - f(0)}{x - 0} = \lim_{x \to 0} \left(x \sin(\tfrac{1}{x})\right) = 0.$$

For $x \neq 0$, the Product and Chain Rules give

$$f'(x) = 2x \sin(\tfrac{1}{x}) - \cos(\tfrac{1}{x}).$$

Now since $\lim_{x \to 0} \left(\cos(\frac{1}{x})\right)$ does not exist, $\lim_{x \to 0} \left(2x \sin(\frac{1}{x}) - \cos(\frac{1}{x})\right)$ does not exist, and so

$$\lim_{x \to 0} f'(x) \neq f'(0).$$

Therefore f' is not continuous at $x = 0$. Even so, by Darboux's Theorem (Theorem 7.2), $f'(x)$ has the Intermediate Value Property on any interval containing $x = 0$. ◇

7.2 Flett's Mean Value Theorem

The following Mean Value – type theorem was discovered by T.M. Flett in 1958 [3]. Our proof follows Trahan's paper [18], which uses Lemma 7.1. (And the g_2 in the proof of Darboux's Theorem (Theorem 7.2) is the g in the proof below.) Another proof can be found in [12]; see also [1].

Theorem 7.4. (Flett's Mean Value Theorem) *Let f be differentiable on $[a,b]$, with $f'(a) = f'(b)$. Then there exists $c \in (a,b)$ such that*

$$f'(c) = \frac{f(c) - f(a)}{c - a}.$$

Proof. Consider the function g defined on $[a,b]$ via

$$g(x) = \begin{cases} f'(a) & \text{if } x = a \\ \frac{f(x)-f(a)}{x-a} & \text{if } x \neq a. \end{cases}$$

Then

$$g'(x) = \frac{(x-a)f'(x) - (f(x) - f(a))}{(x-a)^2} = \frac{1}{x-a}\left(f'(x) - \frac{f(x) - f(a)}{x-a}\right),$$

and so

$$g'(b) = \frac{1}{b-a}\left(f'(b) - \frac{f(b) - f(a)}{b-a}\right).$$

By the definition of g,

$$g(b) - g(a) = \frac{f(b) - f(a)}{b-a} - f'(a).$$

Now since $f'(a) = f'(b)$ by hypothesis, $g'(b)[g(b)-g(a)] < 0$. So by Lemma 7.1 there is $c \in (a,b)$ such that $g'(c) = 0$. That is,

$$f'(c) = \frac{f(c) - f(a)}{c - a},$$

as desired. □

Geometrically, Flett's Mean Value Theorem (Theorem 7.4) says that there exists $c \in (a,b)$ such that the line through $(a, f(a))$ and $(c, f(c))$ coincides with the tangent line at $x = c$. See **Fig. 7.2.**

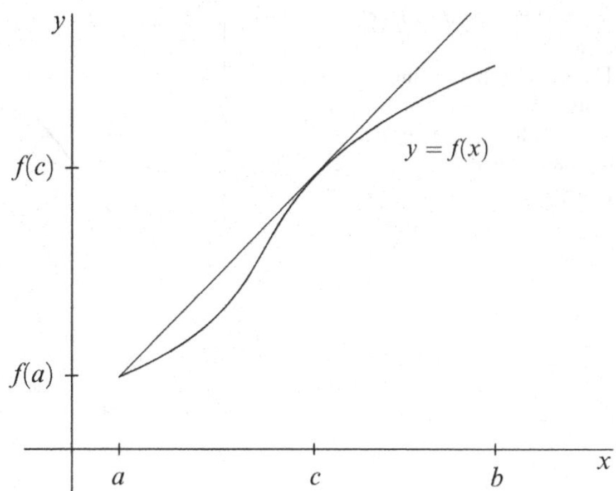

Fig. 7.2 Flett's Mean Value Theorem (Theorem 7.4): The line through $(a, f(a))$ and $(c, f(c))$ coincides with the tangent line at $x = c$

For Mean Value – type theorems having difference quotients other than $\frac{f(b)-f(a)}{b-a}$ (Mean Value Theorem) and $\frac{f(c)-f(a)}{c-a}$ (Flett's Mean Value Theorem) in their conclusions, see [7]. See also Exercise 7.10.

7.3 Pompeiu's Mean Value Theorem

For a function f defined on $[a,b]$, the equation of the line through $(a, f(a))$ and $(b, f(b))$ is

$$y = \frac{f(b) - f(a)}{b - a}(x - a) + f(a).$$

Setting $x = 0$, this line has y-intercept $\frac{bf(a)-af(b)}{b-a}$. (We met this quotient in our proof of the Mean Value Theorem (Theorem 5.2)). For f also differentiable on $[a,b]$, the equation of the tangent line at $c \in [a,b]$ is

$$y = f'(c)(x - c) + f(c).$$

This line has y-intercept $f(c) - cf'(c)$. The following theorem says that as long as $0 \notin [a,b]$, there is $c \in (a,b)$ for which these two y-intercepts coincide. See **Fig. 7.3.** This theorem was discovered by D. Pompeiu in 1946 [11].

Theorem 7.5. (Pompeiu's Mean Value Theorem) *Let $[a,b]$ be an interval not containing 0 and let f be differentiable on $[a,b]$. Then there exists $c \in (a,b)$ such that*

$$\frac{bf(a) - af(b)}{b - a} = f(c) - cf'(c).$$

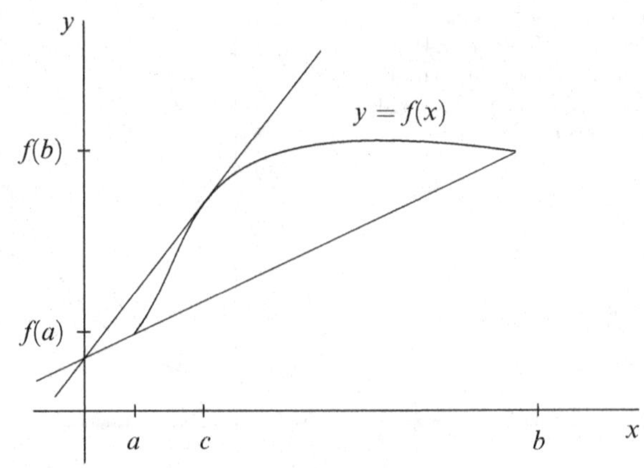

Fig. 7.3 Pompeiu's Mean Value Theorem (Theorem 7.5): The line through $(a, f(a))$ and $(b, f(b))$ has the same y-intercept as the tangent line at $x = c$

Proof. We may suppose that $0 < a < b$. On $[\frac{1}{b}, \frac{1}{a}]$, define the function F by

$$F(t) = tf(\tfrac{1}{t}).$$

Then F is continuous on $[\frac{1}{b}, \frac{1}{a}]$ and differentiable on $(\frac{1}{b}, \frac{1}{a})$. By the Chain and Product Rules,

$$F'(t) = f\left(\tfrac{1}{t}\right) - \frac{1}{t} f'\left(\tfrac{1}{t}\right).$$

Applying the Mean Value Theorem (Theorem 5.2) to F, there is $\xi \in (\frac{1}{b}, \frac{1}{a})$ such that

$$\frac{F(\frac{1}{a}) - F(\frac{1}{b})}{\frac{1}{a} - \frac{1}{b}} = F'(\xi).$$

Finally, setting $c = \frac{1}{\xi}$, this reads

$$\frac{bf(a) - af(b)}{b - a} = f(c) - cf'(c),$$

as desired. □

The proof given here follows [13]. We leave it for Exercise 7.11 to investigate whether the condition $0 \notin [a, b]$ is necessary.

7.4 A Related Result

We close this chapter with another Mean Value – type theorem [6], some variants of which we explore in the exercises. Let f be defined on $[a, b]$. For $c \in [a, b]$ we denote by $C = (c, f(c))$ any point on the graph of f and by

$$M = \left(\tfrac{a+b}{2}, \tfrac{f(a)+f(b)}{2}\right)$$

the midpoint of the chord from $(a, f(a))$ to $(b, f(b))$.

Recall that two lines are perpendicular if their slopes are negative reciprocals of each other. The following result says that for a function f differentiable on (a, b), either M is on the graph of f or there is a C such that the line through M and C is perpendicular to the tangent line at C. See **Fig. 7.4.**

Theorem 7.6. *Let f be continuous on $[a, b]$ and differentiable on (a, b). Then there exists $c \in [a, b]$ such that*

$$f'(c)\left[f(c) - \tfrac{f(a)+f(b)}{2}\right] = -\left[c - \tfrac{a+b}{2}\right].$$

Proof. Define h on $[a, b]$ via

$$h(x) = [x-a]^2 + [f(x) - f(a)]^2 + [x-b]^2 + [f(x) - f(b)]^2.$$

Then h is continuous on $[a, b]$ and differentiable on (a, b) and, as the reader can easily verify, $h(a) = h(b)$. So we apply Rolle's Theorem (Theorem 5.1) to h to conclude that there is $c \in (a, b)$ such that

$$h'(c) = 2(c-a) + 2(f(c) - f(a))f'(c) + 2(c-b) + 2(f(c) - f(b))f'(c) = 0.$$

After some simplification, we get

$$f'(c)[4f(c) - 2f(a) - 2f(b)] = -4c + 2a + 2b.$$

Finally, division by 4 and a little more simplification yields the desired result. □

Exercises

7.1. [16] Here is an extension of the Mean Value Theorem (Theorem 5.2) which contains two functions (but is different from Cauchy's Mean Value Theorem (Theorem 5.11)). **(a)** Apply Rolle's Theorem to

$$F(x) = [f(x) - f(a)](x-b) - [g(x) - g(b)](x-a)$$

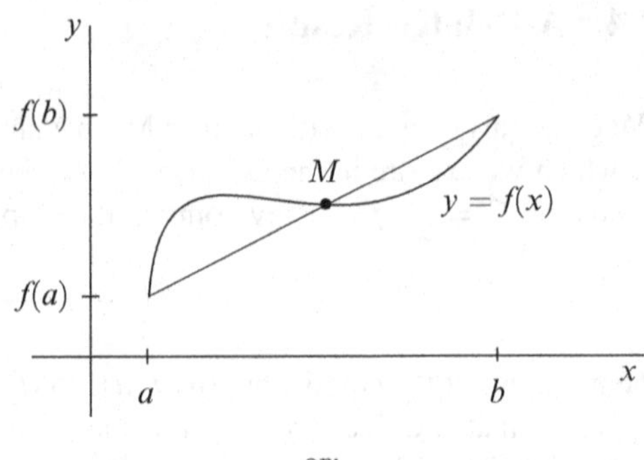

or:

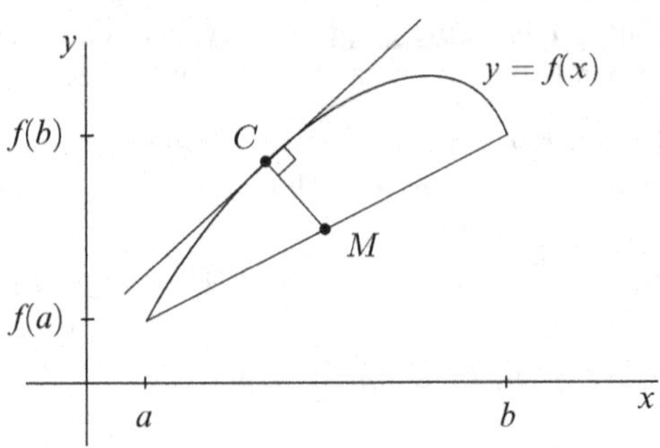

Fig. 7.4 Theorem 7.6: Either $M = (\frac{a+b}{2}, \frac{f(a)+f(b)}{2})$ is on the graph of f or there is $C = (c, f(c))$ such that the line through M and C is perpendicular to the tangent line at C

to prove the following. *Let f and g be continuous on $[a,b]$ and differentiable on (a,b). Then there is $c \in (a,b)$ such that*

$$f'(c)(b-c) + g'(c)(c-a) = [g(b) - g(c)] + [f(c) - f(a)].$$

(b) Verify that if $f = g$, this is the Mean Value Theorem (Theorem 5.2).

7.2. Consider Lemma 7.1, but from the other side. That is: *Let f be continuous on $[a,b]$ and differentiable on $[a,b)$, with $f'(a)[f(b) - f(a)] \leq 0$. Then there exists $c \in [a,b)$ such that $f'(c) = 0$.*
(a) Deduce this result from Lemma 7.1. **(b)** Prove it directly, by suitably adapting the proof of Lemma 7.1. **(c)** Draw a picture which illustrates this result geometrically.

7.3. Here is the proof of Darboux's Theorem (Theorem 7.2) found in most textbooks. Let y be between $f'(a)$ and $f'(b)$, say $f'(a) < y < f'(b)$. Consider the function $g(x) = yx - f(x)$, and show that g' has a zero in $[a,b]$. Hint: Apply the Extreme Value Theorem (Theorem 3.23) to g, and look carefully at the proof of Fermat's Theorem (Theorem 4.13).

7.4. [2] Let f be continuous on $[a,b]$, differentiable on $(a,b]$, with $f'(b) = 0$, and let $K > 0$. Prove, as follows, that there is $c \in (a,b)$ such that

$$f'(c) = K(f(c) - f(a)).$$

(a) Argue that we may take $a = 0$ and $f(a) = 0$.
(b) Looking for a contradiction, suppose that $f'(x) > Kf(x)$ for $x \in (a,b)$. Use $g(x) = e^{-Kx} f(x)$ to show that $f(b) > 0$.
(c) Then by Darboux's Theorem (Theorem 7.2), $f'(b) \geq Kf(b)$, a contradiction.
(d) Show that the $f'(x) < Kf(x)$ case can be handled similarly.

7.5. [9] Here is a curious fact. Consider the functions

$$F(t) = \begin{cases} t^2 \sin(1/t) & \text{if } t \neq 0 \\ 0 & \text{if } t = 0 \end{cases} \quad \text{and} \quad G(t) = \begin{cases} t^2 \cos(1/t) & \text{if } t \neq 0 \\ 0 & \text{if } t = 0\,. \end{cases}$$

Show that $f(t) = F'(t)^2$ and $g(t) = G'(t)^2$ each have the Intermediate Value Property on any interval which contains 0, but $f(t) + g(t)$ does not.

7.6. [14] Let f be differentiable on (a,b) and let $x_1, x_2, \ldots, x_n \in (a,b)$. Prove that there is $c \in (a,b)$ such that $f'(c) = \frac{1}{N} \sum_{j=1}^{N} f'(x_j)$.

7.7. Suppose that you take a ride on a roller coaster. Show that there is a moment during the ride at which your instantaneous speed is equal to your average speed up to that moment.

7.8. Consider Flett's Mean Value Theorem (Theorem 7.4), but from the other side. That is: *Let f be differentiable on $[a,b]$ with $f'(a) = f'(b)$. Then there exists $c \in (a,b)$ such that*

$$\frac{f(b) - f(c)}{b - c} = f'(c).$$

(a) Deduce this result from Flett's Mean Value Theorem.
(b) Prove this directly by suitably adapting the proof of Flett's Mean Value Theorem.
(c) Draw a picture which illustrates geometrically what this says.

7.9. [13] Apply Flett's Mean Value Theorem (Theorem 7.4) to the function

$$g(x) = f(x) - \frac{1}{2}\frac{f'(b) - f'(a)}{b - a}(x - a)^2$$

to obtain a version which does not require $f'(a) = f'(b)$.

7.10. [5, 7, 17]

(a) Apply Rolle's Theorem (Theorem 5.1) to

$$h(x) = (b - x)[f(x) - f(a)]$$

to prove the following: *Let f be continuous on $[a,b]$ and differentiable on (a,b). Then there exists $c \in (a,b)$ such that*

$$f'(c) = \frac{f(c) - f(a)}{b - c}.$$

(b) Show that the triangle formed by the x-axis, the tangent line at $(c, f(c))$, and the line through $(c, f(c))$ and $(b, f(a))$, is isosceles.

7.11. In Pompeiu's Mean Value Theorem (Theorem 7.5), is it *necessary* that the interval $[a,b]$ does not contain 0? Explain.

7.12. **(a)** What does the function h in the proof of Theorem 7.6 represent geometrically?

(b) Prove Theorem 7.6 by instead applying Rolle's Theorem (Theorem 5.1) to

$$g(x) = \left[x - \frac{a+b}{2}\right]^2 + \left[f(x) - \frac{f(a)+f(b)}{2}\right]^2.$$

(c) What does the function g in (b) represent geometrically?

7.13. [6]

(a) Apply Rolle's Theorem (Theorem 5.1) to

$$h(x) = [x-a]^2 - [f(x) - f(a)]^2 + [x-b]^2 - [f(x) - f(b)]^2$$

to prove the following: *Let f be continuous on $[a,b]$ and differentiable on (a,b). Then there exists $c \in (a,b)$ such that*

$$f'(c)\left[f(c) - \tfrac{f(a)+f(b)}{2}\right] = \left[c - \tfrac{a+b}{2}\right].$$

(b) Draw a picture which illustrates geometrically what this says.

(c) Extending the results in (a) and in Theorem 7.6, state and prove results which have conclusions

$$f'(c)\left[f(c) - \tfrac{f(a)+f(b)}{2}\right] = \pm g'(c)\left[g(c) - \tfrac{g(a)+g(b)}{2}\right].$$

7.14. [15] In Exercise 5.39 we saw that applying Rolle's Theorem (Theorem 5.1) to

$$g(x) = [f(x) - f(a)](x-b) - [f(x) - f(b)](x-a)$$

yields a proof of the Mean Value Theorem (Theorem 5.2).

(a) Apply Rolle's Theorem to

$$h(x) = [f(x) - f(a)](x-b) + [f(x) - f(b)](x-a)$$

to prove the following: *Let* f *be continuous on* $[a,b]$ *and differentiable on* (a,b). *Then there exists* $c \in (a,b)$ *such that*

$$f'(c)\left[c - \tfrac{a+b}{2}\right] = -\left[f(c) - \tfrac{f(a)+f(b)}{2}\right].$$

(b) Draw a picture which illustrates geometrically what this says.
(c) Extending the result in (a), state and prove a result which has conclusion

$$f'(c)\left[g(c) - \tfrac{g(a)+g(b)}{2}\right] = -g'(c)\left[f(c) - \tfrac{f(a)+f(b)}{2}\right].$$

7.15. Is there a Mean Value—type theorem similar to those in Exercises 7.13 and 7.14, but having conclusion

$$f'(c)\left[c - \tfrac{a+b}{2}\right] = \left[f(c) - \tfrac{f(a)+f(b)}{2}\right] \ ?$$

If yes, prove it. If no, provide a counterexample.

References

1. Boas, R.P., Jr.: A Primer of Real Functions. Mathematical Association of America, Washington, DC (1981)
2. Flanders, H.: In: Larson, L.L. (ed.) Review of Problem-Solving Through Problems. Springer, New York (1983). Am. Math. Mon. **92**, 676–678 (1985)
3. Flett, T.M.: A mean value theorem. Math. Gaz. **42**, 38–39 (1958)
4. Gillman, L.: Order relations and a proof of L'Hôpital's rule. Coll. Math. J. **28**, 288–292 (1997)
5. Martinez, S.M., Martinez de la Rosa, F.: A generalization of the mean value theorem. Wolfram Demonstrations Project. (http://demonstrations.wolfram.com)
6. Mercer, P.R.: On a mean value theorem. Coll. Math. J. **33**, 46–48 (2002)
7. Meyers, R.E: Some elementary results related to the mean value theorem. Coll. Math. J. **8**, 51–53 (1977)
8. Nadler, S.B., Jr.: A proof of Darboux's theorem. Am. Math. Mon. **117**, 174–175 (2010)
9. Neuser, D.A., Wayment, S.G.: A note on the intermediate value property. Am. Math. Mon. **81**, 995–997 (1974)
10. Olsen, L.: A new proof of Darboux's theorem. Am. Math. Mon. **111**, 713–715 (2004)
11. Pompeiu, D.: Sur une proposition analogue au theoreme des accroissements finis. Mathematica (Cluj.) **22**, 143–146 (1946)
12. Reich, S.: On mean value theorems. Am. Math. Mon. **76**, 70–73 (1969)
13. Sahoo, P.K., Riedel, T.: Mean value theorems and functional equations. World Scientific, Singapore/River Edge (1998)
14. Talman, L.A.: Simpson's rule is exact for cubics. Am. Math. Mon. **113**, 144–155 (2006)
15. Tong, J.: The mean value theorems of Lagrange and Cauchy. Int. J. Math. Educ. Sci. Tech. **30**, 456–458 (1999)
16. Tong, J.: The mean value theorem of Lagrange generalized to involve two functions. Math. Gaz. **84**, 515–516 (2000)
17. Tong, J.: The mean-value theorem generalised to involve two parameters. Math. Gaz. **88**, 538–540 (2004)
18. Trahan, D.H.: A new type of mean value theorem. Math. Mag. **39**, 264–268 (1966)

Chapter 8
Convex Functions and Taylor's Theorem

A smile is a curve that sets everything straight.

– Phyllis Diller

In this chapter we consider the higher derivatives of a function f. These are $f'' = (f')'$, $f^{(3)} = (f'')'$, etc. We extend the Mean Value Theorem to an analogous statement about the second derivative, and this takes us naturally to the notion of convexity. Once there, we meet the very important Jensen's Inequality. Then we extend the Mean Value Theorem to the (n+1)st derivative—this is Taylor's Theorem. We prove that e is irrational and we take a brief look at Taylor series.

8.1 Higher Derivatives

We saw in Sect. 4.1 that

$$f'(x_0) = \lim_{h\to 0} \frac{f(x_0+h) - f(x_0)}{h},$$

whenever this limit exists. It is often useful to consider *higher derivatives* of f, whenever possible:

$$f''(x_0) = \lim_{h\to 0} \frac{f'(x_0+h) - f'(x_0)}{h}, \quad f^{(3)}(x_0) = \lim_{h\to 0} \frac{f''(x_0+h) - f''(x_0)}{h}, \quad \text{etc.}$$

Generally, we write $f^{(0)} = f$ and $f^{(1)} = f'$, then

$$f^{(n)}(x_0) = \lim_{h\to 0} \frac{f^{(n-1)}(x_0+h) - f^{(n-1)}(x_0)}{h} \qquad \text{for } n = 2, 3, 4, \ldots$$

whenever these limits exist. $f^{(n)}(x_0)$ is called the **nth derivative** of f at x_0. As we have already seen with f', it is common to replace x_0 with simply x, when f'', $f^{(3)}$, $f^{(4)}$ etc. are to be thought of as functions.

P.R. Mercer, *More Calculus of a Single Variable*, Undergraduate Texts in Mathematics, DOI 10.1007/978-1-4939-1926-0_8

Example 8.1. Let $n = 1, 2, 3, \ldots$.

(i) For $f(x) = \mathrm{e}^x$ we have $f^{(n)}(x) = \mathrm{e}^x$.
(ii) For $f(x) = \ln(x)$ and $x > 0$, we have $f^{(n)}(x) = (-1)^{n+1}(n-1)!/x^n$.
(iii) If f is a polynomial of degree n then $f^{(n)}$ is constant and $f^{(n+1)} \equiv 0$.
(iv) For $f(x) = \sin(x)$, we have

$$f^{(2n-1)}(x) = (-1)^{n-1}\cos(x), \quad f^{(2n)}(x) = (-1)^n \sin(x),$$

and

$$f^{(n)}(x) = \sin\left(x + \frac{n\pi}{2}\right).$$ ◇

Example 8.2. Let us *assume* (and this is reasonable) that for $n \in \mathbf{N}$,

$$(1+x)^n = \sum_{k=0}^{n} a_k x^k.$$

Then we find a_k's as follows. Taking derivatives up to order j $(0 \le j \le n)$ of each side we get

$$n(n-1)\cdots(n-(j-1))(1+x)^{n-j} = \sum_{k=j}^{n} k(k-1)\cdots(k-(j-1))a_k\, x^{k-j}.$$

If we set $x = 0$ here, the only nonzero term on the right-hand side is that for which $k = j$. Then solving for a_j we get

$$a_j = \frac{n(n-1)\cdots(n-(j-1))}{j(j-1)\cdots(j-(j-1))} = \frac{n!}{(n-j)!j!}.$$

And so we have obtained the **Binomial formula**

$$(1+x)^n = \sum_{k=0}^{n} \binom{n}{k} x^k,$$

where the $\binom{n}{k}$'s are the **binomial coefficients**

$$\binom{n}{k} = \frac{n!}{(n-k)!k!}.$$

We leave it for Exercise 8.3 to find an expression for $(a+b)^n$. See also [19]. ◇

Remark 8.3. Let $k, n \in \mathbf{N}$, with $k \leq n$. The number of ways of selecting a k-element set from a set having n elements is the binomial coefficient $\binom{n}{k}$. The number of ways of arranging k distinct elements is $k!$. The number of arrangements of k elements taken from a set having n elements is $k!\binom{n}{k} = \frac{n!}{(n-k)!}$. ◦

Example 8.4. Here is a neat fact about derivatives which we will use in the next section. It provides a way of computing f'' using f, but not f'. We show that if f is defined on an open interval containing x, and if f'' exists, then

$$f''(x) = \lim_{h\to 0} \frac{f(x+h) - 2f(x) + f(x-h)}{h^2}.$$

With h as the independent variable, we apply L'Hospital's Rule (Theorem 5.13) to get

$$\lim_{h\to 0} \frac{f(x+h) - 2f(x) + f(x-h)}{h^2} = \lim_{h\to 0} \frac{f'(x+h) - f'(x-h)}{2h}.$$

Now if f'' were continuous, we could apply L'Hospital's Rule again to get

$$\lim_{h\to 0} \frac{f''(x+h) + f''(x-h)}{2} = f''(x).$$

But since we only assumed that f'' exists we must take more care. Instead, we write

$$\lim_{h\to 0} \frac{f'(x+h) - f'(x-h)}{2h} = \lim_{h\to 0} \frac{1}{2}\left(\frac{f'(x+h) - f'(x)}{h} + \frac{f'(x-h) - f'(x)}{-h}\right),$$

and this $= \frac{1}{2}(f''(x) + f''(x)) = f''(x)$, just as we wanted to show. ◇

Remark 8.5. The quantity

$$\lim_{h\to 0} \frac{f(x+h) - f(x-h)}{2h} = \lim_{h\to 0} \frac{1}{2}\left(\frac{f(x+h) - f(x)}{h} + \frac{f(x-h) - f(x)}{-h}\right)$$

is the average of the right-hand and left-hand derivatives of f at x. (See Exercise 4.10.) This average is called the **Schwarz derivative**, or the **symmetric derivative**. It may exist even when $f'(x)$ does not: Consider $f(x) = |x|$ at $x = 0$. For Rolle – type theorems and Mean Value – type theorems for the symmetric derivative, see [2]. For Flett's Mean Value Theorem (Theorem 7.4) as regards the symmetric derivative, see [39]. See also Exercise 8.17. ◦

Suppose now that f is differentiable on an open interval I and that $x_0 \in I$. The Mean Value Theorem (Theorem 5.2) says that for each $x \in I$ there is c between x and x_0 such that

$$f(x) - f(x_0) = f'(c)(x - x_0).$$

Therefore $|f'(c)(x - x_0)|$ is the *error* which comes about, in approximating $f(x)$ with the constant function $f(x_0)$. The following result carries this one step further—it provides an expression for the error in approximating $f(x)$ with the linear function

$$L(x) = f(x_0) + f'(x_0)(x - x_0).$$

$L(x)$ mimics $f(x)$, to the extent that $L(x_0) = f(x_0)$, and $L'(x_0) = f'(x_0)$. See **Fig. 8.1**.

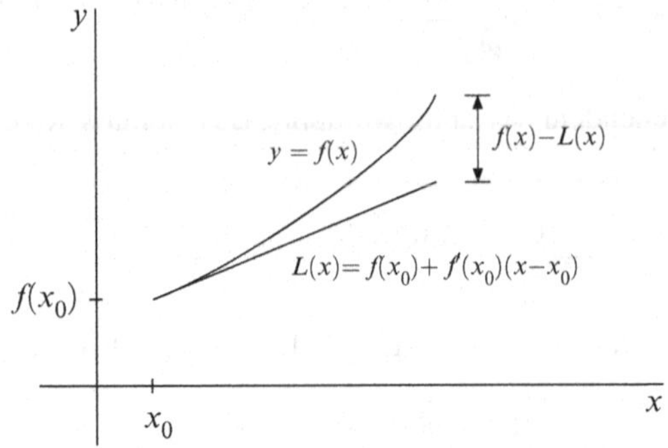

Fig. 8.1 $L(x)$ mimics $f(x)$, to the extent that $L(x_0) = f(x_0)$ and $L'(x_0) = f'(x_0)$

Theorem 8.6. (Mean Value Theorem for the Second Derivative) *Let f be defined on an open interval I, let f'' exist there, and let $x_0 \in I$. Then for each $x \in I$, there is c between x and x_0 such that*

$$f(x) = f(x_0) + f'(x_0)(x - x_0) + \frac{f''(c)}{2}(x - x_0)^2.$$

Proof. Let

$$F(x) = f(x) - f(x_0) - f'(x_0)(x - x_0).$$

Then $F(x_0) = F'(x_0) = 0$. So it is reasonable to compare F with the function

$$G(x) = (x - x_0)^2,$$

which also has $G(x_0) = G'(x_0) = 0$. Applying Cauchy's Mean Value Theorem (Theorem 5.11) two times, there are c_1 and c_2 between x and x_0 such that

$$\frac{F(x)}{G(x)} = \frac{F(x) - F(x_0)}{G(x) - G(x_0)} = \frac{F'(c_1)}{G'(c_1)}$$

$$= \frac{F'(c_1) - F'(x_0)}{G'(c_1) - G'(x_0)} = \frac{F''(c_2)}{G''(c_2)}.$$

That is, $F(x) = \frac{F''(c_2)}{G''(c_2)} G(x)$. Now we observe that $F''(c_2) = f''(c_2)$ and that $G''(c_2) = 2$. This gives

$$F(x) = \frac{f''(c_2)}{2}(x - x_0)^2,$$

as desired (with $c = c_2$). □

So the Mean Value Theorem for the Second Derivative (Theorem 8.6) says that $\left|\frac{f''(c)}{2}(x - x_0)^2\right|$ is the *error* which comes about, in approximating $f(x)$ with the linear function $L(x) = f(x_0) + f'(x_0)(x - x_0)$.

If $f''(x) = 0$ for all $x \in (a, b)$ then by the Mean Value Theorem for the Second Derivative (Theorem 8.6), f must be a linear function. (Compare with Lemma 5.6.)

If $f''(x) \geq 0$ for all $x \in (a, b)$ then applying Lemma 5.6 to $f'' = (f')'$, we see that f' is increasing on (a, b). In freshman calculus, the graph of such a function is usually called **concave upward**. (And $-f$ is called **concave downward**.)

The following is immediate from the Mean Value Theorem for the Second Derivative (Theorem 8.6). It says that the graph of a function which is concave upward lies on or above all of its tangent lines.

Lemma 8.7. *Let f be such that $f'' \geq 0$ on (a, b), and let $x_0 \in (a, b)$. Then for each $x \in (a, b)$,*

$$f(x) \geq f(x_0) + f'(x_0)(x - x_0).$$

Proof. Let $x_0 \in (a, b)$. By the Mean Value Theorem for the Second Derivative (Theorem 8.6), there is c between x and x_0 such that

$$f(x) = f(x_0) + f'(x_0)(x - x_0) + \frac{f''(c)}{2}(x - x_0)^2.$$

Now observing simply that $f''(c) \geq 0$, the proof is complete. □

Example 8.8. $f(x) = e^x$ satisfies $f''(x) > 0$ for all $x \in \mathbf{R}$. The tangent line to $y = f(x)$ at $x = 0$ has equation $y = x + 1$. So by Lemma 8.7, we have again the inequality (6.5):

$$e^x > 1 + x, \quad \text{except at } x = 0 \text{ where we have equality.}$$ ◇

Example 8.9. $f(x) = \ln(x)$ satisfies $f''(x) < 0$ for $x > 0$. The tangent line to $y = f(x)$ at $x = 1$ has equation $y = x - 1$. So by Lemma 8.7, we have again the inequality (6.7):

$$\ln(x) < x - 1, \quad \text{except at } x = 1 \text{ where we have equality.}$$ ◇

Example 8.10. Again, $f(x) = \ln(x)$ satisfies $f''(x) < 0$ for $x > 0$. The tangent line to $y = f(x)$ at $x = \mathrm{e}$ has equation $y = x/\mathrm{e}$. Therefore, by Lemma 8.7, $\ln(x) < x/\mathrm{e}$ for $x > 0$. That is,

$$x^{\mathrm{e}} < \mathrm{e}^x,$$

and in particular,

$$\pi^{\mathrm{e}} < \mathrm{e}^{\pi}.$$

Exercise 8.19 contains a neat proof that $x^{\mathrm{e}} < \mathrm{e}^x$ implies the AGM Inequality (Theorem 2.10). ◇

Also Immediate from the Mean Value Theorem for the Second Derivative (Theorem 8.6) is the **Second Derivative Test**, which we leave for Exercise 8.18. It says that if f'' exists on (a, b) and if for some $c \in (a, b)$ we have $f'(c) = 0$, then $f''(c) > 0$ implies that f has a local minimum at c, and $f''(c) < 0$ implies that f has a local maximum at c. See also Exercise 8.54.

8.2 Convex Functions

Let $x, y \in \mathbf{R}$ with $x < y$. For $t \in [0, 1]$, the expression $(1 - t)x + ty$ is a natural *parameterization* of the interval $[x, y]$. For example, $t = 0$ gives x, $t = 1/3$ gives $\frac{2}{3}x + \frac{1}{3}y$, $t = 1/2$ gives the midpoint $(x + y)/2$, and $t = 1$ gives y. See **Fig. 8.2**.

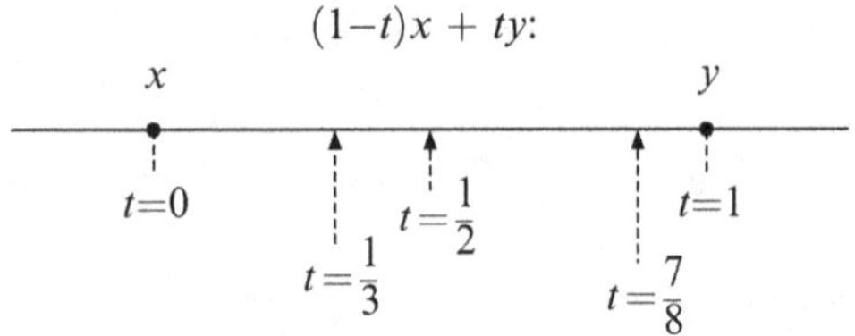

Fig. 8.2 $(1 - t)x + ty$ for a few values of $t \in [0, 1]$

Now let f be a function defined on some interval I. Then f is **convex** on I means that for any $x, y \in I$ with $x < y$,

$$f\big((1-t)x + ty\big) \le (1-t)f(x) + tf(y) \quad \text{for every } t \in (0, 1).$$

And f being **strictly convex** means that the $\le$ above can be replaced with $<$. So geometrically, a strictly convex function is one whose graph lies below all of its chords. See **Fig. 8.3**.

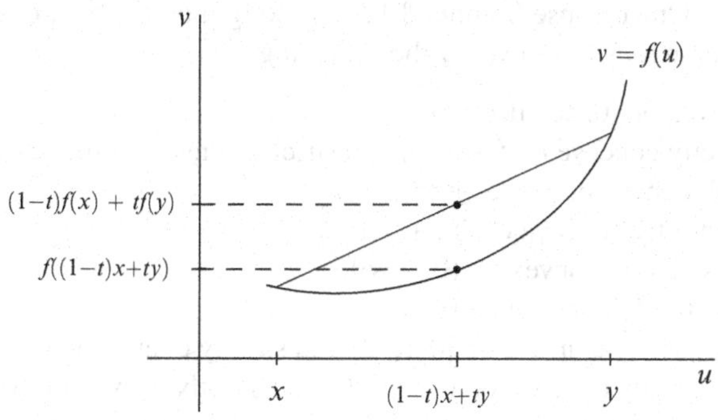

Fig. 8.3 A convex function: $f\big((1-t)x+ty\big) \le (1-t)f(x)+tf(y)$ for every $t \in [0,1]$

The function f is **concave**, or **strictly concave** if these inequalities are reversed. Consequently, f being concave means that $-f$ is convex.

Remark 8.11. Let f be continuous on $[a,b]$ and differentiable on (a,b). Applying Rolle's Theorem (Theorem 5.1) to $g(t) = f\,((1-t)a+tb) - \big((1-t)f(a)+tf(b)\big)$ on $[0,1]$ gives another proof of the Mean Value Theorem (Theorem 5.2). This is essentially the proof outlined in Exercise 5.9; see also Exercise 5.11. ∘

The convexity condition can be tricky to verify, depending on the nature of f. So the following result is often used in practice. It gives a sufficient condition for a function to be convex.

Lemma 8.12. *If $f'' \ge 0$ on (a,b), then f is convex on (a,b). That is, if f is concave upward on (a,b), then f is convex on (a,b).*

Proof. Let $x, y \in (a,b)$. Then $\alpha = (1-t)x + ty \in (a,b)$ for $t \in [0,1]$. By Lemma 8.7,

$$f(x) \ge f(\alpha) + f'(\alpha)(x-\alpha) \quad \text{and} \quad f(y) \ge f(\alpha) + f'(\alpha)(y-\alpha).$$

Therefore

$$\begin{aligned}(1-t)f(x)+tf(y) &\ge (1-t)f(\alpha)+tf(\alpha)+(1-t)f'(\alpha)(x-\alpha)+tf'(\alpha)(y-\alpha)\\ &= f(\alpha)+f'(\alpha)\big[(1-t)x-(1-t)\alpha+ty-t\alpha\big]\\ &= f(\alpha)+f'(\alpha)\big[\alpha-\alpha\big] = f(\alpha).\end{aligned}$$

That is, $(1-t)f(x)+tf(y) \ge f(\alpha) = f\big((1-t)x+ty\big)$, and so f is convex. (We point out that the same argument shows that if $f'' > 0$ then f is *strictly* convex.) □

Example 8.13. One can use Lemma 8.12 (checking that $f'' \geq 0$, or > 0, or ≤ 0, or < 0, as the case may be) to verify the following.

(i) x^r is convex on $[0, \infty)$ if $r > 1$.
(ii) x^3 is strictly concave on $(-\infty, 0)$ and strictly convex on $(0, \infty)$.
(iii) e^x strictly convex on $(-\infty, +\infty)$.
(iv) $\ln(x)$ is strictly concave on $(0, +\infty)$.
(v) $x \ln(x)$ is strictly convex on $(0, +\infty)$.
(vi) $\sin(x)$ is strictly concave on $(0, \pi)$.
(vii) $\cos(x)$ is strictly concave on $[0, \pi/2)$ and strictly convex on $(\pi/2, \pi]$.
(viii) $\tan(x)$ is strictly concave on $(-\pi/2, 0)$ and strictly convex on $(0, \pi/2)$. ◇

Example 8.14. What is known as **Jordan's Inequality** is the left-hand side of the pair of inequalities

$$\frac{2}{\pi}x \leq \sin(x) \leq x \quad \text{for } x \in [0, \pi/2].$$

These are illustrated in **Fig. 8.4**. Here we provide a simple proof using the fact that $f(x) = \sin(x)$ is concave on $[0, \pi/2]$. Here, $f''(x) = -\sin(x) \leq 0$, so by Lemma 8.7 the graph of f lies below its tangent lines on $[0, \pi/2]$ and at $x = 0$ in particular. This is the right-hand inequality. By Lemma 8.12, the graph of f lies above the chord between the points $(0, 0)$ and $(\pi/2, \sin(\pi/2)) = (\pi/2, 1)$. This is the left-hand inequality. ◇

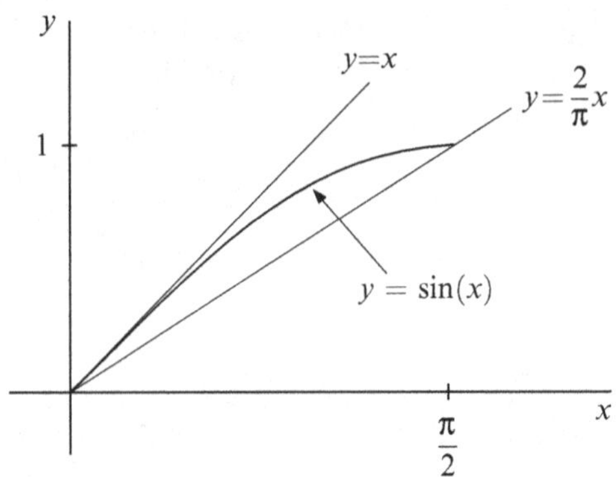

Fig. 8.4 Example 8.14. Jordan's Inequality: $\frac{2}{\pi}x \leq \sin(x) \leq x$ for $x \in [0, \pi/2]$

Example 8.15. We have observed that $\ln(x)$ is strictly concave on $(0, \infty)$. So $-\ln(x)$ is strictly convex there, and the definition of convexity gives

$$(1-t)\ln x + t \ln y < \ln\big((1-t)x + ty\big), \quad \text{for } t \in [0, 1] \text{ and } x, y \in (0, \infty).$$

Applying the exponential function to both sides we obtain

$$x^{(1-t)}y^t < (1-t)x + ty\,.$$

This is the weighted AGM Inequality with n=2 (Corollary 6.16). We have seen that this is equivalent to Young's Inequality (Corollary 6.19):

$$ab \le \frac{a^p}{p} + \frac{b^q}{q} \qquad \left(\text{where } \frac{1}{p} + \frac{1}{q} = 1\right).$$

These results can also be obtained by using the fact that e^x is convex on $(-\infty, +\infty)$. We leave this for Exercise 8.34. ◇

We close this section by showing that the converse of Lemma 8.12 is also true, as long as f'' exists.

Lemma 8.16. *If f is convex on (a,b) and f'' exists, then $f'' \ge 0$. That is, if f is convex on (a,b) and f'' exists there, then f is concave upward on (a,b).*

Proof. Let $x \in (a,b)$ and choose $h > 0$ small enough that $(x-h, x+h) \subset (a,b)$. We write $x = (1-\frac{1}{2})(x-h) + \frac{1}{2}(x+h)$. Then since f is convex,

$$f(x) = f\big((1-\tfrac{1}{2})(x-h) + \tfrac{1}{2}(x+h)\big) \le (1-\tfrac{1}{2})f(x-h) + \tfrac{1}{2}f(x+h).$$

Therefore $f(x-h) - 2f(x) + f(x+h) \ge 0$. Now we saw in Example 8.4 that since f'' exists,

$$f''(x) = \lim_{h\to 0} \frac{f(x+h) - 2f(x) + f(x-h)}{h^2},$$

and so must have $f''(x) \ge 0$, as desired. □

The paper [7] contains a thorough treatment of many of the various geometric characterizations of a convex function.

8.3 Jensen's Inequality

The big theorem in the world of convex functions is due to Danish mathematician J.W. Jensen (1859–1925). Many of the most important results related to convexity follow from Jensen's Inequality. In the definition of convexity, we have

$$f\big((1-t)x + ty\big) \le (1-t)f(x) + tf(y).$$

The idea in Jensen's Inequality is that the x and y can be replaced by any number of points in I, and the $(1-t)x + ty$ can be replaced by any weighted Arithmetic Mean of these points. The proof is an easy consequence of Lemma 8.16 if we assume that f'' exists; we discuss a more general version at the end of this section.

Theorem 8.17. (Jensen's Inequality) *Let f be convex on (a,b) and let f'' exist there. For $n \geq 2$, let $x_1, x_2, \ldots, x_n \in (a,b)$, and let $w_1, w_2, \ldots, w_n$ satisfy $w_j > 0$, with $\sum_{j=1}^{n} w_j = 1$. Then*

$$f\left(\sum_{j=1}^{n} w_j x_j\right) \leq \sum_{j=1}^{n} w_j f(x_j).$$

Proof. Let $A = \sum_{j=1}^{n} w_j x_j$ which is in (a,b), because A is a weighted Arithmetic Mean of the x_j's. Since f is convex and f'' exists, $f'' \geq 0$ by Lemma 8.16. Then by Lemma 8.7, for each x_j we have

$$f(x_j) \geq f(A) + f'(A)(x_j - A).$$

Now multiplying by w_j and summing from 1 to n we get

$$\begin{aligned}
\sum_{j=1}^{n} w_j f(x_j) &\geq \sum_{j=1}^{n} w_j f(A) + \sum_{j=1}^{n} w_j f'(A)(x_j - A) \\
&= \sum_{j=1}^{n} w_j f(A) + f'(A) \sum_{j=1}^{n} w_j (x_j - A) \\
&= \sum_{j=1}^{n} w_j f(A) + f'(A)(0).
\end{aligned}$$

That is, $\sum_{j=1}^{n} w_j f(x_j) \geq \sum_{j=1}^{n} w_j f(A) = f\left(\sum_{j=1}^{n} w_j x_j\right)$, as desired. □

For an interesting geometric explanation of Jensen's Inequality (Theorem 8.17), see [35]. We point out that in the equal weights case $w_1 = w_2 = \cdots = w_n = 1/n$, Jensen's Inequality reads:

$$f\left(\frac{1}{n}\sum_{j=1}^{n} x_j\right) \leq \frac{1}{n}\sum_{j=1}^{n} f(x_j). \tag{8.1}$$

Example 8.18. We showed in Example 8.15 that the concavity of $\ln(x)$ can be used to obtain the weighted AGM Inequality with n = 2 (Corollary 6.16). More generally, the concavity of $\ln(x)$ and Jensen's Inequality (Theorem 8.17) can be used to obtain the full weighted AGM Inequality (Theorem 6.15), as follows. Let $x_1, x_2, \ldots, x_n$ and $w_1, w_2, \ldots, w_n$ be positive numbers with $\sum_{j=1}^{n} w_j = 1$. Let $f(x) = \ln(x)$, which is strictly concave on $(0, \infty)$. Jensen's Inequality ($-\ln(x)$ is strictly convex) gives

$$\sum_{j=1}^{n} w_j \ln(x_j) \leq \ln\left(\sum_{j=1}^{n} w_j x_j\right).$$

The left-hand side here is $\sum_{j=1}^{n} \ln(x_j^{w_j}) = \ln(x_1^{w_1} \cdot x_2^{w_2} \cdots x_n^{w_n})$, and applying the exponential function to both sides we obtain

$$x_1^{w_1} \cdot x_2^{w_2} \cdots x_n^{w_n} \leq \sum_{j=1}^{n} w_j x_j\,,$$

just as we wanted to show. Of course this can also be obtained by using the fact that e^x is convex on $(-\infty, +\infty)$. We leave this for Exercise 8.43. ◇

Example 8.19. Jensen's Inequality (Theorem 8.17) can be used to obtain the Cauchy–Schwarz Inequality (Theorem 2.18), as follows. Let $a_1, a_2, \ldots, a_n$, and $b_1, b_2, \ldots, b_n$ be real numbers. We may assume that $\sum_{k=1}^{n} a_k^2 \neq 0$. Then applying Jensen's Inequality to the convex function $f(x) = x^2$, with $w_j = a_j^2 / \sum_{k=1}^{n} a_k^2$ (so that $\sum_{j=1}^{n} w_j = 1$) and $x_j = a_j b_j / w_j$, we get

$$\left(\sum_{j=1}^{n} w_j x_j\right)^2 \leq \sum_{j=1}^{n} w_j x_j^2\,.$$

After some tidying, this reads

$$\left(\sum_{j=1}^{n} a_j b_j\right)^2 \leq \sum_{j=1}^{n} a_j^2 \sum_{j=1}^{n} b_j^2\,,$$

which is the Cauchy–Schwarz Inequality. ◇

We close this section by showing that Jensen's Inequality (Theorem 8.17) actually holds assuming only that f is convex—that is, without assuming even that f is continuous, much less $f'' \geq 0$. We prove it for the equal weights case (8.1) and leave the more general version for Exercise 8.39. The proof is exactly analogous to Cauchy's proof of the AGM Inequality (Theorem 2.10) which we provided at the end of Sect. 2.2. In fact, it was a careful analysis of Cauchy's proof of the AGM inequality which led Jensen to discover his inequality and thus initiate the study of convex functions [40].

Proof. If $n = 2$ then Jensen's Inequality is simply the convexity condition (with $t = 1/2$). If $n = 4$ we use the condition twice:

$$\begin{aligned} f\left(\frac{1}{4}(x_1 + x_2 + x_3 + x_4)\right) &= f\left(\frac{1}{2}\left(\tfrac{1}{2}[x_1 + x_2] + \tfrac{1}{2}[x_3 + x_4]\right)\right) \\ &\le \frac{1}{2}\left(f\left(\tfrac{1}{2}[x_1 + x_2] + \tfrac{1}{2}[x_3 + x_4]\right)\right) \\ &= \frac{1}{2}\left(f\left(\tfrac{1}{2}\left([x_1 + x_2] + [x_3 + x_4]\right)\right)\right) \\ &\le \frac{1}{2}\frac{1}{2} f\left([x_1 + x_2] + [x_3 + x_4]\right) \\ &= \frac{1}{4} f(x_1 + x_2 + x_3 + x_4). \end{aligned}$$

And for $n = 8$, we would use the $n = 4$ case twice. Etcetera: We could continue this procedure indefinitely, and so we may assume that Jensen's Inequality holds for any n of the form 2^m ($m \ge 0$). For any (other) n, we choose m so large that $2^m > n$. Now writing

$$A = \frac{1}{n}\sum_{j=1}^{n} x_j,$$

we observe that

$$\frac{x_1 + x_2 + \cdots + x_n + (2^m - n)A}{2^m} = A.$$

The numerator of the left-hand side here has 2^m members in the sum and so we can apply what we have proved so far to see that

$$\begin{aligned} f(A) &= f\left(\frac{x_1 + x_2 + \cdots + x_n + (2^m - n)A}{2^m}\right) \\ &\le \frac{1}{2^m}\left(f(x_1) + f(x_2) + \cdots + f(x_n) + (2^m - n)f(A)\right). \end{aligned}$$

That is,

$$f(A) \le \frac{1}{2^m}\big(f(x_1)+f(x_2)+\cdots+f(x_n)-nf(A)\big)+f(A),$$

and so

$$f(A) \le \frac{1}{n}\big(f(x_1)+f(x_2)+\cdots+f(x_n)\big).$$

This is (8.1), as desired. □

8.4 Taylor's Theorem: e Is Irrational

Looking at the Mean Value Theorem (Theorem 5.2) and then the Mean Value Theorem for the Second Derivative (Theorem 8.6) one might ask, "why stop at two derivatives?" Indeed, continuing on to the (n+1)st derivative yields the important theorem below, named for English mathematician Brook Taylor (1685–1731). The proof we provide is just an extension of the proof of the Mean Value Theorem for the Second Derivative. (We shall prove it an entirely different way in Sect. 11.4.)

Theorem 8.20. (Taylor's Theorem) *Let f be such that $f^{(n+1)}$ exists on some open interval I and let $x_0 \in I$. Then for each $x \in I$ there is c between x and x_0 such that*

$$f(x) = \sum_{k=0}^{n} \frac{f^{(k)}(x_0)}{k!}(x-x_0)^k + \frac{f^{(n+1)}(c)}{(n+1)!}(x-x_0)^{n+1}.$$

Proof. Let

$$F(x) = f(x) - \sum_{k=0}^{n} \frac{f^{(k)}(x_0)}{k!}(x-x_0)^k.$$

Then $F(x_0) = F'(x_0) = \cdots = F^{(n)}(x_0) = 0$. So it is reasonable to compare F with the function

$$G(x) = (x-x_0)^{n+1},$$

which also has $G(x_0) = G'(x_0) = \cdots = G^{(n)}(x_0) = 0$. Applying Cauchy's Mean Value Theorem (Theorem 5.11) n+1 times, there are $c_1, c_2, \ldots c_{n+1}$ between x and x_0 such that

$$\frac{F(x)}{G(x)} = \frac{F(x)-F(x_0)}{G(x)-G(x_0)} = \frac{F'(c_1)}{G'(c_1)}$$

$$= \frac{F'(c_1) - F'(x_0)}{G'(c_1) - G'(x_0)} = \frac{F''(c_2)}{G''(c_2)}$$

$$= \frac{F'(c_2) - F'(x_0)}{G'(c_2) - G'(x_0)} = \cdots = \frac{F^{(n+1)}(c_{n+1})}{G^{(n+1)}(c_{n+1})}.$$

That is,

$$F(x) = \frac{F^{(n+1)}(c_{n+1})}{G^{(n+1)}(c_{n+1})} G(x).$$

Now we observe that $F^{(n+1)}(c_{n+1}) = f^{(n+1)}(c_{n+1})$ and that $G^{(n+1)}(c_{n+1}) = (n+1)!$. This gives

$$F(x) = \frac{f^{(n+1)}(c_{n+1})}{(n+1)!}(x - x_0)^{n+1},$$

as desired (with $c = c_{n+1}$.) □

In Taylor's Theorem (Theorem 8.20), the polynomial

$$p_n(x) = \sum_{k=0}^{n} \frac{f^{(k)}(x_0)}{k!}(x - x_0)^k$$

is called the **Taylor polynomial of degree** n**, at** $x = x_0$. The term

$$\frac{f^{(n+1)}(c)}{(n+1)!}(x - x_0)^{n+1}$$

is called the **remainder term.** It gives the *error* which arises, in approximating $f(x)$ with $p_n(x)$. The polynomial $p_n(x)$ mimics $f(x)$ to the extent that

$$f(x_0) = p(x_0), \quad f'(x_0) = p'(x_0), \quad \ldots \quad , f^{(n)}(x_0) = p^{(n)}(x_0).$$

So we might expect that the error should be small if n is large and/or if x is close to x_0. And this expectation seems to be supported by the form of the remainder term. (See also Exercise 8.59.) Of course, having $n = 0$ and $n = 1$ gives the Mean Value Theorem (Theorem 5.2) and the Mean Value Theorem for the Second Derivative (Theorem 8.6) respectively.

Example 8.21. For $f(x) = \mathrm{e}^x$ and for $k = 0, 1, 2, \ldots,$

$$f^{(k)}(x) = \mathrm{e}^x \quad \text{and so} \quad f^{(k)}(0) = 1.$$

So by Taylor's Theorem (Theorem 8.20), with $x_0 = 0$, there is c between 0 and x such that

$$\begin{aligned}
e^x &= \sum_{k=0}^{n} \frac{f^{(k)}(0)}{k!} x^k + \frac{f^{(n+1)}(c)}{(n+1)!} x^{n+1} \\
&= \sum_{k=0}^{n} \frac{x^k}{k!} + \frac{e^c x^{n+1}}{(n+1)!} \\
&= 1 + x + \frac{x^2}{2!} + \frac{x^3}{3!} + \cdots + \frac{x^n}{n!} + \frac{e^c x^{n+1}}{(n+1)!}.
\end{aligned}$$

In particular, if n is odd then

$$e^x \geq 1 + x + \frac{x^2}{2!} + \frac{x^3}{3!} + \cdots + \frac{x^n}{n!},$$

with equality holding only for $x = 0$. This vastly improves inequality (6.5), namely $e^x \geq 1 + x$. (For n even the same inequality holds for $x \geq 0$, and it is reversed for $x \leq 0$.) ◇

Example 8.22. We prove that **e is irrational**. Taking $x = 1$ in Example 8.21, there is c between 0 and 1 such that

$$e = 1 + 1 + \frac{1}{2!} + \frac{1}{3!} + \cdots + \frac{1}{n!} + \frac{e^c}{(n+1)!}.$$

Looking for a contradiction, we suppose that e is rational. That is, $e = a/b$, where a and b are positive integers (since $e > 0$). Then

$$\frac{a}{b} = 1 + 1 + \frac{1}{2!} + \cdots + \frac{1}{n!} + \frac{e^c}{(n+1)!},$$

and so

$$\begin{aligned}
n!\frac{a}{b} &= n!\left(1 + 1 + \frac{1}{2!} + \cdots + \frac{1}{n!}\right) + n!\frac{e^c}{(n+1)!} \\
&= n!\left(1 + 1 + \frac{1}{2!} + \cdots + \frac{1}{n!}\right) + \frac{e^c}{n+1}.
\end{aligned}$$

The first term on the right-hand side is an integer for any n. If $n \geq b$, then the left-hand side is an integer. And if $n > e^c$ then the second term on the right-hand side is between 0 and 1. So choosing $n > \max\{e^c, b\}$ yields a contradiction. Therefore, e must be irrational. ◇

We generalize Example 8.22 considerably in Theorem 12.3 and then Corollary 12.4, showing that e^r is irrational for *any* nonzero rational number r.

Remark 8.23. Euler was the first to prove that e is irrational, in 1737. Saying that e is irrational is the same as saying that e is not the solution to any linear equation $ax + b = 0$ with integer coefficients. The French mathematician J. Liouville (1809–1882) proved around 1844 that e is not a solution to any *quadratic* equation $ax^2 + bx + c = 0$ with integer coefficients. The French mathematician C. Hermite (1822–1901) proved in 1873 that e is not a solution to *any* polynomial equation of *any* degree with integer coefficients. That is, e is not an **algebraic number**; it is a **transcendental number**. ○

Example 8.24. For $x > 0$ and $f(x) = \ln(x)$, and for $k = 1, 2, 3, \ldots,$

$$f^{(k)}(x) = \frac{(-1)^{k+1}(k-1)!}{x^k} \quad \text{and so} \quad f^{(k)}(1) = (-1)^{k+1}(k-1)!.$$

So by Taylor's Theorem (Theorem 8.20), with $x_0 = 1$, there is c between 1 and x such that

$$\begin{aligned}
\ln(x) &= \sum_{k=0}^{n} \frac{f^{(k)}(1)}{k!}(x-1)^k + \frac{f^{(n+1)}(c)}{(n+1)!}(x-1)^{n+1} \\
&= \sum_{k=1}^{n} \frac{(-1)^{k+1}(k-1)!}{k!}(x-1)^k + \frac{(-1)^{n+2}n!}{c^{n+1}(n+1)!}(x-1)^{n+1} \\
&= \sum_{k=1}^{n} \frac{(-1)^{k+1}}{k}(x-1)^k + \frac{(-1)^n}{c^{n+1}}\frac{(x-1)^{n+1}}{n+1} \\
&= (x-1) - \frac{(x-1)^2}{2} + \frac{(x-1)^3}{3} - \frac{(x-1)^4}{4} + \cdots + \frac{(-1)^n}{c^{n+1}}\frac{(x-1)^{n+1}}{n+1}.
\end{aligned}$$

In particular, if n is odd then

$$\ln(x) \leq (x-1) - \frac{(x-1)^2}{2} + \frac{(x-1)^3}{3} - \frac{(x-1)^4}{4} + \cdots + \frac{(x-1)^n}{n},$$

with equality holding only for $x = 1$. This considerably improves (6.7), namely $\ln(x) \leq x - 1$. (If n is even the inequality holds for $x \leq 1$ and it is reversed for $x \geq 1$.) ◇

8.5 Taylor Series

The conclusion of Taylor's Theorem (Theorem 8.20), says that for each $x \in I$ there is c between x and x_0 such that

$$f(x) = \sum_{k=0}^{n} \frac{f^{(k)}(x_0)}{k!}(x - x_0)^k + \frac{f^{(n+1)}(c)}{(n+1)!}(x - x_0)^{n+1}.$$

Now *if* f has derivatives of all orders *and* it so happens that for a given x, it is the case that the remainder term

$$\frac{f^{(n+1)}(c)}{(n+1)!}(x - x_0)^{n+1} \to 0 \qquad \text{as } n \to \infty,$$

then we may reasonably write (for such x):

$$f(x) = \sum_{n=0}^{\infty} \frac{f^{(n)}(x_0)}{n!}(x - x_0)^n.$$

This is called the **Taylor series for f about the point** $x = x_0$. If $x_0 = 0$ it is often called the **Maclaurin series for** f, for Scottish mathematician Colin Maclaurin (1698–1746).

Example 8.25. Again, for $f(x) = \mathrm{e}^x$ and $x_0 = 0$, the remainder term is

$$\frac{f^{(n+1)}(c)}{(n+1)!}(x - x_0)^{n+1} = \frac{\mathrm{e}^c x^{n+1}}{(n+1)!}.$$

We claim that for any given $x \in \mathbf{R}$,

$$\frac{x^N}{N!} \to 0 \quad \text{as} \quad N \to \infty.$$

Then since c is between 0 and x, the remainder term

$$\frac{\mathrm{e}^c x^{n+1}}{(n+1)!} \to 0 \quad \text{as} \quad n \to \infty,$$

and so the Taylor series for $f(x) = \mathrm{e}^x$ about $x_0 = 0$ is

$$\mathrm{e}^x = \sum_{n=0}^{\infty} \frac{f^{(n)}(0)}{n!}x^n = \sum_{n=0}^{\infty} \frac{x^n}{n!} = 1 + x + \frac{1}{2!}x^2 + \frac{1}{3!}x^3 + \frac{1}{4!}x^4 + \cdots \quad \text{for all } x \in \mathbf{R}.$$

Now to verify the claim. Suppose first that $x > 0$ and take M to be the greatest integer $\leq x$. That is, $M \leq x < M+1$. Then for any $N > M$,

$$\frac{x^N}{N!} = \frac{x}{N}\frac{x}{N-1}\cdots\frac{x}{M+1}\frac{x}{M}\frac{x}{M-1}\cdots\frac{x}{2}\frac{x}{1}$$

$$< \frac{x}{N}\frac{x}{M}\frac{x}{M-1}\cdots\frac{x}{2}\frac{x}{1} = \frac{x}{N}\frac{x^M}{M!},$$

which clearly $\to 0$ as $N \to \infty$. We leave the (very similar) proof for $x < 0$ to Exercise 8.62. So the claim is verified. ◇

From Example 8.25 we see that in particular (taking $x = 1$):

$$\mathrm{e} = \sum_{n=0}^{\infty} \frac{1}{n!} = 1 + 1 + \frac{1}{2} + \frac{1}{3!} + \frac{1}{4!} + \cdots .$$

See [50] for a neat geometric argument, based on this series, which shows that e is irrational.

Example 8.26. For $f(x) = \ln(x)$ and $x_0 = 1$, the remainder term is

$$\frac{f^{(n+1)}(c)}{(n+1)!}(x-x_0)^{n+1} = \frac{(-1)^n}{c^{n+1}}\frac{(x-1)^{n+1}}{n+1} = \left(\frac{x-1}{c}\right)^{n+1}\frac{(-1)^n}{n+1}.$$

The reader may verify that for $1/2 \leq x \leq 2$ and c between x and 1,

$$-1 \leq \frac{x-1}{c} \leq 1.$$

(The cases $1/2 \leq x \leq c \leq 1$ and $1 \leq c \leq x \leq 2$ should be considered separately.) Therefore

$$\left(\frac{x-1}{c}\right)^{n+1}\frac{(-1)^n}{n+1} \to 0 \quad \text{as} \quad n \to \infty.$$

So, for $1/2 \leq x \leq 2$, the Taylor series for $f(x) = \ln(x)$ about $x_0 = 1$ is

$$\ln(x) = \sum_{n=1}^{\infty} \frac{(-1)^{n+1}}{n}(x-1)^n$$

$$= (x-1) - \frac{(x-1)^2}{2} + \frac{(x-1)^3}{3} - \frac{(x-1)^4}{4} + \cdots .$$

(We shall see in Example 10.10 that this equality in fact holds for $0 < x \leq 2$.) In particular, taking $x = 2$:

$$\ln(2) = \sum_{n=1}^{\infty} \frac{(-1)^{n+1}}{n} = 1 - \frac{1}{2} + \frac{1}{3} - \frac{1}{4} + \cdots .$$

So we have again found the sum of the Alternating Harmonic series. ◇

We leave it for Exercises 8.65 and 8.66 to verify that for each $x \in \mathbf{R}$, we have the Maclaurin series

$$\sin(x) = \sum_{n=0}^{\infty} \frac{(-1)^n x^{2n+1}}{(2n+1)!} = x - \frac{1}{3!}x^3 + \frac{1}{5!}x^5 - \frac{1}{7!}x^7 + \cdots ,$$

and

$$\cos(x) = \sum_{n=0}^{\infty} \frac{(-1)^n x^{2n}}{(2n)!} = 1 - \frac{1}{2!}x^2 + \frac{1}{4!}x^4 - \frac{1}{6!}x^6 + \cdots .$$

(Recall that sine is an odd function, and cosine is an even function...)

Exercises

8.1. [18]

(a) Show that $\dfrac{(fg)''}{fg} = \dfrac{f''}{f} + \dfrac{g''}{g} + 2\dfrac{f'}{f}\dfrac{g'}{g}$.

(b) Show that $\dfrac{(f/g)''}{f/g} = \dfrac{f''}{f} - \dfrac{g''}{g} - 2\dfrac{(f/g)'}{f/g}\dfrac{g'}{g}$.

8.2. [52] This is an extension of Exercise 5.4.

(a) Let f be continuous on $[a, b]$ and differentiable on (a, b) with $f(a) = f(b) = 0$. Show that there is $c \in (a, b)$ such that $f'(c) = f(c)$. Hint: Consider $g(x) = \mathrm{e}^{-x} f(x)$.

(b) Let f be continuous on $[a, b]$, differentiable on (a, b), and $f^{(k)}(a) = f^{(k)}(b) = 0$ for $k = 0, 1, 2, \ldots, n$. Show that there is $c \in (a, b)$ such that $f^{(n+1)}(c) = f(c)$.

8.3. **(a)** Show that

$$\frac{1}{2^n} \sum_{k=0}^{n} \binom{n}{k} = 1 \quad \text{and} \quad \sum_{k=0}^{n} (-1)^k \binom{n}{k} = 0.$$

(b) Use the Binomial formula $(1+x)^n = \sum_{k=0}^{n} \binom{n}{k} x^k$ to find an expression for $(a+b)^n$.

8.4. (e.g., [5]) Here's another proof that the sequence $\{(1+\frac{1}{n})^n\}$ is increasing and bounded above (and hence converges).

(a) Apply the Binomial formula $(1+x)^n = \sum_{k=0}^{n} \binom{n}{k} x^k$ to $\left(1+\frac{1}{n}\right)^n$, then simplify to get

$$\left(1+\frac{1}{n}\right)^n = 1+1+\frac{1}{2!}\left(1-\frac{1}{n}\right)+\frac{1}{3!}\left(1-\frac{1}{n}\right)\left(1-\frac{2}{n}\right)$$
$$+\cdots+\frac{1}{n!}\left(1-\frac{1}{n}\right)\left(1-\frac{2}{n}\right)\cdots\left(1-\frac{n-1}{n}\right).$$

(b) Do the same for $\left(1+\frac{1}{n+1}\right)^{n+1}$ then conclude that $\left(1+\frac{1}{n}\right)^n < \left(1+\frac{1}{n+1}\right)^{n+1}$.

(c) Show that

$$\left(1+\frac{1}{n}\right)^n < 1+1+\frac{1}{2!}+\frac{1}{3!}+\cdots+\frac{1}{n!} < 1+1+\frac{1}{2}+\frac{1}{2^2}+\cdots+\frac{1}{2^{n-1}} < 3.$$

8.5. Let $h(x) = f(x)g(x)$. Here we obtain **Leibniz's formula**

$$h^{(n)}(x) = \sum_{k=0}^{n} \binom{n}{k} f^{(k)}(x) g^{(n-k)}(x),$$

where $\binom{n}{k}$ is the binomial coefficient $\binom{n}{k} = \frac{n!}{k!(n-k)!}$.

(a) Argue that it is reasonable to assume that

$$h^{(n)}(x) = \sum_{k=0}^{n} a_k f^{(k)}(x) g^{(n-k)}(x).$$

(b) Set $f(x) = x^p$ and $g(x) = x^q$ with $p+q=n$, to show that $n! = a_k k!(n-k)!$. Now solve for a_k.

8.6. [37]

(a) Show that

$$2^{n+1}-2 = \sum_{k=1}^{n} \binom{n+1}{k}.$$

(b) Use the AGM Inequality (Theorem 2.10) to show that

$$(2^{n+1}-2)^n \geq \frac{(n(n+1)!)^n}{\prod_{k=1}^{n}(k!)^2},$$

and that equality holds if and only if $n = 1$ or $n = 2$.

8.7. [20] Here's a slick way of verifying the trigonometric identities

$$\sin(x+y) = \sin(x)\cos(y) + \sin(y)\cos(x) \quad \text{and}$$
$$\cos(x+y) = \cos(x)\cos(y) - \sin(x)\sin(y).$$

(a) Fix y and set

$$f(x) = \sin(x+y) - \big(\sin(x)\cos(y) + \sin(y)\cos(x)\big),$$

then show that $f' + f'' = 0$.

(b) Set

$$g(x) = \big(f'(x)\big)^2 + \big(f(x)\big)^2,$$

and conclude that $g' = 0$.

(c) So g is constant—find the constant.

(d) What does this say about f and f' ?

8.8. [8]

(a) Show that $y = \sin(x)$ and $y = \cos(x)$ each satisfy the differential equation

$$y'' + y = 0.$$

(b) Show, as follows, that *any* solution to this differential equation is of the form $y = c_1 \sin(x) + c_2 \cos(x)$, where c_1 and c_2 are constants. Set

$$p(x) = y\cos(x) - y'\sin(x)$$
$$q(x) = y\sin(x) + y'\cos(x).$$

Show that $p'(x) \equiv q'(x) \equiv 0$, so that $p(x)$ and $q(x)$ are each constant. Now eliminate y' in the pair of equations above.

8.9. [24] For $x > 0$ and $n = 1, 2, \ldots$ let $f(x) = x^n \ln(x)$.

(a) Show that

$$\frac{1}{n!} f^{(n)}(x) = \frac{1}{n} + \frac{1}{n+1} + \cdots + \frac{1}{3} + \frac{1}{2} + 1 + \ln(x).$$

(b) Conclude that Euler's constant $\gamma = \lim_{n\to\infty} \frac{1}{n!} f^{(n)}(\frac{1}{n})$.

8.10. [27] Let

$$p(x) = a_n x^n + a_{n-1}x^{n-1} + \cdots + a_1 x + a_0$$

be a polynomial of degree n, with roots $x_1, x_2, \ldots, x_n$.

(a) Show that for $x \neq x_j$,

$$\frac{p'(x)}{p(x)} = \sum_{j=1}^{n} \frac{1}{x - x_j}.$$

(b) Show that if each root x_j is real, then

$$(p'(x))^2 - p(x)p''(x) \geq \frac{(p'(x))^2}{n}, \quad \text{or} \quad (n-1)(p'(x))^2 - np(x)p''(x) \geq 0.$$

Hint: Differentiate the result in (a) and apply the Cauchy–Schwarz Inequality (Theorem 2.18) to the result.

Notice that if $x = 0$ this reduces to simply $(n-1)a_1^2 - 2na_2a_0 \geq 0$. Therefore, if $(n-1)a_1^2 - 2na_2a_0 < 0$, then p must have at least one, and hence at least two, complex roots. So if p is a quadratic we get the familiar discriminant condition $a_1^2 - 4a_2a_0 \geq 0$ for real roots.

8.11. Prove, using Cauchy's Mean Value Theorem (Theorem 5.11) instead of L'Hospital's Rule, that if f'' exists then

$$f''(x) = \lim_{h \to 0} \frac{f(x+h) - 2f(x) + f(x-h)}{h^2}.$$

8.12. **(a)** Show that if f'' is continuous, then

$$f''(x) = \lim_{h \to 0} \frac{f(x+3h) + 3f(x-h)}{6h^2}.$$

(b) Show that this still holds even if we assume only that f'' exists.

8.13. **(a)** Show that if $f^{(3)}$ is continuous, then

$$f^{(3)}(x) = \lim_{h \to 0} \frac{3f(x+2h) - 10f(x+h) - 6f(x-h) - f(x-2h)}{2h^3}.$$

(b) Find a, b, c such that if $f^{(3)}$ is continuous, then

$$f^{(3)}(x) = \lim_{h \to 0} \frac{af(x+3h) + bf(x+2h) + cf(x-h) - f(x)}{h^3}.$$

8.14. [29] In Exercise 1.12 we showed that

$$\left|f(x+h)h - 2hf(x) + hf(x-h)\right|$$

is twice the area $A(x,h)$ of the triangle determined by $(x-h, f(x-h))$, $(x, f(x))$, and $(x+h, f(x+h))$. Conclude that if f'' exists, then $f''(x) = \lim_{h\to 0} \dfrac{2A(x,h)}{h^3}$.

8.15. [5] Suppose that f is continuous on $[a,b]$ and that f'' exists on (a,b). Suppose that the chord between the points $(a, f(a))$ and $(b, f(b))$ intersects the graph of f at $(x_0, f(x_0))$, where $a < x_0 < b$. Prove that there is a point $c \in (a,b)$ such that $f''(c) = 0$. Hint: Begin by applying the Mean Value Theorem (Theorem 5.2) on $[a, x_0]$ and on $[x_0, b]$.

8.16. [32] Here's another proof of the Mean Value Theorem for the Second Derivative (Theorem 8.6). Let

$$F(t) = f(t) - f(x_0) - f'(x_0)(t - x_0) + K(t - x_0)^2,$$

where K is chosen so that $F(x) = 0$.
(a) Verify that

$$K = \frac{f(x) - f(x_0) - f'(x_0)(x - x_0)}{(x - x_0)^2}.$$

(b) Apply Rolle's Theorem (Theorem 5.1) to F on $[x, x_0]$ to show there is $c \in (x, x_0)$ such that $F'(c) = 0$.
(c) Now apply Rolle's Theorem to F' on $[x, c]$.

8.17. [33] Write

$$f^{[1]}(x) = \lim_{h\to 0} \frac{f(x+h) - f(x-h)}{2h}$$

for the Schwarz derivative, or the symmetric derivative. The **second Schwarz derivative**, or the **second symmetric derivative** is:

$$f^{[2]}(x) = \lim_{h\to 0} \frac{f(x+h) - 2f(x) + f(x-h)}{h^2}.$$

It is easy to see (e.g., Exercise 4.10) that if $f'(x)$ exists, then $f'(x) = f^{[1]}(x)$ and we know that if $f' \equiv 0$ then f is constant.

(a) If $f^{[1]} = 0$ then is f necessarily constant?
(b) In Example 8.4 we showed that if $f''(x)$ exists, then $f''(x) = f^{[2]}(x)$ and we know that if $f'' \equiv 0$ then f is a linear function. Show that if $f^{[2]}(x) = 0$ for

all $x \in (a,b)$ then f is a linear function on $[a,b]$, as follows. If f is linear, it must look like $\frac{f(b)-f(a)}{b-a}(x-a)+f(a)$. So consider

$$F(x) = F_n(x) = f(x) - \left(\frac{f(b)-f(a)}{b-a}(x-a)+f(a)\right) + \frac{(x-a)(x-b)}{n}.$$

Now use $F^{[2]}(x) = 2/n$ to show that $F(x) \leq 0$.

(c) Consider

$$G(x) = G_n(x) = \frac{f(b)-f(a)}{b-a}(x-a)+f(a)-f(x)+\frac{(x-a)(x-b)}{n},$$

and show that $G(x) \leq 0$.

(d) Combine (b) and (c) and let $n \to \infty$.

8.18. This is the **Second Derivative Test**. Suppose that f'' exists on (a,b) and that for some $c \in (a,b)$, we have $f'(c) = 0$. Show that if $f''(c) > 0$ then f has a local minimum at c. Show that if $f''(c) < 0$ then f has a local maximum at c. What if $f''(c) = 0$? What if $f''(c)$ does not exist? Draw generic pictures which illustrate these cases.

8.19. [44] We saw in Example 8.10 that $x^{\mathrm{e}} < \mathrm{e}^x$ for $x \neq \mathrm{e}$. Use this to prove the AGM Inequality (Theorem 2.10) as follows. Set $x = \mathrm{e}a_j/G$ for $j = 1,2,\ldots n$, then multiply.

8.20. [14] Suppose that $f > 0$ and has two derivatives on **R**. Show that there is $x_0 \in \mathbf{R}$ such that $f''(x_0) \geq 0$.

8.21. [13] Let f be a function with continuous second derivative on **R**. Show that if $\lim\limits_{n\to\infty} f(x) = 0$ and f'' is bounded, then $\lim\limits_{n\to\infty} f'(x) = 0$.

8.22. [12] Extend Lemma 8.7 as follows. Show that if f is convex and differentiable on (a,b) with $x_0 \in (a,b)$ then for $x \in (a,b)$,

$$f(x) \geq f(x_0) + f'(x_0)(x-x_0).$$

Hint: Show that the convexity condition can be manipulated to obtain (for $t \neq 0$ and $x \neq x_0$)

$$\frac{f(t(x-x_0)+x_0)-f(x_0)}{t(x-x_0)}(x-x_0) \leq f(x)-f(x_0),$$

then let $t \to 0$, and hence $t(x-x_0) \to 0$.

8.23. [12] Show that the converse of the Exercise 8.22 holds: Suppose that f is differentiable on (a,b) and for each $x, x_0 \in (a,b)$,

$$f(x) \geq f(x_0) + f'(x_0)(x-x_0).$$

Then f is convex on (a,b).

8.24. [10]

(a) Suppose that f is convex on $[a,b]$, f'' exists, and that $f < 1$. Show that $1/(1-f)$ is convex.

(b) Let f be such that f'' is continuous on $[a,b]$. Show that

$$f(x) - m\frac{x^2}{2} \quad \text{and} \quad M\frac{x^2}{2} - f(x)$$

are each convex, where $m = \min_{x\in[a,b]} \{f''(x)\}$ and $M = \max_{x\in[a,b]} \{f''(x)\}$ (m and M exist by the Extreme Value Theorem (Theorem 3.23).)

8.25. Here is another proof of Lemma 8.12, that if $f'' \geq 0$ on (a,b) then f is convex on (a,b).

(a) Since $f'' \geq 0$, f' is increasing. Conclude that for $x < c < y$,

$$\frac{f(c) - f(x)}{c - x} \leq \frac{f(y) - f(c)}{y - c}.$$

(b) Therefore

$$f(c) \leq \frac{(y-c)f(x) + (c-x)f(y)}{y - x}.$$

(c) Now translate this to a statement involving t and $1-t$.

8.26. **(a)** Suppose that f is convex on $[a,b]$ and let $a \leq x_1 < x_2 < x_3 \leq b$. In the definition of convexity set $x = x_1$, $y = x_3$, and $x_2 = (1-t)x_1 + tx_3$ to show that

$$(x_3 - x_2)f(x_1) + (x_3 - x_1)f(x_2) + (x_2 - x_1)f(x_3) \geq 0.$$

(b) Let $x_1 < x_2 < x_3$. In Exercise 1.12 we saw that the area A of the triangle T with vertices (x_1, y_1), (x_2, y_2), (x_3, y_3) is

$$A = \frac{1}{2}\left|x_1(y_2 - y_3) + x_3(y_1 - y_2) + x_2(y_3 - y_1)\right|,$$

for $x_1 < x_2 < x_3$. Show that if f is convex on $[a,b]$ then

$$\frac{1}{2}\left[x_1(f(x_2) - f(x_3)) + x_3(f(x_1) - f(x_2)) + x_2(f(x_3) - f(x_1))\right] \geq 0$$

whenever $a \leq x_1 < x_2 < x_3 \leq b$.

(c) Draw a picture which shows what (b) says.

8.27. [31] Let $a, b > 0$, with $a + b = 1$. Exercise 2.18 asks to show that

$$\left(a + \frac{1}{a}\right)^2 + \left(b + \frac{1}{b}\right)^2 \geq \frac{25}{2}.$$

Here's another way. **(a)** Show that $f(x) = (x + 1/x)^2$ is convex on $(0, 1)$, and so

$$\frac{f(a) + f(b)}{2} \geq f\left(\frac{a + b}{2}\right).$$

(b) Let $\alpha > 0$. Prove the more general inequality

$$\left(a + \frac{1}{a}\right)^\alpha + \left(b + \frac{1}{b}\right)^\alpha \geq \frac{5^\alpha}{2^{\alpha-1}}.$$

8.28. [3] Let $0 < a \leq b$. Show that

$$\frac{a + b}{2} \leq \left(a^a b^b\right)^{\frac{1}{a+b}}.$$

Hint: Set $x = a/b$ and show that

$$\ln\left(\frac{x + 1}{2}\right) \leq \frac{x}{x + 1} \ln(x).$$

8.29. Show that for $x \in (0, \pi/4)$,

$$x < \tan(x) < \frac{4}{\pi}x.$$

8.30. [25,45]

(a) Show that for $x \in (0, \pi/2)$, we have $\cos(x) > 1 - 2x/\pi$.
(b) Conclude that for such x,

$$\tan(x) < \frac{\pi x}{\pi - 2x}.$$

(c) Show that for $x \in (0, \pi/6)$,

$$\frac{2x}{\pi - 2x} < \sin(x).$$

8.31. [34] Let $-\pi/2 < \theta_j < \pi/2$ for $j = 1, 2, \ldots, n$. Show that

$$\left(\cos(\theta_1)\cos(\theta_2)\cdots\cos(\theta_n)\right)^{1/n} \leq \cos\left(\frac{\theta_1+\theta_2+\cdots+\theta_n}{n}\right).$$

Hint: Consider $f(x) = \ln(\cos(x))$.

8.32. [47] Let $x, y \in (0, \pi)$, with $x \neq y$. Show that

$$\sin\left(\sqrt{xy}\right) > \sqrt{\sin(x)\sin(y)}.$$

Hint: Consider $f(x) = \ln(\sin(e^x))$. Can you show that

$$\sin\left(x^p y^q\right) > (\sin(x))^p\,(\sin(y))^q\,,$$

with $p, q > 1$ satisfying $1/p + 1/q = 1$?

8.33. [1]

(a) Let f be convex on $[a, c]$ and let $a < b < c$. Show that

$$f(a - b + c) \leq f(a) - f(b) + f(c).$$

Hint: Write $b = (1 - t)a + tc$ and observe that $a - b + c = ta + (1 - t)c$.

(b) Draw a picture which illustrates this inequality.

8.34. Use the fact that e^x is convex on $(-\infty, +\infty)$ to prove the weighted AGM Inequality with $n = 2$ (Corollary 6.16). Deduce Young's Inequality (Corollary 6.19).

8.35. A function $f : [a, b] \to \mathbf{R}$ is **logarithmically convex** if for every $x, y \in [a, b]$,

$$f((1 - t)x + ty) \leq f(x)^{1-t} f(y)^t \qquad \text{for every } t \in [0, 1].$$

(a) Show that for $f > 0$, f being logarithmically convex is equivalent to $\ln(f)$ being convex.

(b) Show that if f is logarithmically convex then f convex.

(c) Show that for $a, b \in [0, \pi/4]$,

$$\sqrt{ab} \leq \tan\left(\frac{a + b}{2}\right) \leq \frac{\tan a + \tan b}{2}$$

(d) Find a function which is convex but not logarithmically convex.

8.36. [15, 16] Let $a_1, a_2, \ldots, a_n$ be positive numbers and denote by G their Geometric Mean. Prove the AGM Inequality (Theorem 2.10) as follows.

(a) Show that

$$g(x) = \frac{G}{n} \sum_{j=1}^{n} \left(\frac{a_j}{G}\right)^x \qquad \text{is convex.}$$

(b) Verify that $g'(0) = 0$ and use this to conclude that $g(0) \le g(1)$.
(c) Can you extend this to prove the weighted AGM Inequality (Theorem 6.15)?

8.37. [30, 46] Fill in the details of another proof of Jensen's Inequality (Theorem 8.17), as follows. Set $A = \sum_{j=1}^{n} w_j x_j$, and for $0 \le t \le 1$, let

$$g(t) = \sum_{j=1}^{n} w_j f((1-t)x_j + tA).$$

(a) Show that $g'' \ge 0$. **(b)** Conclude that $g'(t) \le g'(1)$. **(c)** Compute $g''(1)$ then conclude that $g(0) \ge g(1)$. **(d)** Write down what $g(0) \ge g(1)$ says.

8.38. Fill in the details of another proof of Jensen's Inequality (Theorem 8.17) which is closely related to the proof in Exercise 8.37. Set $A = \sum_{j=1}^{n} w_j x_j$, and for $0 \le t \le 1$, let

$$g(t) = \sum_{j=1}^{n} w_j f((1-t)x_j + tA).$$

(a) Apply the Mean Value Theorem for the Second Derivative (Theorem 8.6) to get $g(0) = g(1) + g'(1)(0-1) + \frac{g''(c)}{2}(0-1)^2$.
(b) Compute each of $g(0)$, $g(1)$, $g'(1)$, and $g''(c)$ to see what (a) says.
(c) Now use the convexity of f.

8.39. Fill in the details, as follows, of a proof of Jensen's Inequality (Theorem 8.17) which does not assume that f'' exists. The $n = 2$ case is the definition of convexity. Now write

$$f\left(\sum_{j=1}^{n} w_j x_j\right) = f\left((1-w_n)\sum_{j=1}^{n-1} \frac{w_j}{1-w_n} x_j + w_n x_n\right)$$

and proceed by induction. (Compare with Exercise 6.30.)

8.40. [43]

(a) Show that

$$\frac{1}{n}\sum_{j=1}^{n} \sin(x_j) \le \sin\left(\frac{1}{n}\sum_{j=1}^{n} x_j\right) \qquad \text{for } x_j \in (0, \pi)$$

and

$$\frac{1}{n}\sum_{j=1}^{n} \tan(x_j) \ge \tan\left(\frac{1}{n}\sum_{j=1}^{n} x_j\right) \qquad \text{for } x_j \in (0, \pi/2).$$

(b) Find and prove a similar inequality for $\cos(x)$, on $(-\pi/2, \pi/2)$.
(c) Show that a triangle with interior angles A, B, C satisfies

$$\sin(A) + \sin(B) + \sin(C) \leq \frac{3\sqrt{3}}{2} \quad \text{and} \quad \sin(A)\sin(B)\sin(C) \leq \frac{3\sqrt{3}}{8}.$$

8.41. Apply Jensen's Inequality (Theorem 8.17) to $f(x) = x\ln(x)$, with $n = 3$, to prove that for $a, b, c > 0$ we have $a^a b^b c^c \geq \left(\frac{a+b+c}{3}\right)^{a+b+c}$.

8.42. [41] Use Jensen's Inequality (Theorem 8.17) and the AGM Inequality (Theorem 2.10) to show that for $0 < a_1 \leq a_2 \leq \cdots \leq a_n$,

$$\sum_{j=1}^{n} a_j^{n+1} \geq a_1 a_2 \cdots a_n \sum_{j=1}^{n} a_j .$$

(We saw another way to do this in Exercise 2.56.)

8.43. Use Jensen's Inequality (Theorem 8.17) and the fact that e^x is convex on $(-\infty, +\infty)$ to prove the weighted AGM Inequality (Theorem 6.15).

8.44. [26, 51] Let $w_1, \ldots, w_n > 0$ satisfy $\sum_{j=1}^{n} w_j = 1$. Let $0 < x_1 < \cdots < x_n < 1/2$, $y_j = 1 - x_j$,

$$A_1 = \sum_{j=1}^{n} w_j x_j, \quad \text{and} \quad A_2 = \sum_{j=1}^{n} w_j y_j = 1 - A_1 .$$

Apply Jensen's Inequality (Theorem 8.17) to $f(1-x) - f(x)$ to on $(0, 1)$ to prove **Levinson's Inequality**:

$$f^{(3)} \geq 0 \text{ on } (0,1) \;\Rightarrow\; f(A_2) - f(A_1) \leq \sum_{j=1}^{n} w_j f(y_j) - \sum_{j=1}^{n} w_j f(x_j) .$$

8.45. [17]. Let $\alpha_1, \alpha_2, \alpha_3$ be the measures (in radians) of the angles in an acute triangle. **(a)** Show that

$$\frac{\alpha_1}{\alpha_2\alpha_3}\tan(\alpha_1) + \frac{\alpha_2}{\alpha_1\alpha_3}\tan(\alpha_2) + \frac{\alpha_3}{\alpha_1\alpha_2}\tan(\alpha_3) \geq 3\sqrt{3}.$$

(b) Show that

$$\tfrac{\alpha_1}{\alpha_2\alpha_3}(3 + \tan^2(\alpha_1))^{1/4} + \tfrac{\alpha_2}{\alpha_1\alpha_3}(3 + \tan^2(\alpha_2))^{1/4} + \tfrac{\alpha_3}{\alpha_1\alpha_2}(3 + \tan^2(\alpha_3))^{1/4} \geq 3\sqrt{3}.$$

Hint for both: Chebyshev's Inequality (Exercise 2.54), AGM Inequality (Theorem 2.10), and Jensen's Inequality (Theorem 8.17).

8.46. [49] The conclusion of Jensen's Inequality (Theorem 8.17) on $[a, b]$ reads

$$0 \le \sum_{j=1}^{n} w_j f(x_j) - f\Big(\sum_{j=1}^{n} w_j x_j\Big).$$

Prove the upper bound

$$\sum_{j=1}^{n} w_j f(x_j) - f\Big(\sum_{j=1}^{n} w_j x_j\Big) \le \max_{0 \le t \le 1} \big[(1-t)f(a) + tf(b) - f\big((1-t)a + tb\big)\big].$$

Hint: In $\sum_{j=1}^{n} w_j f(x_j) - f(\sum_{j=1}^{n} w_j x_j)$, begin by writing $x_j = (1-t_j)a + t_j b$ and use the fact that f is convex.

8.47. [22] For $j = 1, 2, \ldots, n$, let $0 \le x_j \le 1$ and $w_j > 0$ with $\sum_{j=1}^{n} w_j = 1$. Show that

$$\frac{w_1}{1+x_1} + \frac{w_2}{1+x_2} + \cdots + \frac{w_n}{1+x_n} \le \frac{1}{1 + x_1^{w_1} x_2^{w_2} \cdots x_n^{w_n}}.$$

Hint: First dispense with cases in which any $x_j = 0$. Then show that

$$f(t) = \frac{1}{1+e^t}$$

is concave, and apply Jensen's Inequality (Theorem 8.17), with $t_j = \ln(x_j)$.

8.48. Let $x_1, x_2, \cdots, x_n$ be positive numbers. For $r \ne 0$ their **Power Mean** M_r is:

$$M_r = \left(\frac{1}{n}\sum_{j=1}^{n} x_j^r\right)^{1/r}.$$

(a) Verify, for example, that M_1 is the Arithmetic Mean, M_{-1} is the Harmonic Mean, and M_2 is the Root Mean Square.
(b) Use Jensen's Inequality (Theorem 8.17) to show that if $s < r$, then $M_s < M_r$.
(c) Show that it is reasonable, for the sake of continuity, to define M_0 = the Geometric Mean $G = (x_1 \cdot x_2 \cdots x_n)^{1/n}$.
(d) What are reasonable definitions of $M_{-\infty}$ and M_{∞} ?
(e) How would you define the *weighted* Power Means?

8.49. [10] Let f be convex on $[a,b]$, and let $a = x_1 \le x_2 \le \cdots \le x_n = b$. Show that

$$\frac{1}{n}\sum_{j=1}^{n} f(x_j) \le \frac{b - \frac{1}{n}\sum_{j=1}^{n} f(x_j)}{b-a} f(a) + \frac{\frac{1}{n}\sum_{j=1}^{n} f(x_j) - a}{b-a} f(b).$$

8.50. We in Example 8.19 how Jensen's Inequality (Theorem 8.17) can be used to prove the Cauchy–Schwarz Inequality (Theorem 2.18). Use Jensen's Inequality to prove Hölder's Inequality (Lemma 6.18) : *Let* $a_1, \ldots, a_n > 0$ *and* $b_1, \ldots, b_n > 0$, *and let* $p, q > 1$ *satisfy* $\frac{1}{p} + \frac{1}{q} = 1$. *Then*

$$\sum_{j=1}^{n} a_j\, b_j \;\le\; \left(\sum_{j=1}^{n} a_j^p\right)^{1/p} \left(\sum_{j=1}^{n} b_j^q\right)^{1/q}.$$

8.51. [10, 42] The conclusion of the Mean Value Theorem (Theorem 5.2) is: there exists $c \in (a,b)$ such that $f'(c) = \frac{f(b)-f(a)}{b-a}$. Assume for this problem that $f'' > 0$.

(a) Show that the c above is unique.

(b) For $a > 0$, the number

$$c = (f')^{-1}\left(\tfrac{f(b)-f(a)}{b-a}\right)$$

is called the **Lagrangian Mean** of a and b. So that the Lagrangian Mean is continuous, what should we define as the Lagrangian Mean of a and b, if $a = b$?

(c) Compute the Lagrangian Mean for $f(x) = x^2$, for $f(x) = 1/x$, and for a few other functions of your choice. Try $f(x) = x^r$ and let $r \to 0$.

8.52. Extend Exercise 8.16 above to prove Taylor's Theorem (Theorem 8.20). This is essentially the proof to be found in most textbooks. Yet another proof can be found in [5].

8.53. Let n be a positive integer. Prove the **Binomial formula**

$$(1+x)^n = \sum_{k=0}^{n} \binom{n}{k} x^k,$$

where the coefficient of x^k is the **binomial coefficient**

$$\binom{n}{k} = \frac{n!}{(n-k)!k!} = \frac{n(n-1)(n-2)\cdots(n-k+1)}{k(k-1)(k-2)\cdots 2\cdot 1}.$$

by using Taylor's Theorem (Theorem 8.20), as follows.

(a) Let $f(x) = (1+x)^n$ and verify that for $k \le n$,

$$f^{(k)}(x) = n(n-1)\cdots(n-k+1)(1+x)^{n-k}.$$

(b) Conclude that

$$f^{(k)}(0) = \begin{cases} n(n-1)\cdots(n-k+1) & \text{if } k \le n \\ 0 & \text{if } k > n. \end{cases}$$

(c) Now apply Taylor's Theorem with $x_0 = 0$.

8.54. Here we extend the **Second Derivative Test** from Exercise 8.18. Suppose that each of f, f', f'', …, $f^{(n+1)}$ is continuous on an open interval I containing c, that

$$0 = f'(c) = f''(c) = \cdots = f^{(n)}(c), \quad \text{but that } f^{(n+1)}(c) \neq 0.$$

Show that if $f^{(n+1)}(c) > 0$ and n is even, then c yields a local minimum for f. Can you summarize the other possibilities—for example $f^{(n+1)}(c) < 0$ and n odd?

8.55. Use the Taylor polynomial of degree n and corresponding remainder for e^x with $x_0 = 0$, to show that $n! > \left(\frac{n}{e}\right)^n$. We did this another way in Exercise 2.27. In Exercise 2.17 we saw that $n! < \left(\frac{n+1}{2}\right)^n$.

8.56. [11] Suppose that $f''(x)$ exists for all x and that $p, q > 1$ satisfy $1/p + 1/q = 1$. Show that if

$$\frac{f(x) - f(y)}{x - y} = f'(px + qy)$$

for all $x, y \in \mathbf{R}$, then f is either linear (for $p \neq q$) or a quadratic.

8.57. [28] Suppose that $f^{(3)}(x)$ exists for all $x \in \mathbf{R}$. Show that

$$f(x+y) = f(x-y) + y\left(f'(x+y) + f'(x-y)\right)$$

for all $x, y \in \mathbf{R}$ if and only if f is a quadratic.

8.58. [38] Let p be a polynomial of degree n. Show that

$$\sum_{k=0}^{n} \frac{p^{(k)}(0)}{(k+1)!} x^{k+1} = \sum_{k=0}^{n} (-1)^k \frac{p^{(k)}(x)}{(k+1)!} x^{k+1}.$$

Hint: Show that the left and right hand sides differ by a constant, then show that the constant is zero.

8.59. Under the hypotheses of Taylor's Theorem (Theorem 8.20),

$$\begin{aligned} f(x) &= \sum_{k=0}^{n} \frac{f^{(k)}(x_0)}{k!}(x - x_0)^k + \frac{f^{(n+1)}(c)}{(n+1)!}(x - x_0)^{n+1} \\ &= p_n(x) + \frac{f^{(n+1)}(c)}{(n+1)!}(x - x_0)^{n+1}. \end{aligned}$$

Show that

$$\lim_{x \to x_0} \frac{f(x) - p_n(x)}{(x - x_0)^n} = 0.$$

8.60. [21] Taylor's Theorem (Theorem 8.20) is often used to prove things about functions which satisfy various specific conditions. Here's an example. Let f be a function with continuous third derivative on $[0, 1]$. Suppose that $f(0) = f'(0) = f''(0) = f'(1) = f''(1) = 0$ and $f(1) = 1$. Show that there exists $x \in [0, 1]$ such that $f^{(3)}(x) \geq 24$.

8.61. [31] Taylor's Theorem (Theorem 8.20) is often used to prove inequalities for functions which satisfy various general conditions. Here's an example which is essentially due to German mathematician E. Landau (1877–1938): *Let f, f', and f'' be continuous on $[0, 2]$, with $|f(x)| \leq 1$ and $|f''(x)| \leq 1$ there. Then $|f'(x)| \leq 2$ for all $x \in [0, 2]$.* Fill in the details of the following proof.

(a) Show that by Taylor's Theorem we have, for some $c_1, c_2 \in (0, 2)$,

$$\begin{aligned} f(0) &= f(x) + f'(x)(0 - x) + \frac{f''(c_1)}{2}(0 - x)^2 \\ &= f(x) - f'(x)x + \frac{f''(c_1)}{2}x^2 \end{aligned}$$

and

$$\begin{aligned} f(2) &= f(x) + f'(x)(2 - x) + \frac{f''(c_2)}{2}(2 - x)^2 \\ &= f(x) + 2f'(x) - f'(x)x + \frac{f''(c_2)}{2}(2 - x)^2. \end{aligned}$$

(b) Subtract the first equation from the second to get

$$2f'(x) = f(2) - f(0) - \frac{f''(c_2)}{2}(2 - x)^2 + \frac{f''(c_1)}{2}x^2.$$

(c) Show then, by the hypotheses and the triangle inequality, that

$$|2f'(x)| \leq 1 + 1 + \frac{1}{2}(2 - x)^2 + \frac{1}{2}x^2 = 4 - x(2 - x) \leq 4.$$

(d) Consider the function $f(x) = \frac{1}{2}x^2 - 1$ on $[0, 2]$ to show that equality can occur.

8.62. Show that for each $x \in \mathbf{R}$,

$$\frac{x^N}{N!} \to 0 \quad \text{as} \quad N \to \infty.$$

We showed this for $x > 0$, in Example 8.25. So only $x \leq 0$ needs consideration.

8.63. **(a)** Use the Taylor polynomial of degree n and corresponding remainder, for $f(x) = \ln(x)$ and $x_0 = 1$ (see Example 8.24) to obtain the Taylor polynomial of degree n and corresponding remainder, for $f(x) = \ln(1 + x)$ and $x_0 = 0$.

(b) Show that for $-1/2 \leq x \leq 1$,

$$\ln(1 + x) = x - \frac{x^2}{2} + \frac{x^3}{3} - \frac{x^4}{4} + \cdots.$$

8.64. **(a)** Compute the Taylor polynomial of degree n at $x_0 = 1$ and corresponding remainder, for $f(x) = 1/x$.

(b) Find the Taylor series for $f(x) = 1/x$ at $x_0 = 1$ and show that it converges for $1/2 \leq x \leq 2$.

8.65. **(a)** Compute the Taylor polynomial of degree n at $x_0 = 0$ and corresponding remainder, for $f(x) = \cos(x)$.

(b) Show that for $x \in \mathbf{R}$,

$$1 - \frac{x^2}{2} \leq \cos(x) \leq 1.$$

(In particular, $\lim_{x \to 0} (\cos(x)) = 1$.)

(c) Show that for $x \in \mathbf{R}$,

$$1 - \frac{x^2}{2} \leq \cos(x) \leq 1 - \frac{x^2}{2} + \frac{x^4}{24}.$$

(d) Show that for $x \in \mathbf{R}$,

$$\cos(x) = \sum_{n=0}^{\infty} \frac{(-1)^n x^{2n}}{(2n)!} = 1 - \frac{1}{2}x^2 + \frac{1}{4!}x^4 - \frac{1}{6!}x^6 + \cdots.$$

8.66. **(a)** Compute the Taylor polynomial of degree n at $x_0 = 0$ and corresponding remainder, for $f(x) = \sin(x)$.

(b) Show that for $x \geq 0$,

$$x - \frac{x^3}{6} \leq \sin(x) \leq x.$$

(c) Show that the inequalities in (b) are reversed for $x \le 0$. (So $\lim_{x\to 0} \frac{\sin(x)}{x} = 1$.)

(d) Show that for $x \in \mathbf{R}$,

$$\sin(x) = \sum_{n=0}^{\infty} \frac{(-1)^n x^{2n+1}}{(2n+1)!} = x - \frac{1}{3!}x^3 + \frac{1}{5!}x^5 - \frac{1}{7!}x^7 + \cdots .$$

8.67. [23]

(a) Apply the Mean Value Theorem (Theorem 5.2) to $f(t) = \sin(t)$ on $[0, x]$, with $x < \pi/2$, to show that

$$\sin(x) \ge \frac{x}{\sqrt{1+x^2}}.$$

(b) How does this compare with Jordan's Inequality $\sin(x) \ge \frac{2}{\pi}x$ from Example 8.14 ?

(c) How does this compare with the $\sin(x) \ge x - \frac{x^3}{6}$ from Exercise 8.66 ?

8.68. (a) Compute the Taylor polynomial of degree n at $x_0 = 0$ and corresponding remainder, for

$$\cosh(x) = \frac{e^x + e^{-x}}{2}.$$

(b) Find (with justification) the Maclaurin series for $\cosh(x)$.

(c) Compute the Taylor polynomial of degree n at $x_0 = 0$ and corresponding remainder, for

$$\sinh(x) = \frac{e^x - e^{-x}}{2}.$$

(d) Find (with justification) the Maclaurin series for $\sinh(x)$.

8.69. [48] Show that

$$\sum_{n=0}^{\infty} \frac{1}{n!(n^4 + n^2 + 1)} = \frac{e}{2}.$$

8.70. Define

$$f(x) = \begin{cases} e^{-1/x^2} & \text{if } x \ne 0 \\ 0 & \text{if } x = 0. \end{cases}$$

Show that f has derivatives of all orders at $x_0 = 0$, but that

$$f(x) \neq \sum_{n=0}^{\infty} \frac{f^{(n)}(0)}{n!} x^n,$$

unless $x = 0$. (This function is not *analytic* except at zero. A function is **analytic** wherever it equals its Taylor series.)

8.71. Newton's method (Sir Isaac Newton, English (1642–1727); no introduction necessary) is a method for approximating a root c of a function f, i.e., a number c for which $f(c) = 0$. It begins with an initial guess x_0, then the iteration scheme

$$x_{n+1} = x_n - \frac{f(x_n)}{f'(x_n)} \qquad \text{for } n = 0, 1, 2, 3, \ldots.$$

As long as $f'(c) \neq 0$ and x_0 is close to c, the scheme converges to c. (See, for example, [5] or [12] for details).

(a) Show that x_{n+1} is the x intercept of the tangent line to $y = f(x)$ at $x = x_n$. That is, x_{n+1} is where the Taylor polynomial p_1 of degree 1 at $x = x_n$ has a zero.

(b) Show that if we take instead x_{n+1} as a zero the Taylor polynomial p_2 of degree 2 then we get the expression

$$x_{n+1} = x_n - \frac{f(x_n)}{f'(x_n) + \frac{f''(x_n)}{2}(x_{n+1} - x_n)}.$$

(c) [36] Solve this for x_{n+1} to get another iteration scheme.

(d) [9] A different approach from (c) is to use Newton's method to approximate the x_{n+1} on the right-hand side. Show that this leads to the iteration scheme

$$x_{n+1} = x_n - \frac{2f(x_n)f'(x_n)}{2f'(x_n)f'(x_n) - f(x_n)f''(x_n)}.$$

This scheme is known as **Halley's method**, named for English mathematician and astronomer Edmond Halley (1656–1742). (Yes, this is the same Halley as the comet: Halley's comet can be seen from Earth with the naked eye every 75 years or so. It is due to next come around in 2061.)

(e) [6] Show that Newton's method applied to

$$g(x) = \frac{f(x)}{\sqrt{f'(x)}}$$

yields Halley's method.

(f) Show that Newton's method is fixed point iteration (see Exercise 5.36) applied to

$$g(x) = x - \frac{f(x)}{f'(x)}.$$

(g) Show that Halley's method is fixed point iteration (see Exercise 5.36) applied to

$$g(x) = x - \frac{2f(x)f'(x)}{2f'(x)f'(x) - f(x)f''(x)}.$$

See [4] for an interesting historical account of iteration methods.

References

1. Andreescu, T., Enescu, B.: Mathematical Olympiad Treasures. Birkhauser, Boston (2003)
2. Aull, C.E.: The first symmetric derivative. Am. Math. Mon. **74**, 708–711 (1967)
3. Ayoub, A.B., Amdeberhan, T.: Problem 705. Coll. Math. J. **33**, 246 (2002)
4. Bailey, D.F.: A historical survey of solution by functional iteration. Math. Mag. **62**, 155–166 (1989)
5. Bartle, R.G., Sherbert, D.R.: Introduction to Real Analysis, 2nd edn. Wiley, New York (1992)
6. Bateman, H.: Halley's methods for solving equations. Am. Math. Mon. **45**, 11–17 (1938)
7. Belfi, V.A.: Convexity in elementary calculus: some geometric equivalences. Am. Math. Mon. **15**, 37–41 (1984)
8. Brenner, J.L.: An elementary approach to $y'' = -y$. Am. Math. Mon. **95**, 344 (1988)
9. Brown, G.H., Jr.: On Halley's variation of Newton's method. Am. Math. Mon. **84**, 726–728 (1977)
10. Bullen, P.S.: Handbook of Means and Their Inequalities. Kluwer Academic, Dordrecht/Boston (2003)
11. Callahan, F.P., Coen, R.: Problem E1803. Am. Math. Mon. **74**, 82 (1967)
12. Carlson, R.: A Concrete Approach to Real Analysis. Chapman & Hall/CRC, Boca Raton (2006)
13. Chico Problem Group, Cal. State – Chico and Weiner, J.: Problem 213. Coll. Math. J. **14**, 356–357 (1983)
14. Clark, J.: Derivative sign patterns. Coll. Math. J. **42**, 379–382 (2011)
15. Daykin, D.E., Eliezer, C.J.: Generalizations of the A.M. and G.M. inequality. Math. Mag. **40**, 247–250 (1967)
16. Daykin, D.E., Eliezer, C.J.: Elementary proofs of basic inequalities. Am. Math. Mon. **76**, 543–546 (1969)
17. Díaz-Barrero, J.J., Brase, R.: Problem 11385. Am. Math. Mon. **117**, 285 (2010)
18. Eggleton, R., Kustov, V.: The product and quotient rules revisited. Coll. Math. J. **42**, 323–326 (2011)
19. Friedberg, S.H.: The power rule and the binomial formula. Coll. Math. J. **20**, 322 (1989)
20. Geist, R.: Response to Query 4. Coll. Math. J. **11**, 126 (1980)
21. Klamkin, M.S.: Problem Q913. Math. Mag. **74**, 325, 330 (2001)
22. Klamkin, M.S., Beesing, M.: Problem E2480. Am. Math. Mon. **82**, 670–671 (1975)
23. Kroopnick, A.J.: A lower bound for $\sin(x)$. Math. Gazette **81**, 88–89 (1997)
24. Littlejohn, L., Ahuja, M., Girardeau, C.: Problem 122. Coll. Math. J. **11**, 63 (1980)
25. Laforgia, A., Weston, S.R.: Problem 1215. Math. Mag. **59**, 119 (1986)

26. Levinson, N.: Generalization of an inequality of Ky Fan. J. Math. Anal. Appl. **8**, 133–134 (1964)
27. Love, J.B., Blundon, W.J., Philipp, S.: Problem E1532. Am. Math. Mon. **70**, 443 (1963)
28. Magnotta, F., Gerber, L.: Problem E2280. Am. Math. Mon. **79**, 181 (1972)
29. Mascioni, V.: An area approach to the second derivative. Coll. Math. J. **38**, 378–380 (2007)
30. Mercer, A.Mc.D.: Short proofs of Jensen's and Levinson's inequalities. Math. Gazette **94**, 492–494 (2010)
31. Mitrinovic, D.S.: Elementary Inequalities Noordhoff Ltd., Groningen (1964)
32. Moakes, A.J.: The mean value theorems. Math. Gazette **45**, 141–142 (1961)
33. Natanson, I.P.: Theory of Functions of a Real Variable, vol. 2. Frederick Ungar Pub. Co., New York (1961)
34. Nievergelt, Y., Nakhash, A.: Problem 10940. Am. Math. Mon. **110**, 546–547 (2003)
35. Needham, T.: A visual explanation of Jensen's inequality. Am. Math. Mon. **100**, 768–771 (1993)
36. Parker, F.D.: Taylor's theorem and Newton's method. Am. Math. Mon. **66**, 51 (1959)
37. Patruno, G., Vowe, M.: Problem 359. Coll. Math. J. **20**, 344–345 (1989)
38. Rolland, P., Jr., Van de Vyle, C.: Problem E1087. Am. Math. Mon. **74**, 84–85 (1967)
39. Reich, S.: On mean value theorems. Am. Math. Mon. **76**, 70–73 (1969)
40. Roy, R.: Sources in the Development of Mathematics. Cambridge University Press, Cambridge/New York (2011)
41. Sadoveanu, I., Vowe, M., Wagner, R.J.: Problem 458. Coll. Math. J. **23**, 344–345 (1992)
42. Sahoo, P.K., Riedel, T.: Mean Value Theorems and Functional Equations. World Scientific, Singapore/River Edge (1998)
43. Schaumberger, N.: Another proof of Jensen's inequality. Coll. Math. J. **20**, 57–58 (1989)
44. Schaumberger, N.: The AM-GM inequality via $x^{1/x}$. Coll. Math. J. **20**, 320 (1989)
45. Schaumberger, N.: Problem Q806. Math. Mag. **66**, 193, 201 (1993)
46. Schaumberger, N., Kabak, B.: Proof of Jensen's inequality. Coll. Math. J. **20**, 57–58 (1989)
47. Schaumberger, N., Farnsworth, D., Levine, E.: Problem 289. Coll. Math. J. **17**, 362–364 (1986)
48. Senum, G.I., Bang, S.J.: Problem E3352. Am. Math. Mon. **98**, 369–370 (1991)
49. Simic, S.: On an upper bound for Jensen's inequality. J. Ineq. Pure Appl. Math. **10**, article 60 (2009)
50. Sondow, J.: A geometric proof that e is irrational and a new measure of its irrationality. Am. Math. Mon. **113**, 637–641 (2003)
51. Witkowski, A.: Another proof of Levinson inequality. ajmaa.org/RGMIA/papers/v12n2/ (2009)
52. Zhang, G.Q., Roberts, B.: Problem E3214. Am. Math. Mon. **96**, 739–740 (1980)

Chapter 9
Integration of Continuous Functions

It has long been an axiom of mine that the little things are infinitely more important.

– Sherlock Holmes, in *A Case of Identity*,
by Sir Arthur Conan Doyle

A function's range is a collection of values and so we might expect that it should have an *average* value, as long as the function is reasonably well behaved. A fence or a wall for example, no matter how long or how irregular in height, should have an average height.

In Sect. 3.4 we considered the average value

$$\frac{1}{N}\sum_{j=1}^{N} f(x_j)$$

of a continuous function $f : [a,b] \to \mathbf{R}$ evaluated at N sample points $x_1, x_2, \ldots, x_N$ from $[a,b]$. By choosing these sample points in a systematic way and then letting $N \to \infty$, we define the *average value of f over the interval* $[a,b]$. This naturally gives rise to the notion of area under a curve and the definite integral. Then, since the definite integral is defined in terms of sums, we see that many properties of sums give rise to properties of definite integrals—and vice-versa. For example, we obtain integral analogues for many of the inequalities from Chaps. 2 and 6.

9.1 The Average Value of a Continuous Function

Consider the closed interval $[a,b]$ and let $N \in \mathbf{N}$. Choose the points

$$a = x_0 < x_1 < x_2 < \cdots < x_{N-1} < x_N = b$$

P.R. Mercer, *More Calculus of a Single Variable*, Undergraduate Texts in Mathematics, DOI 10.1007/978-1-4939-1926-0_9

according to

$$x_j = a + j\tfrac{b-a}{N} \quad \text{for} \quad j = 0, 1, 2, \ldots, N.$$

Then the x_j's are *equally spaced*: each is distance $\frac{b-a}{N}$ from its closest neighbor(s).
Now however, we consider only N of the form

$$N = 2^n \quad \text{for } n = 0, 1, 2, 3, \ldots, \qquad \text{so that} \quad N = 1, 2, 4, 8, 16, 32, \ldots.$$

This way, points x_j of each *partition* $P_n = \{x_0, x_1, x_2, \ldots, x_{N-1}, x_N\}$ of $[a, b]$ are also points of any partition arising from a larger n. As such, each partition (after $n = 0$) is called a *refinement* of every previous partition. See **Fig. 9.1**.

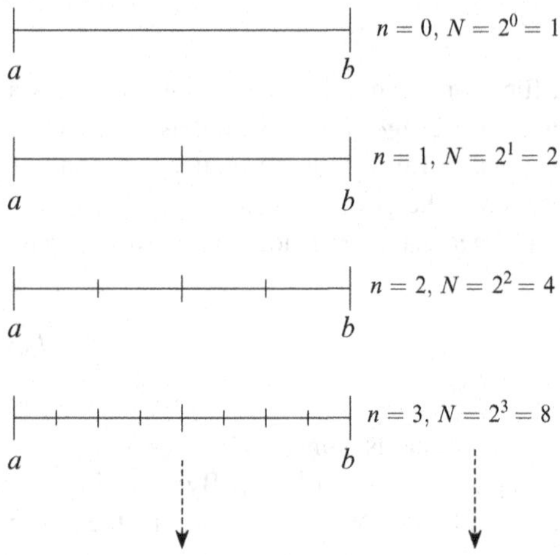

Fig. 9.1 Each partition P_n of $[a, b]$ gives $N = 2^n$ subintervals of $[a, b]$, for $n = 0, 1, 2, \ldots$

Finally, we denote by x_j^* any particular point of each subinterval $[x_{j-1}, x_j]$:

$$x_j^* \in [x_{j-1}, x_j] \quad \text{for} \quad j = 1, 2, \ldots, N.$$

That is,

$$x_1^* \in [x_0, x_1], \quad x_2^* \in [x_1, x_2], \quad \ldots \ , \quad x_N^* \in [x_{N-1}, x_N].$$

With all of this notation in place, it is a very important fact that if f is continuous on $[a, b]$, then $\lim\limits_{N\to\infty} \left(\frac{1}{N}\sum_{j=1}^{N} f(x_j^*)\right)$ exists and is independent of the choices for x_j^*. This limit is the **average value of** f **over** $[a, b]$ and we denote it by

$$A_f\big([a, b]\big).$$

We merely *state* this fact as a theorem below. Its proof requires some rather deep ideas that would take us somewhat off course, so we leave it for Appendix A.

Theorem 9.1. *Let f be continuous on $[a,b]$. With the notation as above (in particular $N = 2^n$),*

$$A_f\big([a,b]\big) = \lim_{N\to\infty}\left(\frac{1}{N}\sum_{j=1}^{N} f(x_j^*)\right) \quad \text{exists,}$$

and is independent of the choices $x_j^* \in [x_{j-1}, x_j]$.

Proof. See the Appendix, Sect. A.3. □

Example 9.2. For a constant function $f(x) \equiv C$, we should expect that the average value $A_f\big([a,b]\big) = C$. Indeed, for any choice of the x_j^*'s,

$$A_f\big([a,b]\big) = \lim_{N\to\infty}\left(\frac{1}{N}\sum_{j=1}^{N} f(x_j^*)\right) = \lim_{N\to\infty}\left(\frac{1}{N}\sum_{j=1}^{N} C\right) = C\,. \qquad \diamond$$

In practice, since x_j^* can be any particular point of each interval $[x_{j-1}, x_j]$, a convenient choice is usually made—like for example $x_j^* = x_{j-1}$, or $x_j^* = x_j$, or $x_j^* =$ the midpoint: $x_j^* = \frac{x_{j-1}+x_j}{2}$. We shall make such choices in next few examples. We shall also make use of the formulas (for N being a natural number)

$$\sum_{j=1}^{N} j = \frac{N(N+1)}{2} \quad \text{and} \quad \sum_{j=1}^{N} j^2 = \frac{N(N+1)(2N+1)}{6}.$$

These were verified in Exercises 2.17 and 5.39 (and again in Exercise 9.1).

Example 9.3. We compute $A_f\big([a,b]\big)$ for $f(x) = x^2$, and $[a,b] = [0,1]$. Here we have $x_0 = a = 0$,

$$x_j = a + j\tfrac{b-a}{N} = 0 + j\tfrac{1-0}{N} = \tfrac{j}{N}, \quad \text{and we take } x_j^* = x_j \text{ for } j = 1, 2, \ldots, N.$$

Then

$$\frac{1}{N}\sum_{j=1}^{N} f(x_j^*) = \frac{1}{N}\sum_{j=1}^{N}\left(\frac{j}{N}\right)^2 = \frac{1}{N^3}\sum_{j=1}^{N} j^2.$$

Therefore,

$$A_f\big([0,1]\big) = \lim_{N\to\infty}\left(\frac{1}{N}\sum_{j=1}^{N} f(x_j^*)\right) = \lim_{N\to\infty}\left(\frac{1}{N^3}\frac{N(N+1)(2N+1)}{6}\right) = \frac{1}{3}\,.$$

◇

Example 9.4. Extending Example 9.3, we compute $A_f([a,b])$ for $f(x) = x^2$, and for any interval $[a,b]$. Here $x_0 = a$,

$$x_j = a + j\tfrac{b-a}{N}, \text{ and we take } x_j^* = x_j \text{ for } j = 1,2,\ldots,N.$$

Then

$$\begin{aligned}
\frac{1}{N}\sum_{j=1}^{N} f(x_j^*) &= \frac{1}{N}\sum_{j=1}^{N}\left(a + j\frac{b-a}{N}\right)^2 \\
&= \frac{1}{N}\left(\sum_{j=1}^{N} a^2 + 2a\frac{b-a}{N}\sum_{j=1}^{N} j + \frac{(b-a)^2}{N^2}\sum_{j=1}^{N} j^2\right) \\
&= \frac{1}{N}\left(Na^2 + 2a\frac{b-a}{N}\frac{N(N+1)}{2} + \frac{(b-a)^2}{N^2}\frac{N(N+1)(2N+1)}{6}\right).
\end{aligned}$$

Therefore,

$$A_f([a,b]) = \lim_{N\to\infty}\left(\frac{1}{N}\sum_{j=1}^{N} f(x_j^*)\right) = a^2 + a(b-a) + \frac{(b-a)^2}{3} = \frac{b^2+ab+a^2}{3}.$$

◇

Example 9.5. [10] We compute $A_f([a,b])$ for $f(x) = \mathrm{e}^x$. Here $x_0 = a$,

$$x_j = a + j\tfrac{b-a}{N}, \text{ and we take } x_j^* = x_{j-1} \text{ for } j = 1,2,\ldots,N.$$

Then

$$\frac{1}{N}\sum_{j=1}^{N} f(x_j^*) = \frac{1}{N}\sum_{j=0}^{N-1} f(x_j) = \frac{1}{N}\sum_{j=0}^{N-1} \mathrm{e}^{a+j(b-a)/N} = \frac{\mathrm{e}^a}{N}\sum_{j=0}^{N-1}\left(\mathrm{e}^{(b-a)/N}\right)^j.$$

Now for $R \neq 1$ and natural numbers $N \geq 2$, the following identity can be found by doing long division on the right-hand side, or simply verified by cross multiplication:

$$1 + R + R^2 + R^3 + \cdots + R^{N-1} = \frac{1-R^N}{1-R}.$$

So taking $R = \mathrm{e}^{(b-a)/N} \neq 1$ here, we get

$$\frac{1}{N}\sum_{j=1}^{N} f(x_j^*) = \frac{\mathrm{e}^a}{N}\sum_{j=0}^{N-1}(\mathrm{e}^{(b-a)/N})^j = \frac{\mathrm{e}^a}{N}\frac{1-\left(\mathrm{e}^{(b-a)/N}\right)^N}{1-\mathrm{e}^{(b-a)/N}}$$

$$= \frac{e^a}{N} \frac{1 - e^{b-a}}{1 - e^{(b-a)/N}} = \frac{\frac{b-a}{N}}{e^{(b-a)/N} - 1} \frac{e^b - e^a}{b - a}.$$

Now because the derivative of e^x at $x = 0$ is 1 (or use L'Hospital's Rule (Theorem 5.13)), we get

$$\lim_{N\to\infty} \frac{\frac{b-a}{N}}{e^{(b-a)/N} - 1} = 1.$$

Therefore

$$A_f([a,b]) = \lim_{N\to\infty} \left(\frac{1}{N} \sum_{j=1}^{N} f(x_j^*) \right) = \frac{e^b - e^a}{b - a}. \qquad \diamond$$

Remark 9.6. In a very similar fashion, for $f(x) = e^{-x}$ on $[a,b]$,

$$A_f([a,b]) = \frac{e^{-a} - e^{-b}}{b - a} = \frac{1/e^a - 1/e^b}{b - a}.$$

This is the content of Exercise 9.3. ∘

Example 9.7. [6] We compute $A_f([a,b])$ for $f(x) = \sin(x)$. Here $x_0 = a$,

$$x_j = a + j\tfrac{b-a}{N}, \text{ and we take } x_j^* = x_j \text{ for } j = 1, 2, \ldots, N.$$

Then

$$\frac{1}{N} \sum_{j=1}^{N} f(x_j^*) = \frac{1}{N} \sum_{j=1}^{N} \sin\left(a + j\frac{b-a}{N}\right).$$

Now we use the trigonometric identity

$$2\sin(A)\sin(B) = \cos(A - B) - \cos(A + B),$$

with $A = a + j\frac{b-a}{N}$ and $B = \frac{1}{2}\frac{b-a}{N}$, to get

$$2\sin\left(a + j\tfrac{b-a}{N}\right)\sin\left(\tfrac{1}{2}\tfrac{b-a}{N}\right) = \cos\left(a - \tfrac{(2j-1)(b-a)}{2N}\right) - \cos\left(a + \tfrac{(2j+1)(b-a)}{2N}\right).$$

Then summing from $j = 1$ to N we get lots of cancellation (the sum *telescopes*):

$$2\sum_{j=1}^{N} \sin\left(a + j\tfrac{b-a}{N}\right)\sin\left(\tfrac{1}{2}\tfrac{b-a}{N}\right) = \cos\left(a - \tfrac{b-a}{2N}\right) - \cos\left(a + \tfrac{(2N+1)(b-a)}{2N}\right).$$

Therefore,

$$\frac{1}{N}\sum_{j=1}^{N}\sin\left(a+j\tfrac{b-a}{N}\right)=\frac{1}{2N}\frac{1}{\sin\left(\frac{1}{2}\frac{b-a}{N}\right)}\left[\cos\left(a-\tfrac{b-a}{2N}\right)-\cos\left(a+\tfrac{(2N+1)(b-a)}{2N}\right)\right]$$

$$=\frac{1}{b-a}\frac{\frac{b-a}{2N}}{\sin\left(\frac{b-a}{2N}\right)}\left[\cos\left(a-\tfrac{b-a}{2N}\right)-\cos\left(a+\tfrac{(2N+1)(b-a)}{2N}\right)\right].$$

Now because the derivative of $\sin(x)$ at $x = 0$ is 1 (or use L'Hospital's Rule (Theorem 5.13)), we have

$$\lim_{N\to\infty}\frac{\frac{b-a}{2N}}{\sin\left(\frac{b-a}{2N}\right)}=1.$$

And because the cosine function is continuous we get finally, for $f(x) = \sin(x)$:

$$A_f\big([a,b]\big)=\lim_{N\to\infty}\left(\frac{1}{N}\sum_{j=1}^{N}f(x_j^*)\right)=\frac{1}{b-a}\big[\cos(a)-\cos(b)\big].\qquad\diamond$$

Remark 9.8. For $f(x) = \cos(x)$ on $[a, b]$,

$$A_f\big([a,b]\big)=\frac{1}{b-a}\big[\sin(b)-\sin(a)\big].$$

This is the content of Exercise 9.4. ∘

If f is continuous on $[a, b]$ then $\min\limits_{x\in[a,b]}\{f(x)\}$ and $\max\limits_{x\in[a,b]}\{f(x)\}$ exist by the Extreme Value Theorem (Theorem 3.23). Obviously

$$\min_{x\in[a,b]}\{f(x)\}\ \le f(x)\ \le \max_{x\in[a,b]}\{f(x)\},$$

and therefore, essentially by Example 9.2,

$$\min_{x\in[a,b]}\{f(x)\}\ \le\ A_f\big([a,b]\big)\le \max_{x\in[a,b]}\{f(x)\}.\tag{9.1}$$

Now recall from Sect. 2.2 that the average, or Arithmetic Mean, $A = \frac{1}{n}\sum\limits_{j=1}^{n} a_j$ of the n numbers $a_1, a_2, \ldots, a_n$ is called a *mean* simply because it satisfies

$$\min_{1\le j\le n}\{a_j\}\le A\le \max_{1\le j\le n}\{a_j\}.$$

So in view of (9.1), calling $A_f([a,b])$ an *average* is natural. This is the analogue, for functions, of the Arithmetic Mean. But somewhat more than (9.1) is true, as follows.

Lemma 9.9. *Let f and g be continuous on $[a,b]$, with $f \leq g$ there. Then*

$$A_f([a,b]) \leq A_g([a,b]).$$

Proof. This is Exercise 9.5. □

The following important result is the analogue, for functions, of the Average Value Theorem for Sums (Theorem 3.19). It says that the average value of a continuous function on a closed interval is actually *attained* by the function.

Theorem 9.10. (Average Value Theorem) *Let f be continuous on $[a,b]$. Then there is $c \in [a,b]$ such that*

$$A_f([a,b]) = f(c).$$

Proof. By the Extreme Value Theorem (Theorem 3.23), there exist numbers $x_m, x_M \in [a,b]$ such that

$$f(x_m) \leq f(x) \leq f(x_M) \quad \text{for every } x \in [a,b].$$

So by (9.1),

$$f(x_m) \leq A_f([a,b]) \leq f(x_M).$$

Therefore, by the Intermediate Value Theorem (Theorem 3.17) there is c between x_m and x_M (and so $c \in [a,b]$) such that

$$f(c) = A_f([a,b]),$$

as desired. □

9.2 The Definite Integral

In the sum $\frac{1}{N}\sum_{j=1}^{N} f(x_j^*)$, if we let $\Delta x_N = \frac{b-a}{N}$ then we get

$$\frac{1}{N}\sum_{j=1}^{N} f(x_j^*) = \frac{1}{b-a}\sum_{j=1}^{N} f(x_j^*)\Delta x_N.$$

So it is customary to denote the average value of the continuous function f on $[a,b]$ by

$$A_f([a,b]) = \frac{1}{b-a}\int_a^b f(x)\,dx.$$

This notation serves vaguely as a reminder of where it comes from: As $N \to \infty$, the idea is that

$$\sum_{j=1}^{N} \to \int_a^b, \qquad f(x_j^*) \to f(x), \qquad \text{and} \qquad \Delta x_N \to dx.$$

Here,

$$\int_a^b f(x)\,dx$$

is called the **definite integral** (or simply the **integral**) of f from a to b.

In case we need to interchange the roles of a and b, we *define*

$$\int_b^a f(x)\,dx = -\int_a^b f(x)\,dx, \tag{9.2}$$

which is consistent with the set-up: for $b < a$, we have $x_0 = b$ and $x_N = a$, and $\Delta x_N < 0$. And notice that taking $a = b$ in (9.2), we get $\int_a^a f(x)\,dx = -\int_a^a f(x)\,dx$, so that (as we should expect):

$$\int_a^a f(x)\,dx = 0.$$

With this notation in place, Lemma 9.9 reads, for f and g continuous on $[a,b]$:

$$f \le g \;\Rightarrow\; \int_a^b f(x)\,dx \le \int_a^b g(x)\,dx. \tag{9.3}$$

This says that the definite integral is a *positive operator*. This simple but very important property of the definite integral is sometimes taken for granted. This property is not shared, for example, by the derivative: The reader should agree that it is *not* the case that $f(x) \le g(x) \Rightarrow f'(x) \le g'(x)$.

Remark 9.11. The expression

$$\sum_{j=1}^{N} f(x_j^*)\Delta x_N$$

is called a **Riemann sum**, after the great German mathematician Bernhard Riemann (1826–1866). For N large,

$$\sum_{j=1}^{N} f(x_j^*)\Delta x_N \cong \int_a^b f(x)\,dx.$$

This is made more precise in the Appendix (Theorem A.9 of Sect. A.3). ◦

Remark 9.12. In any sum, the *index of summation* plays no essential role. For example,

$$\sum_{j=1}^{N} f(x_j^*)\Delta x_N = \sum_{j=1}^{N} f(t_j^*)\Delta t_N = \sum_{j=1}^{N} f(u_j^*)\Delta u_N \quad \text{etc.}$$

In the same way, the *variable of integration* in a definite integral plays no essential role. It might be x, or just as well be t, or u, or virtually anything else:

$$\int_a^b f(x)\,dx = \int_a^b f(t)\,dt = \int_a^b f(u)\,du \quad \text{etc.}$$ ◦

Since integrals are defined in terms of sums, we can often use a property of sums to deduce a property of integrals. For example, the property

$$\sum_{j=1}^{N} \big(\alpha f(x_j) + \beta g(x_j)\big) = \alpha \sum_{j=1}^{N} f(x_j) + \beta \sum_{j=1}^{N} g(x_j) \quad \text{for } \alpha, \beta \in \mathbf{R}$$

easily gives rise to the following.

Lemma 9.13. *Let f and g be continuous on $[a, b]$. Then for any $\alpha, \beta \in \mathbf{R}$,*

$$\int_a^b \big(\alpha f(x) + \beta g(x)\big)\,dx = \alpha \int_a^b f(x)\,dx + \beta \int_a^b g(x)\,dx.$$

Proof. This is Exercise 9.6. □

Lemma 9.13 says that the definite integral is a *linear operator*. The derivative is also a linear operator—we saw in Sect. 4.2 that the derivative obeys what we called the Linear Combination Rule.

Observe now that the conclusion of the Average Value Theorem (Theorem 9.10) reads

$$f(c) = \frac{1}{b-a}\int_a^b f(x)\,dx = \frac{\int_a^b f(x)1\,dx}{\int_a^b 1\,dx}.$$

If we replace the 1's in the numerator and the denominator of the right-hand side above with a suitable continuous function g then we get the more general Mean Value Theorem for Integrals below. It is the analogue, for functions, of the Mean Value Theorem for Sums (Theorem 3.22).

Theorem 9.14. (Mean Value Theorem for Integrals) *Let f and g be continuous on $[a,b]$ and suppose that g does not change signs on $[a,b]$, and that $g(x) \not\equiv 0$. Then there is $c \in [a,b]$ such that*

$$f(c) = \frac{\int_a^b f(x)g(x)\,dx}{\int_a^b g(x)\,dx}.$$

Proof. We may assume that $g(x) \geq 0$ on $[a,b]$ for otherwise, we would consider $-g(x)$. By the Extreme Value Theorem (Theorem 3.23), there are $x_m, x_M \in [a,b]$ such that

$$f(x_m) \leq f(x) \leq f(x_M) \quad \text{for every } x \in [a,b].$$

Multiplying through by $g(x)$, we get

$$f(x_m)g(x) \leq f(x)g(x) \leq f(x_M)g(x) \quad \text{for every } x \in [a,b].$$

Then integrating and using (9.3) and Lemma 9.13 we obtain

$$f(x_m)\int_a^b g(x)\,dx \;\leq\; \int_a^b f(x)g(x)\,dx \;\leq\; f(x_M)\int_a^b g(x)\,dx.$$

Now since $g(x) \not\equiv 0$, there exists $x_0 \in [a,b]$ such that $g(x_0) > 0$. Then by Lemma 3.4 there is a closed interval J containing x_0 such that $g(x) > 0$ for $x \in J$. Therefore $\int_a^b g(x)\,dx > 0$, and we may divide through to obtain

$$f(x_m) \le \frac{\int_a^b f(x)g(x)\,dx}{\int_a^b g(x)\,dx} \le f(x_M).$$

Then by the Intermediate Value Theorem (Theorem 3.17) there exists c between x_m and x_M (and so $c \in [a,b]$) such that

$$f(c) = \frac{\int_a^b f(x)g(x)\,dx}{\int_a^b g(x)\,dx},$$

as desired. □

In the context of the Mean Value Theorem for Integrals (Theorem 9.14), for $g(x) > 0$ on $[a,b]$ one often sets

$$w(x) = \frac{g(x)}{\int_a^b g(t)\,dt}.$$

Then w is continuous, $w(x) > 0$ for $x \in [a,b]$, and $\int_a^b w(x)\,dx = 1$. Here, $w(x)$ is naturally called a **weight function**, and

$$\int_a^b w(x)f(x)\,dx$$

is the analogue, for functions, of the weighted Arithmetic Mean. Then the conclusion of the Mean Value Theorem for Integrals (Theorem 9.14) reads

$$f(c) = \int_a^b w(x)f(x)\,dx.$$

9.3 The Definite Integral as Area

For f continuous on $[a, b]$, the conclusion of the Average Value Theorem (Theorem 9.10) reads:

$$f(c) = \frac{1}{b-a} \int_a^b f(x)\, dx\,.$$

Therefore $f(c)(b-a)$ is the average value of f over $[a, b]$, multiplied by the length of $[a, b]$. So if f is also nonnegative we adopt this (very naturally) as the *definition* of the **area** between the graph of f and the x-axis, from $x = a$ to $x = b$.

That is,

$$\int_a^b f(x)\, dx = \text{the area between the graph of } f \text{ and the x-axis,}$$

$$\text{from } x = a \text{ to } x = b. \quad \text{(See Fig. 9.2.)}$$

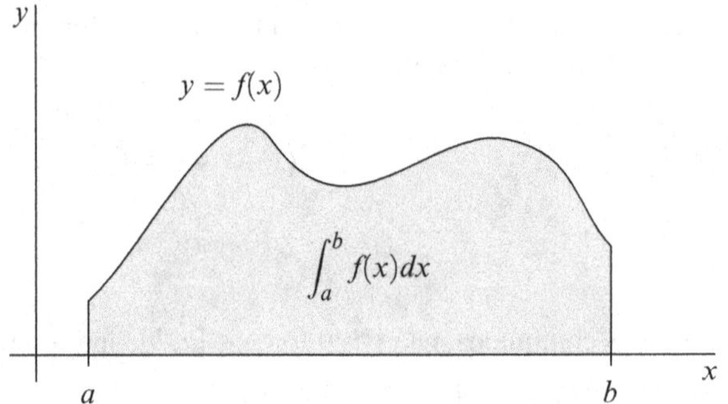

Fig. 9.2 The area of the shaded region is $\int_a^b f(x)\, dx$

So for example, if f defines the varying height of a wall running straight along the ground from a to b, then the area of the wall's face is

$$\int_a^b f(x)\, dx.$$

The average height of the wall is

$$\frac{1}{b-a} \int_a^b f(x)\, dx,$$

and the Average Value Theorem (Theorem 9.10) says that there is at least one place along the ground over which the wall is exactly its average height.

For any continuous function f, not just nonnegative ones, $\int_a^b f(x)\,dx$ is the **signed area** between the graph of f and the x-axis, from $x = a$ to $x = b$. The area is *signed* because function values below the x-axis give negative contributions in the sums which ultimately define the integral. The **total area** between the graph of f and the x-axis, from $x = a$ to $x = b$ is then

$$\int_a^b |f(x)|\,dx.$$

The following is the analogue, for functions, of the triangle inequality (Lemma 1.1). It appears reasonable upon drawing a picture. Not surprisingly, its proof comes from the triangle inequality.

Lemma 9.15. *Let f be continuous on $[a,b]$. Then*

$$\left|\int_a^b f(x)\,dx\right| \le \int_a^b |f(x)|\,dx.$$

Proof. This is Exercise 9.10. □

Example 9.16. We saw in Example 9.2 that $\int_a^b C\,dx = C(b-a)$. For $a < b$ and $C > 0$, this is the area of the rectangle with base $[a,b]$ and height C. ◇

Example 9.17. For $r > 0$ the graph of $f(x) = \sqrt{r^2 - x^2}$ on $[-r, r]$ is the top half of the circle with radius r, centered at the origin. Therefore

$$\int_{-r}^{r} \sqrt{r^2 - x^2}\,dx = \frac{\pi r^2}{2}.$$ ◇

Example 9.18. By interpreting the definite integral as a signed area (and knowing the formula for the area of a trapezoid), one can verify that

$$\int_a^b x\,dx = \frac{a+b}{2}(b-a) = \frac{1}{2}\left(b^2 - a^2\right).$$ ◇

Example 9.19. Let $f(x) = x^2$. We saw in Example 9.4 that

$$A_f\big([a,b]\big) = \frac{b^2 + ab + a^2}{3}.$$

Therefore the signed area between the graph of $f(x) = x^2$ and the x-axis, from $x = a$ to $x = b$, is

$$\int_a^b x^2\,dx = (b-a)A_f([a,b]) = (b-a)\frac{b^2+ab+a^2}{3} = \frac{1}{3}(b^3-a^3). \qquad \diamond$$

Let $P,\ Q,\ R \in \mathbf{R}$. Then by Examples 9.16, 9.18 and 9.19, and Lemma 9.13,

$$\int_a^b (Px^2+Qx+R)\,dx = P\int_a^b x^2\,dx \ + Q\int_a^b x\,dx \ + R\int_a^b 1\,dx$$

$$= \frac{P}{3}(b^3-a^3) + \frac{Q}{2}(b^2-a^2) + R(b-a).$$

Example 9.20. We saw in Example 9.5 that for $f(x) = \mathrm{e}^x$,

$$A_f([a,b]) = \frac{\mathrm{e}^b - \mathrm{e}^a}{b-a}.$$

Therefore the area between the graph of $f(x) = \mathrm{e}^x$ and the x-axis, from $x = a$ to $x = b$, is

$$\int_a^b \mathrm{e}^x\,dx = \mathrm{e}^b - \mathrm{e}^a. \qquad \diamond$$

Example 9.21. We saw in Example 9.7 that for $f(x) = \sin(x)$,

$$A_f([a,b]) = \frac{\cos(a)-\cos(b)}{b-a}.$$

Therefore the signed area between the graph of $f(x) = \sin(x)$ and the x-axis, from $x = a$ to $x = b$, is

$$\int_a^b \sin(x)\,dx = \cos(a) - \cos(b).$$

So, for example,

$$\int_0^{3\pi/2} \sin(x)\,dx = \cos(0) - \cos(3\pi/2) = 1.$$

And using the symmetry of the sine function,

$$\int_0^{3\pi/2} |\sin(x)|\,dx = 3\int_0^{\pi/2} \sin(x)\,dx = 3\big(\cos(0) - \cos(\pi/2)\big) = 3. \qquad \diamond$$

9.4 Some Applications

The following lemma seems perfectly reasonable but its proof is surprisingly tricky, so we leave it for Appendix A. It enables us to consider the definite integral of certain functions which are not continuous. See **Fig. 9.3**.

Lemma 9.22. *Let f be continuous on $[a,b]$ and let $c \in (a,b)$. Then*

$$\int_a^b f(x)\,dx = \int_a^c f(x)\,dx + \int_c^b f(x)\,dx\,.$$

Proof. See the Appendix, Sect. A.3. □

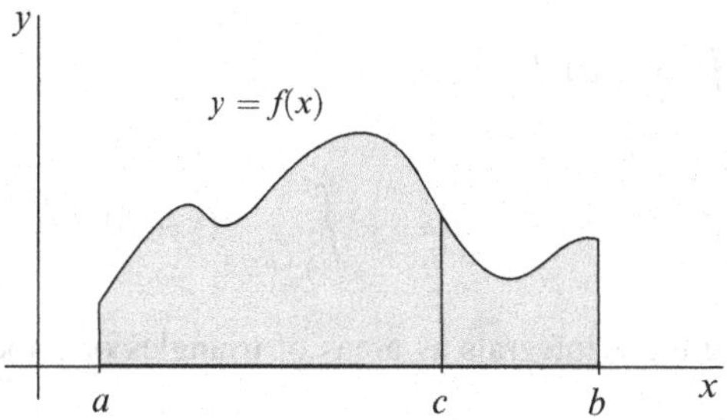

Fig. 9.3 Lemma 9.22: $\int_a^b f(x)\,dx = \int_a^c f(x)\,dx + \int_c^b f(x)\,dx$

Example 9.23. Consider the function

$$f(x) = \begin{cases} x - a & \text{if } x \in \left[a, \frac{a+b}{2}\right] \\ x - b & \text{if } x \in \left(\frac{a+b}{2}, b\right]. \end{cases}$$

See the graph of f in **Fig. 9.4**.

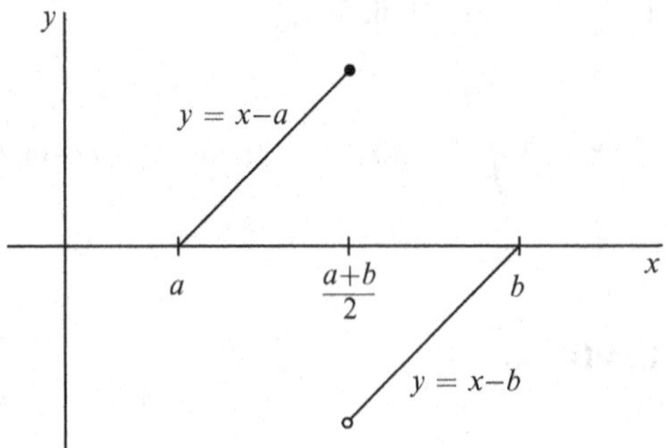

Fig. 9.4 Example 9.23. $f(x) = x - a$ on $[a, \frac{a+b}{2}]$, and $f(x) = x - b$ on $(\frac{a+b}{2}, b]$

Now f is not continuous on $[a, b]$, but in view of Lemma 9.22 (with $c = (a + b)/2$ there), we write

$$\int_a^b f(x)\,dx = \int_a^{(a+b)/2} f(x)\,dx + \int_{(a+b)/2}^b f(x)\,dx = 0.$$

Also in view of Lemma 9.22 (for any $t \in [a, b]$), we write

$$\int_a^t f(x)\,dx = \begin{cases} \displaystyle\int_a^t (x-a)\,dx & \text{if } t \in [a, \frac{a+b}{2}] \\ \displaystyle\int_a^{(a+b)/2} (x-a)\,dx + \int_{(a+b)/2}^t (x-b)\,dx & \text{if } t \in (\frac{a+b}{2}, b]. \end{cases}$$

Then by interpreting these integrals as areas of triangles or trapezoids (depending on t), we get

$$\int_a^t f(x)\,dx = \begin{cases} \frac{1}{2}(t-a)^2 & \text{if } t \in [a, \frac{a+b}{2}] \\ \frac{1}{2}(\frac{b-a}{2})^2 + \frac{1}{2}\left[(t-b)^2 - (\frac{b-a}{2})^2\right] & \text{if } t \in (\frac{a+b}{2}, b]. \end{cases}$$

Here, as the reader may check, we get $\displaystyle\int_a^b f(x)\,dx = 0$ as we should. Also, the reader may verify that $\displaystyle\int_a^b |f(x)|\,dx = \frac{(b-a)^2}{4}$. ◇

Here is an application of the definite integral wherein a property of integrals yields a useful property of sums, and vice-versa.

Theorem 9.24. (The Integral Test) *Let f be continuous, positive, and decreasing on $[1, \infty)$. Then $\sum_{n=1}^{\infty} f(n)$ converges if and only if $\int_1^{\infty} f(x)\,dx = \lim_{N\to\infty} \int_1^N f(x)\,dx$ exists.*

Proof. Since f is decreasing, $f(b) \le f(x) \le f(a)$ for $1 \le a \le x \le b$, and so

$$f(b)[b-a] \le \int_a^b f(x)\,dx \le f(a)[b-a].$$

Therefore $f(n+1) \le \int_n^{n+1} f(x)\,dx \le f(n)$ for each integer $n \ge 1$. See **Fig. 9.5**.

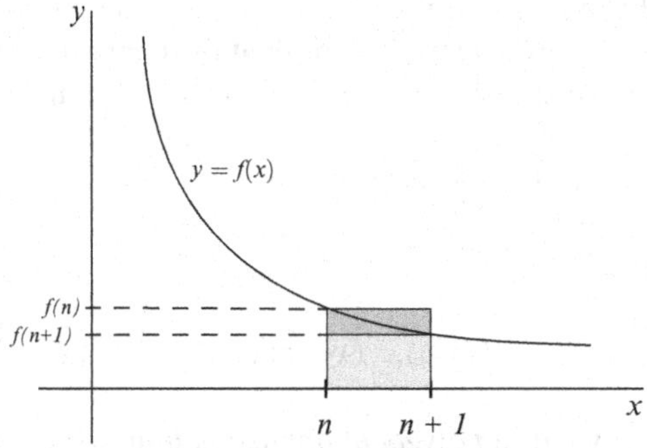

Fig. 9.5 In the proof of the Integral Test (Theorem 9.24), f is decreasing on $[1, \infty)$. Therefore we have $f(n+1) \le \int_n^{n+1} f(x)\,dx \le f(n)$ for integers $n \ge 1$

Then

$$\sum_{n=1}^{N-1} f(n+1) \le \sum_{n=1}^{N-1} \int_n^{n+1} f(x)\,dx \le \sum_{n=1}^{N-1} f(n),$$

which, by applying Lemma 9.22 $N-1$ times, reads

$$\sum_{n=1}^{N-1} f(n+1) \le \int_1^N f(x)\,dx \le \sum_{n=1}^{N-1} f(n).$$

Now since f is positive, the right-hand inequality shows that if $\sum_{n=1}^{\infty} f(n)$ exists, then $\int_1^{\infty} f(x)\,dx$ exists. And the left-hand inequality shows that if $\int_1^{\infty} f(x)\,dx$ exists, then $\sum_{n=1}^{\infty} f(n+1)$ exists; therefore so does $\sum_{n=1}^{\infty} f(n)$. $\square$

Example 9.25. We showed in Sect. 6.7 that a **p-series**

$$\sum_{n=1}^{\infty} \frac{1}{n^p}$$

converges for $p > 1$ and diverges to $+\infty$ for $p \leq 1$. In Example 10.5 we shall show that for $p \neq 1$:

$$\int_1^N \frac{1}{x^p}\,dx = \frac{1}{1-p}\left(\frac{N}{N^p} - 1\right), \qquad \text{and for } p = 1: \quad \int_1^N \frac{1}{x}\,dx = \ln(N).$$

Therefore (as we shall conclude in Example 10.6) the convergence/divergence of a p-series also follows from the Integral Test (Theorem 9.24). ◇

For another application of the definite integral, we suppose that f and f' are continuous on $[a,b]$. Then we *define* the **length of the curve** described by $y = f(x)$ from $x = a$ to $x = b$, as follows. (But see also [9, 21].) Again, let

$$a = x_0 \text{ and } x_j = a + j\tfrac{b-a}{N} = a + j\Delta x_N \text{ for } j = 0, 1, 2, \ldots, N = 2^n.$$

Joining the successive points

$$(x_0, f(x_0)),\quad (x_1, f(x_1)),\quad (x_2, f(x_2)),\quad \ldots\quad, (x_N, f(x_N))$$

with N line segments yields a *polygonal approximation* to the graph of $y = f(x)$ over $[a,b]$. By the Pythagorean Theorem, each segment has length

$$\sqrt{(x_j - x_{j-1})^2 + \left(f(x_j) - f(x_{j-1})\right)^2}.$$

So (see **Fig. 9.6**) the total length of these segments is

$$\sum_{j=1}^{N} \sqrt{(x_j - x_{j-1})^2 + \left(f(x_j) - f(x_{j-1})\right)^2}.$$

And taking very N large it seems that the total length of these segments would provide a pretty good approximation to what we think would be the length of the curve $y = f(x)$ from $x = a$ to $x = b$.

Now applying the Mean Value Theorem (Theorem 5.2) on each interval $[x_{j-1}, x_j]$, there is $x_j^* \in (x_{j-1}, x_j)$ such that

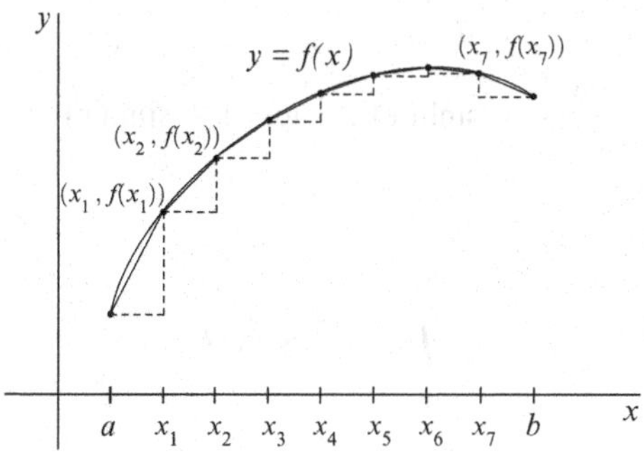

Fig. 9.6 A polygonal approximation for $y = f(x)$. Here $n = 3$, so there are $N = 2^3 = 8$ subintervals

$$\begin{aligned}\sum_{j=1}^{N} \sqrt{(x_j - x_{j-1})^2 + \big(f(x_j) - f(x_{j-1})\big)^2} &= \sum_{j=1}^{N} \sqrt{(x_j - x_{j-1})^2 + f'(x_j^*)^2 (x_j - x_{j-1})^2} \\ &= \sum_{j=1}^{N} \sqrt{1 + f'(x_j^*)^2}\,(x_j - x_{j-1}) \\ &= \sum_{j=1}^{N} \sqrt{1 + f'(x_j^*)^2}\,\Delta x_N .\end{aligned}$$

Since f' is continuous, so is $\sqrt{1 + (f')^2}$ and therefore

$$\lim_{N\to\infty} \sum_{j=1}^{N} \sqrt{(x_j - x_{j-1})^2 + \big(f(x_j) - f(x_{j-1})\big)^2} = \int_a^b \sqrt{1 + f'(x)^2}\,dx \quad \text{exists.}$$

This is the **length of the curve** $y = f(x)$ over $[a, b]$. It is typically denoted by the letter s. For example, if f defines the varying height of a fence that runs straight along the ground from a to b, then

$$s = \int_a^b \sqrt{1 + f'(x)^2}\,dx$$

is the length of the top of the fence.

Example 9.26. We compute the length of the curve

$$f(x) = \cosh(x) = \frac{e^x + e^{-x}}{2}, \quad \text{from} \quad x = 0 \quad \text{to} \quad x = 1.$$

Here we have

$$f'(x) = \frac{e^x - e^{-x}}{2} = \sinh(x) \quad \text{and} \quad 1 + \sinh(x)^2 = \cosh(x)^2.$$

Therefore

$$\begin{aligned} s &= \int_0^1 \sqrt{1 + f'(x)^2}\, dx \\ &= \int_0^1 \cosh(x)\, dx. \end{aligned}$$

Then using Remark 9.6 and Lemma 9.13, we get

$$s = \frac{1}{2}\left(e - \frac{1}{e}\right). \qquad \diamond$$

9.5 Famous Inequalities for the Definite Integral

Continuing the theme that properties of sums can give rise to corresponding properties of integrals, we extend some of the famous inequalities for sums to obtain analogous inequalities for integrals. And quite often, their proofs are really no more difficult.

In Sect. 6.4 we obtained Hölder's Inequality (Lemma 6.18) using the weighted AGM Inequality with n = 2 (Corollary 6.16). In an entirely similar way we obtain the following.

Theorem 9.27. (Hölder's Integral Inequality) *Let f and g be continuous and nonnegative on $[a, b]$ and let $p, q > 1$ satisfy $\frac{1}{p} + \frac{1}{q} = 1$. Then*

$$\int_a^b f(x)g(x)\, dx \le \left(\int_a^b f(x)^p\, dx\right)^{1/p} \left(\int_a^b g(x)^q\, dx\right)^{1/q}.$$

Proof. If $f(x) \equiv 0$ or $g(x) \equiv 0$, then the desired inequality holds, with equality. Otherwise, take $t = \frac{1}{p}$ in the weighted AGM Inequality with $n = 2$ (Corollary 6.16), with

$$a = \frac{f(x)^p}{\int_a^b f(x)^p\,dx} \quad \text{and} \quad b = \frac{g(x)^q}{\int_a^b g(x)^q\,dx}, \quad \text{to get}$$

$$\frac{f(x)}{\left(\int_a^b f(x)^p\,dx\right)^{1/p}} \frac{g(x)}{\left(\int_a^b g(x)^q\,dx\right)^{1/q}} \le \frac{f(x)^p}{p\int_a^b f(x)^p\,dx} + \frac{g(x)^q}{q\int_a^b g(x)^q\,dx}.$$

Then integrating from a to b we obtain

$$\frac{\int_a^b f(x)g(x)\,dx}{\left(\int_a^b f(x)^p\,dx\right)^{1/p}\left(\int_a^b g(x)^q\,dx\right)^{1/q}} \le \frac{\int_a^b f(x)^p\,dx}{p\int_a^b f(x)^p\,dx} + \frac{\int_a^b g(x)^q\,dx}{q\int_a^b g(x)^q\,dx} = \frac{1}{p} + \frac{1}{q} = 1,$$

as desired. □

The case of $p = q = 2$ in Hölder's Integral Inequality (Theorem 9.27) gives the integral analogue of the Cauchy–Schwarz Inequality (Theorem 2.18) as follows. (For a sharpened version of this result, see [33].)

Corollary 9.28. (Cauchy–Schwarz Integral Inequality) *Let f and g be continuous on $[a,b]$. Then*

$$\left(\int_a^b f(x)g(x)\,dx\right)^2 \le \int_a^b f(x)^2\,dx \int_a^b g(x)^2\,dx.$$ □

Now let us recall Jensen's Inequality (Theorem 8.17): *Let $x_1, x_2, \ldots, x_n \in [a,b]$ and let $w_1, w_2, \ldots, w_n$ satisfy $w_j > 0$, with $\sum_{j=1}^{n} w_j = 1$. Let φ satisfy $\varphi'' \ge 0$ on $[a,b]$ (so that φ is convex there). Then*

$$\varphi\left(\sum_{j=1}^{n} w_j x_j\right) \le \sum_{j=1}^{n} w_j \varphi(x_j).$$

There is also an integral analogue for Jensen's Inequality, as follows.

Theorem 9.29. (Jensen's Integral Inequality) *Let f be continuous on $[a,b]$ and let $w \ge 0$ be continuous on $[a,b]$, with $\int_a^b w(x)\,dx = 1$. Let φ satisfy $\varphi'' \ge 0$ on the range of f (so that φ is convex there). Then*

$$\varphi\left(\int_a^b w(x) f(x)\, dx\right) \le \int_a^b w(x)\varphi(f(x))\, dx\,.$$

Proof. Let A be in the range of f. Since $\varphi'' \ge 0$, the tangent line to φ at $x = A$ is on or below the graph of φ, by Lemma 8.7. That is, for any t for which φ is defined,

$$\varphi(t) \ge \varphi(A) + \varphi'(A)\,(t - A)\,.$$

Now let $A = \int_a^b w(x) f(x)\, dx$, which is indeed in the range of f, by the Mean Value Theorem for Integrals (Theorem 9.14), and write $t = f(x)$. Then

$$\varphi(f(x)) \ge \varphi(A) + \varphi'(A)\,(f(x) - A)\,.$$

Multiplying through by $w(x) \ge 0$,

$$w(x)\varphi(f(x)) \ge w(x)\varphi(A) + w(x)\varphi'(A)\,(f(x) - A)\,.$$

Now integrating from a to b, we get

$$\begin{aligned}\int_a^b w(x)\varphi(f(x))\, dx &\ge \varphi(A)\int_a^b w(x)\, dx + \varphi'(A)\left(\int_a^b w(x) f(x)\, dx - A\int_a^b w(x)\, dx\right)\\ &= \varphi(A) + \varphi'(A)\,(A - A) = \varphi\left(\int_a^b w(x) f(x)\, dx\right),\end{aligned}$$

as desired. □

Example 9.30. We saw in Example 8.19 how Jensen's Inequality (Theorem 8.17) can be used to obtain the Cauchy–Schwarz Inequality (Theorem 2.18). Here, in an entirely similar way, we use Jensen's Integral Inequality (Theorem 9.29) to obtain the Cauchy–Schwarz Integral Inequality (Corollary 9.28). Let p and q be continuous on $[a, b]$ with p not identically zero. We shall apply Jensen's Integral Inequality (Theorem 9.29) to the convex function $\varphi(x) = x^2$, with

$$w(x) = \frac{p(x)^2}{\int_a^b p(t)^2\, dt}, \qquad \text{so that} \qquad \int_a^b w(x)\, dx = 1.$$

Jensen's Integral Inequality is

$$\left(\int_a^b w(x) f(x)\, dx \right)^2 \le \int_a^b w(x) f(x)^2\, dx.$$

Now setting $f(x) = \dfrac{p(x)q(x)}{w(x)}$, this reads

$$\left(\int_a^b p(x)q(x)\, dx \right)^2 \le \int_a^b \frac{p(x)^2 q(x)^2}{p(x)^2} \left(\int_a^b p(x)^2\, dx \right) dx = \int_a^b p(x)^2\, dx \int_a^b q(x)^2\, dx,$$

which is the Cauchy–Schwarz Integral Inequality (Corollary 9.28). ◇

We saw at the end of Sect. 8.3 with Cauchy's proof of Jensen's Inequality (Theorem 8.17) and again in Exercise 8.39, that Jensen's Inequality continues to hold even if f is only assumed to be convex and continuous, i.e., not requiring that $f'' \ge 0$, or even that f' exists. Likewise, Jensen's Integral Inequality (Theorem 9.29) holds under less restrictive conditions than the ones we have imposed on the function φ. But φ must still be convex; see Exercise 9.35.

For the remainder of this section we focus on the important **special case of Jensen's Integral Inequality** (Theorem 9.29) in which $w(x) \equiv \frac{1}{b-a}$. That is, for f continuous and ϕ convex on the range of f :

$$\varphi\left(\frac{1}{b-a} \int_a^b f(x)\, dx \right) \le \frac{1}{b-a} \int_a^b \varphi(f(x))\, dx.$$

Example 9.31. **(i)** Let $\varphi(x) = x^r$, with $r \ge 1$. Then for f continuous on $[a, b]$, with $a \ge 0$, Jensen's Inequality gives

$$\left(\frac{1}{b-a} \int_a^b f(x)\, dx \right)^r \le \frac{1}{b-a} \int_a^b f(x)^r\, dx.$$

For $r = 2$ this is a special case (i.e., $g(x) \equiv 1$) of the Cauchy–Schwarz Integral Inequality (Corollary 9.28).

(ii) Let $\varphi(x) = \ln(x)$, which is concave on $(0, +\infty)$. For f continuous on $[a, b]$ and $f > 0$ there, Jensen's Inequality gives

$$\ln\left(\frac{1}{b-a} \int_a^b f(x)\, dx \right) \ge \frac{1}{b-a} \int_a^b \ln(f(x))\, dx.$$

(iii) Let $\varphi(x) = \mathrm{e}^x$, which is convex on $(-\infty, +\infty)$. For f continuous on $[a, b]$, Jensen's Inequality gives

$$\mathrm{e}^{\frac{1}{b-a}\int_a^b f(x)\,dx} \leq \frac{1}{b-a}\int_a^b \mathrm{e}^{f(x)}\,dx. \qquad \diamond$$

Example 9.32. Let f be continuous and positive on $[a, b]$. Replacing the f with $\ln(f)$ in Example 9.31 (iii) gives

$$\mathrm{e}^{\frac{1}{b-a}\int_a^b \ln(f(x))\,dx} \leq \frac{1}{b-a}\int_a^b f(x)\,dx. \tag{9.4}$$

This is the **AGM Inequality for Functions**—it is the analogue, for continuous functions, of the weighted AGM Inequality (Theorem 6.15). The left-hand side above is the **Geometric Mean** of f, and the right-hand side is of course the Average Value, or the the **Arithmetic Mean** of f.

Here is a very simple example which shows how the AGM Inequality for Functions can reduce to the weighted AGM Inequality (Theorem 6.15). Let

$$w_1, w_2, w_3 > 0 \quad \text{with} \quad \sum_{j=1}^{3} w_j = 1.$$

And for $x \in [a, b] = [0, 1]$, consider the function

$$f(x) = \begin{cases} a_1 & \text{if } \ 0 \leq x \leq w_1 \\ a_2 & \text{if } \ w_1 < x \leq w_1 + w_2 \\ a_3 & \text{if } \ w_1 + w_2 < x \leq 1\,. \end{cases}$$

A graph of f is shown in **Fig. 9.7**.

Using Lemma 9.22 two times, the left-hand side of (9.4), i.e., the Geometric Mean of f, is

$$\mathrm{e}^{\frac{1}{b-a}\int_a^b \ln(f(x))\,dx} = \mathrm{e}^{w_1\ln(a_1)+w_2\ln(a_2)+w_3\ln(a_3)} = a_1^{w_1}\,a_2^{w_2}\,a_3^{w_3}\,.$$

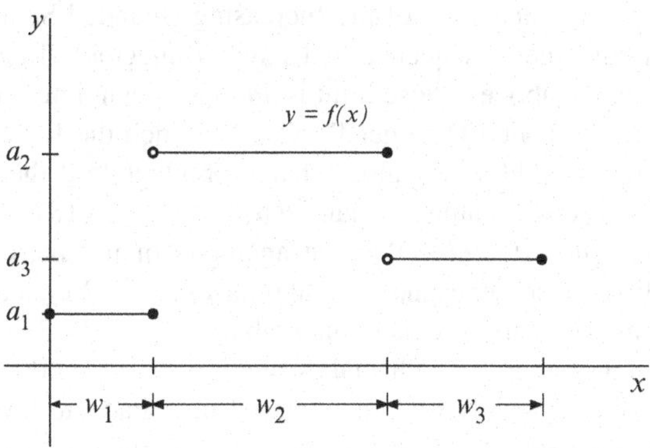

Fig. 9.7 For Example 9.32. Here, $n = 3$. Each a_j is assigned the weight w_j, and $\sum_{j=1}^{3} w_j = 1$

The right-hand side of (9.4), i.e., the Average Value of f, is

$$\frac{1}{b-a}\int_a^b f(x)\,dx = w_1a_1 + w_2a_2 + w_3a_3.$$

Together then, (9.4) gives the weighted AGM Inequality (Theorem 6.15) with $n = 3$:

$$a_1^{w_1}\, a_2^{w_2}\, a_3^{w_3} \le \sum_{j=1}^{3} w_j a_j.$$

The reader should agree that this procedure could be carried out for any number of weights. That is, for any $w_1, w_2, \ldots, w_n > 0$ with $\sum_{j=1}^{n} w_j = 1$. ◇

9.6 Epilogue

The (definite) integral that we have considered is called the **Riemann integral**. We defined it only for *continuous* functions (and we were to able extend it somewhat, using Lemma 9.22). This is usually quite adequate for calculus. But there are functions which are not continuous, and not covered by Lemma 9.22—in fact, very nasty ones—which nevertheless possess a Riemann integral. A nice approach to the Riemann integral, to which ours can be viewed as a precursor, can be found in [29].

Recall from Chap. 1 that the set **Q** was not (for us) a large enough place in which to work: the Increasing Bounded Sequence Property does not hold within **Q**. In the same way, the collection of *Riemann integrable* functions is not large enough, in the sense that there lacks an analogue of the Increasing Bounded Sequence Property. (That is, one can construct a sequence of Riemann integrable functions which is increasing and bounded above, whose limit is *not* a Riemann integrable function.)

So in Chap. 1 we extended **Q** to get **R**, a set in which the Increasing Bounded Sequence Property *does* hold. The analogue here, for integrals, is the **Lebesgue integral**, developed by French mathematician Henri Lebesgue (1875–1941). Enough functions are *Lebesgue integrable* that the analogue of the Increasing Bounded Property *does* hold, and the Lebesgue integral reduces to the Riemann integral when applied to functions which are Riemann integrable.

It is the Lebesgue integral which makes many areas of modern mathematical analysis possible. It is typically first encountered in a graduate level real analysis course. An excellent account of the historical development of the Riemann and Lebesgue integrals, and many other topics from calculus, can be found in [8].

Exercises

9.1. [6] Show, as follows, that

$$S = \sum_{j=1}^{N} j^2 = \frac{N(N+1)(2N+1)}{6}.$$

(a) Verify that $T = \sum_{j=1}^{N} j = \frac{N(N+1)}{2}$.

(b) In the identity

$$(k+1)^3 - k^3 = 3k^2 + 3k + 1,$$

set $k = 0, 1, 2, \ldots, N$ and add each of these together to get

$$(N+1)^3 = 3S + 3T + N + 1.$$

(c) Now solve for S and use (a).

9.2. **(a)** Show that

$$\sum_{j=1}^{N} j^3 = \frac{N^2(N+1)^2}{4} = \left(\sum_{j=1}^{N} j\right)^2.$$

(b) Find the average value of $f(x) = x^3$ over $[0, 1]$.

9.3. Find the average value of $f(x) = \mathrm{e}^{-x}$ over $[a, b]$.

9.4. [18] Find the average value of $f(x) = \cos(x)$ over $[a, b]$.

9.5. Prove Lemma 9.9: *Let f and g be continuous on $[a, b]$, with $f \le g$. Then*

$$A_f([a,b]) \le A_g([a,b]).$$

9.6. Prove Lemma 9.13: *Let f and g be continuous on $[a, b]$. Then for any $\alpha, \beta \in \mathbf{R}$,*

$$\int_a^b (\alpha f(x) + \beta g(x))\, dx = \alpha \int_a^b f(x)\, dx + \beta \int_a^b g(x)\, dx.$$

9.7. [1] Let f and g be continuous and nonnegative on $[0, 1]$, with

$$\int_0^1 e^{-xf(x)}\, dx \ge \int_0^1 e^{-xg(x)}\, dx.$$

Show that

$$\int_0^1 xg(x)e^{-xf(x)}\, dx \ge \int_0^1 xf(x)e^{-xg(x)}\, dx.$$

9.8. Let f be continuous on $[a, b]$. **(a)** Show that $\left(\frac{1}{b-a}\int_a^b f(x)^2\, dx\right)^{1/2}$ is a *mean*. Which mean of n numbers, that we have met before, does this generalize? **(b)** Show that there is $c \in [a, b]$ such that

$$f(c) = \left(\frac{1}{b-a}\int_a^b f(x)^2\, dx\right)^{1/2}.$$

9.9. [22] Let f be a continuous function on $[0, 1]$ with $\int_a^b f(x)\, dx = 1$ and let $n \in \mathbf{N}$.

(a) Apply the Average Value Theorem (Theorem 9.10) on each of $[(k-1)/n,\ k/n]$ for $k = 1, 2, \ldots n$ to show that there are distinct $c_1, c_2, \ldots, c_n \in (a, b)$ such that

$$f(c_1) + f(c_2) + \cdots + f(c_n) = n.$$

(b) Show that there are distinct $c_1, c_2, \ldots, c_n \in (a,b)$ such that

$$\frac{1}{f(c_1)} + \frac{1}{f(c_2)} + \cdots + \frac{1}{f(c_n)} = n.$$

9.10. Prove Lemma 9.15: *Let f be continuous on $[a,b]$. Then*

$$\left| \int_a^b f(x)\,dx \right| \le \int_a^b |f(x)|\,dx.$$

9.11. [15] Let $a, b, c > 0$ with $b < c$. Evaluate the following definite integrals by *only* interpreting the definite integral as an area:

(a) $\displaystyle\int_{-a}^{b} |x|\,dx$ **(b)** $\displaystyle\int_{-a}^{b} |x-c|\,dx$ **(c)** $\displaystyle\int_{-a}^{b} \big||x-c|-|x+c|\big|\,dx.$

9.12. [3] Evaluate $\int_0^2 \sqrt{8-x^2}\,dx$ by *only* interpreting the definite integral as an area.

9.13. [7] Let f be nonnegative and continuous on $[0,1]$. Set

$$P_n = \prod_{k=1}^{n} \left(1 + \frac{1}{n} f\left(\frac{k}{n}\right)\right).$$

Show that as $n \to \infty$,

$$P_n \to \mathrm{e}^{\int_0^1 f(x)\,dx}.$$

Hint: Verify, then use, $0 \le x - \ln(1+x) \le x^2/2$.

9.14. Assume the hypotheses of the Integral Test (Theorem 9.24). Show that if $\sum_{n=1}^{\infty} f(n) = S$ exists, then

$$\sum_{n=1}^{N} f(n) + \int_{N+1}^{\infty} f(x)\,dx \le S \le \sum_{n=1}^{N} f(n) + \int_{N}^{\infty} f(x)\,dx.$$

9.15. Assume the hypotheses of the Integral Test (Theorem 9.24).

(a) Show that

$$0 \le \sum_{n=1}^{N} f(n) \le \int_1^N f(x)\,dx + f(1).$$

(b) Show that the sequence $\left\{ \sum_{n=1}^{N} f(n) - \int_1^N f(x)\,dx \right\}$ is convergent.

9.16. [16] Let f and f' be continuous on $[a,b]$ with $|f'(x)| \le M$ there. Show that

$$\left| \int_a^{(a+b)/2} f(x)\,dx - \int_{(a+b)/2}^{b} f(x)\,dx \right| \le M \left(\frac{b-a}{2} \right)^2 .$$

Hint: Write

$$\left| \int_a^{(a+b)/2} f(x)\,dx - \int_{(a+b)/2}^{b} f(x)\,dx \right| = \left| \int_a^{(a+b)/2} \left(f(x) - f(\tfrac{a+b}{2})\right)\,dx \right.$$
$$\left. + \int_{(a+b)/2}^{b} \left(f(\tfrac{a+b}{2}) - f(x)\right)\,dx \right| ,$$

apply the triangle inequality, use the Mean Value Theorem (Theorem 5.2) on each piece, and bring the absolute value signs inside using Lemma 9.15.

9.17. [19] Let f and g be continuous on $[a,b]$, with $m \le f \le M$ and $\int_a^b g(x)\,dx = 0$.

(a) Write $\int_a^b f(x)g(x)\,dx = \frac{1}{2}\int_a^b (2f(x) - M - m)g(x)\,dx$ to show that

$$\left| \int_a^b f(x)g(x)\,dx \right| \le \tfrac{M-m}{2} \int_a^b |g(x)|\,dx .$$

(b) Prove this another way, by writing

$$\int_a^b f(x)g(x)\,dx = \int_a^b \left(f(x) - \frac{M+m}{2} \right) g(x)\,dx.$$

9.18. Let f be continuous on $[a,b]$ and suppose that $\int_a^b f(x)g(x)\,dx = 0$ for every continuous function g on $[a,b]$. Show that $f(x) \equiv 0$ on $[a,b]$.

9.19. [31]

(a) Suppose that f and g are continuous on $[a,b]$. Show that there is $c \in [a,b]$ such that

$$\int_a^c f(x)\,dx + \int_c^b g(x)\,dx = f(c)(b-c) + g(c)(c-a).$$

(b) Draw a picture to show what this says.
(c) Show that this generalizes the Average Value Theorem (Theorem 9.10).

9.20. [2] Suppose that f is continuous on $[a,b]$, and that there are $m, M \in [a,b]$ such that

$$\frac{1}{2}f(m)(b-a) \le \int_a^b f(x)\,dx \le \frac{1}{2}f(M)(b-a).$$

Show that one of the following holds, for some $c \in [a,b]$:

$$\int_a^b f(x)\,dx = f(c)(c-a), \quad \text{or} \quad \int_a^b f(x)\,dx = f(c)(b-c).$$

Hint: Consider

$$G(t) = \int_a^b f(x)\,dx - f(t)(t-a) \quad \text{and} \quad H(t) = \int_a^b f(x)\,dx - f(t)(b-t).$$

Notice that

$$G(t) + H(t) = 2\int_a^b f(x)\,dx - f(t)(b-a),$$

and show that there is $c \in [a,b]$ such that $G(c) = -H(c)$.

9.21. [27] In the interval $[a,b]$, let $a = x_0 < x_1 < x_2 < \cdots < x_{n-1} < x_n = b$, with the x_j's not necessarily equally spaced. Fix $t \in [0,1]$ and let

$$c_j = x_{j-1} + t(x_j - x_{j-1}) \quad \text{for } j = 1, 2, \ldots, n\,.$$

So each c_j is the same proportion along each subinterval $[x_{j-1}, x_j]$. Let

$$r(t) = \sum_{j=1}^{n} f(c_j)[x_j - x_{j-1}] = \sum_{j=1}^{n} f\big(x_{j-1} + t(x_j - x_{j-1})\big)[x_j - x_{j-1}].$$

Show that there is $c \in [0,1]$ such that $r(c) = \int_a^b f(x)\,dx$. (If $n = 1$ this reduces to the Average Value Theorem (Theorem 9.10).)

9.22. Let f be continuous on $[a,b]$ and let $c \in (a,b)$. Show that the average value of f on $[a,b]$ is a weighted average of the average value of f on $[a,c]$ and the average value of f on $[c,b]$.

9.23. Let f be increasing and continuous on $[a,b]$, with f' is continuous. Show that

$$\int_a^b \sqrt{1+f'(x)^2}\,dx \le (b-a)^2 + (f(b)-f(a))^2.$$

What is this saying geometrically?

9.24. [5]] Show that

$$\int_0^\alpha \sqrt{1+(\cos(x))^2}\,dx > \sqrt{\alpha^2+(\sin(\alpha))^2}, \quad \text{for } 0 < \alpha \le \pi/2.$$

9.25. [32] Suppose that f is continuous on $[a,b]$. Let $a = x_0$ and $x_j = a + j\frac{b-a}{N}$ for $j = 0,1,2,\ldots,N = 2^n$. Denote by s the length of the curve described by $y = f(x)$, from $x = a$ to $x = b$.

(a) Draw a picture which illustrates the approximation

$$s = \int_a^b \sqrt{1+f'(x)^2}\,dx \cong \sum_{j=1}^N \sqrt{\left(x_j - x_{j-1}\right)^2 + \left(f'(x_j^*)(x_j - x_{j-1})\right)^2},$$

where $x_j^* = \frac{x_{j-1}+x_j}{2}$ is the midpoint of $[x_{j-1}, x_j]$.

(b) Show that if f' is continuous, then

$$\lim_{n\to\infty} \sum_{j=1}^N \sqrt{\left(x_j - x_{j-1}\right)^2 + \left(f'(x_j^*)(x_j - x_{j-1})\right)^2} = \int_a^b \sqrt{1+f'(x)^2}\,dx.$$

9.26. [13]

(a) Prove the following Mean Value – type theorem for the length of a curve. *Let f and f' be continuous on $[a,b]$. Denote by $t(x)$ the length of the tangent line to $y = f(x)$ at x, between the vertical lines $x = a$ and $x = b$. Then there is $c \in [a,b]$ such that*

$$t(c) = \int_a^b \sqrt{1+f'(x)^2}\,dx.$$

(b) Draw a picture which shows what this is saying geometrically.

9.27. [14] Let f and g be continuous on $[a,b]$ with $f \geq 0$, and $g \geq 0$ and decreasing. Show that

$$\frac{\int_a^b xf(x)g(x)\,dx}{\int_a^b f(x)g(x)\,dx} \leq \frac{\int_a^b xf(x)\,dx}{\int_a^b f(x)\,dx}.$$

9.28. [23] Let f be differentiable on $[a,b]$ with $f(a) = f(b) = 0$. Show that as long as $f(x) \not\equiv 0$, there is $c \in (a,b)$ such that

$$\left|f'(c)\right| > \tfrac{4}{(b-a)^2} \int_a^b f(x)\,dx.$$

Hint: Apply Mean Value Theorem (Theorem 5.2) to f on $\left[a, \frac{a+b}{2}\right]$ and on $\left[\frac{a+b}{2}, b\right]$, integrate each, then add.

9.29. Let f be differentiable on $[a,b]$ with $M = \max\limits_{x \in [a,b]} \left|f'(x)\right|$. Show that

$$\left|\int_a^b f(x)\,dx - \tfrac{f(a)+f(b)}{2}(b-a)\right| \leq M\left(\tfrac{b-a}{2}\right)^2.$$

What is this saying geometrically?

Hint: Apply the Mean Value Theorem (Theorem 5.2) to f on $[a,x] \subset \left[a, \frac{a+b}{2}\right]$ and on $\left[\frac{a+b}{2}, x\right] \subset \left[\frac{a+b}{2}, b\right]$, integrate each expression, then add them together.

9.30. [23] Let f be defined on $[-r,r]$, with f'' continuous ($r > 0$). Show that there is $c \in (-r,r)$ such that $f''(c) = \frac{3}{r^3}\displaystyle\int_{-r}^{r} \left(f(x) - f(0)\right)dx$.

Hint: Start with the Mean Value Theorem for the Second Derivative (Theorem 8.6), then integrate, then use the Mean Value Theorem for Integrals (Theorem 9.14).

9.31. [12] Let f be continuous and increasing on $[0,1]$. **(a)** Show that for any positive integer n,

$$\frac{1}{n}\sum_{k=1}^{n} f\left(\frac{k}{n}\right) \geq \int_0^1 f(x)\,dx.$$

(b) If f is also convex, then these sums *decrease* to $\displaystyle\int_0^1 f(x)\,dx$. Verify and fill in the details of the following proof. Since f is convex, for $x, y \in [0,1]$,

$$(1-t)f(x)+tf(y)\geq f\big((1-t)x+ty\big)\quad\text{for every } t\in[0,1].$$

Set $x=\frac{k}{n}$, $y=\frac{k-1}{n}$, and $t=\frac{k-1}{n}$. Then we have

$$\left(1-\frac{k-1}{n}\right)f\left(\frac{k}{n}\right)+\frac{k-1}{n}f\left(\frac{k-1}{n}\right)\geq f\left((1-\frac{k-1}{n})\frac{k}{n}+\frac{k-1}{n}\frac{k-1}{n}\right)$$
$$=f\left(\frac{kn^2-k+n+1}{n^2(n+1)}\right)$$
$$\geq f\left(\frac{k}{n+1}\right).$$

Now sum from $k=1$ to n.

(c) Show that the assumption that f is increasing is necessary.

(d) Show that if f is concave the sums also decrease to $\int_0^1 f(x)\,dx$.

9.32. Prove the Cauchy–Schwarz Integral Inequality (Corollary 9.28) by suitably modifying H. Schwarz's proof, from Sect. 2.3, of the Cauchy–Schwarz Inequality (Theorem 2.18) for sums. (It was in fact in the context of integrals that Schwarz's proof first appeared [30].)

9.33. [4] Apply Schwarz's idea as in Exercise 9.32 to $\int_a^b \left(f(x)+t\right)^2dx$.

What do you get?

9.34. [25, 26]

(a) Take $f(x)=1/x$ and $g\equiv 1$, then $f(x)=1/\sqrt{x}$ and $g\equiv 1$, in the Cauchy–Schwarz Integral Inequality (Corollary 9.28) to give another proof of Lemma 6.20:

$$G<L<A,$$

where G, L, and A are respectively, the Geometric, Logarithmic, and Arithmetic Means of a and b.

(b) Manipulate the latter case carefully, to show that in fact $L<\frac{A+G}{2}<A$.

9.35. **(a)** Show that Jensen's Integral Inequality (Theorem 9.29) still holds for φ convex, but φ is only assumed to be differentiable. Hint: Look at Exercise 8.22.

(b) Can you show that Jensen's Integral Inequality (Theorem 9.29) still holds for φ convex, but φ is only assumed to be continuous?

9.36. Fill in the details of another proof of the Cauchy–Schwarz Integral Inequality (Corollary 9.28) which is similar to Schwarz's. (The sum version of this is the content of Exercise 2.38). First dispense with the case in which $g(x)\equiv 0$. Observe that for any real number t,

$$\int_a^b (f(x) - tg(x))^2 dx \geq 0.$$

Now expand this and set

$$t = \frac{-\int_a^b f(x)g(x)\,dx}{\int_a^b g(x)^2\,dx}.$$

(This is the t at which the quadratic $\int_a^b (f(x) - tg(x))^2 dx$ attains its minimum.)

9.37. Here is an ostensibly different proof of Hölder's Integral Inequality (Theorem 9.27). (The sum version of this is the content of Exercise 6.34).

(a) First dispense with the cases $\int_a^b f(x)^p\,dx = 0$ or $\int_a^b g(x)^q\,dx = 0$.

(b) In Young's Inequality (Corollary 6.19) $ab \leq \dfrac{a^p}{p} + \dfrac{b^q}{q}$, set

$$a = f(x)\left(\int_a^b f(x)^p\,dx\right)^{-1/p} \quad \text{and} \quad b = g(x)\left(\int_a^b g(x)^q\,dx\right)^{-1/q},$$

then integrate from a to b.

9.38. Find necessary and sufficient conditions for equality to hold in **(a)** the Cauchy–Schwarz Integral Inequality (Corollary 9.28), and **(b)** Hölder's Integral Inequality (Theorem 9.27).

9.39. Suppose that f and g are continuous on $[0, 1]$ with $\int_0^1 g(x)\,dx = 0$.

(a) Show that

$$\left(\int_0^1 f(x)g(x)\,dx\right)^2 \leq \left(\int_0^1 f(x)^2\,dx - \left(\int_0^1 f(x)\,dx\right)^2\right)\int_0^1 g(x)^2\,dx.$$

(b) How would this read on $[a, b]$ instead of $[0, 1]$?
(The sum version of this exercise is the content of Exercise 2.51.)

9.40. [11] We used the weighted AGM Inequality with n = 2 (Corollary 6.16) to obtain Hölder's Integral Inequality (Theorem 9.27). Use the full weighted AGM

Inequality (Theorem 6.15) to obtain the following extension of Hölder's Integral Inequality. *Let $f_1, f_2, \ldots, f_n$ be continuous and nonnegative on $[a,b]$ and let $p_1, p_2, \ldots, p_n > 1$ satisfy $\frac{1}{p_1} + \frac{1}{p_2} + \cdots + \frac{1}{p_n} = 1$. Then*

$$\int_a^b \prod_{j=1}^n f_j(x)\,dx \le \prod_{j=1}^n \left(\int_a^b f_j(x)^{p_j}\,dx \right)^{1/p_j}$$

(The sum version of this is the content of Exercise 6.37.)

9.41. Use Hölder's Integral Inequality (Theorem 9.27) to prove **Minkowski's Integral Inequality**: *Let f and g be continuous on $[a,b]$ and let $p > 0$. Then*

$$\left(\int_a^b (f(x) + g(x))^p dx \right)^{1/p} \le \left(\int_a^b f(x)^p dx \right)^{1/p} + \left(\int_a^b g(x)^p dx \right)^{1/p}$$

(The sum version of this for $p = 2$ is the content of Exercise 2.53 and the sum version of this for $p > 0$ is the content of Exercise 6.39.)

Hint: Write

$$\big(f(x) + g(x)\big)^p = f(x)\big(f(x) + g(x)\big)^{p-1} + g(x)\big(f(x) + g(x)\big)^{p-1},$$

then integrate, then apply Hölder's Integral Inequality (Theorem 9.27) to each piece.

Note: If $p = 2$, then Minkowski's Integral Inequality is a sort of a triangle inequality for integrals:

$$\sqrt{\int_a^b \big(f(x) + g(x)\big)^2 dx} \le \sqrt{\int_a^b f(x)^2\,dx} + \sqrt{\int_a^b g(x)^2\,dx}.$$

9.42. [17]

(a) Show that $(pq - rs)^2 \ge (p^2 - r^2)(q^2 - s^2)$.

(b) Let f and g be continuous on $[a,b]$. Use (a) and the Cauchy–Schwarz Integral Inequality (Corollary 9.28) to show that for any $p, q \in \mathbf{R}$,

$$\left(pq - \int_a^b f(x)g(x)\,dx \right)^2 \ge \left(p^2 - \int_a^b f(x)^2\,dx \right)\left(q^2 - \int_a^b g(x)^2\,dx \right).$$

9.43. [24] Show that if f is continuous and increasing on $[a,b]$, then

$$\frac{1}{b-a} \int_a^b (x - c)\,f(x)\,dx \ge f(c)\big(\tfrac{a+b}{2} - c\big) \quad \text{for any } c \in [a,b].$$

9.44. The Cauchy–Schwarz Integral Inequality (Corollary 9.28) gives an upper bound for

$$\int_a^b f(x)g(x)\,dx.$$

Chebyshev's Integral Inequality gives a lower bound under certain circumstances: *Let f and g be continuous on $[a,b]$ with either both increasing or both decreasing. Then*

$$\frac{1}{b-a}\int_a^b f(x)\,dx \cdot \frac{1}{b-a}\int_a^b g(x)\,dx \le \frac{1}{b-a}\int_a^b f(x)g(x)\,dx.$$

And the inequality is reversed if f and g have opposite monotonicity. (The sum version of this is the content of Exercise 2.54.) Fill in the details of the following proof of Chebyshev's Integral Inequality. If f and g have the same monotonicity, then

$$\big(f(x)-f(c)\big)\big(g(x)-g(c)\big) \ge 0 \quad \text{for any } c \in [a,b].$$

So by Lemma 9.9,

$$\int_a^b \big(f(x)-f(c)\big)\big(g(x)-g(c)\big)\,dx \ \ge\ 0.$$

Now take $c \in [a,b]$ as given for f by the Average Value Theorem (Theorem 9.10): $f(c) = \frac{1}{b-a}\int_a^b f(x)\,dx$ and expand the left-hand side.

9.45. [28] Suppose that f is positive and has two continuous derivatives on $[a,b]$, with each of f and $1/f$ convex. Use Chebyshev's Integral Inequality (Exercise 9.44) to show that

$$\int_a^b \left(\frac{f'(x)}{f(x)}\right)^2 dx \ \le\ \frac{1}{(b-a)}\,\frac{(f(b)-f(a))^2}{f(a)f(b)}.$$

9.46. **(a)** In Exercise 9.44 is Chebyshev's Integral Inequality. Prove the **weighted version of Chebyshev's Integral Inequality**: *Let F and G be continuous on $[a,b]$ with either both increasing or both decreasing. Let $w > 0$ be continuous on $[a,b]$. Then*

$$\int_a^b w(x)F(x)\,dx \cdot \int_a^b w(x)G(x)\,dx \le \int_a^b w(x)\,dx \int_a^b w(x)F(x)G(x)\,dx.$$

And the inequality is reversed if f and g have opposite monotonicity. (A sum version of this is the content of Exercise 6.40.)

(b) Use this to prove the Cauchy–Schwarz Integral Inequality (Corollary 9.28). Hint: Set $w = G^2$, and $F = G = f/g$. (A sum version of this is also contained in Exercise 6.40.)

9.47. Let f be continuous on $[a, b]$ with $m \le f \le M$, and set $A = \frac{1}{b-a}\int_a^b f(x)\,dx$.

(a) Verify that

$$\frac{1}{b-a}\int_a^b (f(x) - A)^2\,dx = (M - A)(A - m) - \frac{1}{b-a}\int_a^b (M - f(x))\,(f(x) - m)\,dx.$$

(b) Conclude that $\frac{1}{b-a}\int_a^b (f(x) - A)^2\,dx \le (M - A)(A - m)$.

(c) Show that the inequality in (b) is better than—that is, is a *refinement* of—the integral version of **Popoviciu's Inequality**:

$$\frac{1}{b-a}\int_a^b (f(x) - A)^2\,dx \le \frac{1}{4}(M - m)^2.$$

Hint: Show that for $q < Q$, the quadratic $(Q-x)(x-q)$ is maximized precisely when $x = \frac{1}{2}(Q + q)$. (A sum version of this is the content of Exercise 2.57.)

9.48. [20]

(a) Look carefully at Exercise 9.47. Extend the ideas there to prove **Grüss's Integral Inequality**: *Let f, g be continuous on $[a, b]$, with $m \le f \le M$ and $\gamma \le g \le \Gamma$. Then*

$$\left|\frac{1}{b-a}\int_a^b f(x)g(x)\,dx - \frac{1}{b-a}\int_a^b f(x)\,dx \cdot \frac{1}{b-a}\int_a^b g(x)\,dx\right| \le \frac{1}{4}(M - m)(\Gamma - \gamma).$$

Hint: Let $A = \frac{1}{b-a}\int_a^b f(x)\,dx$, $B = \frac{1}{b-a}\int_a^b g(x)\,dx$ and begin by applying the Cauchy–Schwarz Integral Inequality (Corollary 9.28) to $\left(\int_a^b (f(x)-A)(g(x)-B)\right)^2$.

(b) Consider the functions $f(x) = g(x) = sign(x-\frac{a+b}{2})$ to show that the constant $1/4$ above cannot be replaced with anything smaller. That is, it is *sharp*. (A sum version of this is the content of Exercise 2.58.)

9.49. Use Jensen's Integral Inequality (Theorem 9.29), with an appropriate choice of weight function, to prove Hölder's Integral Inequality (Theorem 9.27). (A sum version of this is the content of Exercise 8.50.)

9.50. Extend Example 9.32: State and prove a *weighted* AGM Inequality for Functions, thus generalizing the AGM Inequality for Functions.

9.51. Let f be continuous on $[a, b]$. For $r > 0$ define the **Power Mean** M_r by

$$M_r = \left(\frac{1}{b-a}\int_a^b f(x)^r\,dx\right)^{1/r}.$$

For example, M_1 is the average value of f and M_2 is called the **Root Mean Square**.

(a) Apply Jensen's Integral Inequality (Theorem 9.29) with $\varphi(x) = x^{r/s}$ to show that $M_s < M_r$ if $s < r$.
(b) What is a reasonable way to define M_0?
(c) What is a reasonable definition of M_∞?
(d) How might you define *weighted* Power Means?

(A sum version of this exercise is the content of Exercise 8.48 and another approach is the content of Exercise 10.18.)

9.52. [20] Let f and g be continuous on $[0, 1]$, with f decreasing and $0 \le g \le 1$. Let $\lambda = \int_0^1 g(x)\,dx$. **Steffensen's Inequalities** are:

$$\int_{1-\lambda}^{1} f(x)\,dx \le \int_0^1 f(x)g(x)\,dx \le \int_0^{\lambda} f(x)\,dx.$$

(a) By considering $1 - g(x)$, show that the left-hand inequality follows from the right-hand inequality.

(b) Prove the right-hand inequality. Hint: Verify that

$$\int_0^\lambda f(x)\,dx - \int_0^1 f(x)g(x)\,dx = \int_0^\lambda (1 - g(x))f(x)\,dx - \int_\lambda^1 f(x)g(x)\,dx$$

$$\geq f(\lambda)\int_0^\lambda (1 - g(x))\,dx - \int_\lambda^1 f(x)g(x)\,dx,$$

then show that this is ≥ 0.

(c) Prove the left-hand inequality directly—that is, without using (a).

(d) How would Steffensen's Inequalities read on $[a, b]$?

References

1. Aliprantis, C.D., Wilkins, E.J., Jr.: Problem 10586. Am. Math. Mon. **105**, 960 (1998)
2. Baker, C.W.: Mean value type theorems of integral calculus. Coll. Math J. **10**, 35–37 (1979)
3. Briggs, D.J. Simpson and another circle. Math. Gaz. **62**, 123 (1978)
4. Chambers, L.G.: An inequality concerning means. Math. Gaz. **82**, 456–459 (1998)
5. Cheng, B.N.: Problem Q811. Math. Mag. **66**, 339–344 (1993)
6. Courant, R.: Differential and Integral Calculus, 2nd edn. Wiley Classics Library, Hoboken (1988)
7. Díaz-Barrero, J.L., Viteam, T.: Problem 1892. Math. Mag. **86**, 149–150 (2013)
8. Dunham, W.: The Calculus Gallery. Princeton University Press, Princeton (2005)
9. Goldberg, K.P.: Does the formula for arc length measure arc length? In: Page, W. (ed.) Two-Year College Mathematics Readings, pp. 107–110. Mathematical Association of America, Washington DC (1981)
10. Gordon, S.P.: Riemann sums and the exponential function. Coll. Math. J. **25**, 38–39 (1994)
11. Hardy, G.H., Littlewood, J.E., Polya, G.: Inequalities. Cambridge University Press, Cambridge/New York (1967)
12. Jichang, K.: Some extensions and refinements of the Minc-Sathre inequality. Math. Gaz. **83**, 123–127 (1999)
13. Kaucher, J.: A theorem on arc length. Math. Mag. **42**, 132–133 (1969)
14. Kestelman, H.: A centroidal inequality. Math. Gaz. **46**, 140–141 (1962)
15. Kumar, A.: Definite integration via areas. Math. Gaz. **86**, 95–99 (2002)
16. Lupu, C., Lupu, T., Curtis, C., Yang, Y.: Problem 927. Coll. Math. J. **42**, 236–237 (2011)
17. Lyusternik, L.A.: Convex Figures and Polyhedra. D.C. Heath & Co., Boston (1966)
18. Matthews, J.H., Shultz, H.S.: Riemann integral of cos(x). Coll. Math. J. **20**, 237 (1989)
19. Mercer, P.R.: Error estimates for numerical integration rules. Coll. Math. J. **36**, 27–34 (2005)
20. Mitrinovic, D.S.: Analytic Inequalities. Springer, Berlin/New York (1970)
21. Page, W.: The formula for arc length does measure arc length. In: Page, W. (ed.) Two-Year College Mathematics Readings, pp. 111–114. Mathematical Association of America, Washington DC (1981)
22. Plaza, Á., Rodríguez, C., García, T.M., Suárez, P.: Problem 1867. Math. Mag. **85**, 152–143 (2012)
23. Polya, G., Szegő, G.: Problems and Theorems in Analysis I, p. 80. Springer, Berlin/Heidelberg/New York (2004)

24. Sandor, J.: Some simple integral inequalities. Octogon Math. Mag. **16**, 925–933 (2008)
25. Sandor, J.: Bouniakowsky and the logarithmic mean inequalities. RGMIA Res. Rep. Coll. **17**, article 5 (2014)
26. Sandor, J.: A note on the logarithmic mean. RGMIA Res. Rep. Coll. **17**, article 6 (2014)
27. Sayrafiezadeh, M.: A generalization of the mean value theorem for integrals. Coll. Math. J. **26**, 223–224 (1995)
28. Sieffert, H.J., Lau, K.W., Levine, E.: Problem 1284. Math. Mag. **61**, 322–323 (1988)
29. Sklar, A.: On a sequence approach to integration. Am. Math. Mon. **67**, 897–900 (1960)
30. Steele, J.M.: The Cauchy-Schwarz Master Class. Mathematical Association of America/Cambridge University Press, Washington DC - Cambridge/New York (2004)
31. Tong, J.: A generalization of the mean value theorem for integrals. Coll. Math. J. **33**, 408–409 (2002)
32. White, J.B.: A note on arc length. Math. Mag. **43**, 44 (1970)
33. Xiang, J.X.: A note on the Cauchy-Schwarz inequality. Am. Math. Mon. **120**, 456–459 (2013)

Chapter 10
The Fundamental Theorem of Calculus

There is nothing more thrilling than getting a new song – when you think of the piece that will fit the puzzle.

—James Taylor, in *Rolling Stone* #1062, October 2008

In Chap. 9 we evaluated a few definite integrals by using Riemann sums, a method that is generally not easy, even for simple integrands. Then we evaluated a few more, by interpreting the definite integral as a certain familiar area. For example:

$$\int_a^b (Mx + B)\,dx = \frac{M}{2}(b^2 - a^2) + B(b-a) \quad \text{and} \quad \int_{-r}^{r} \sqrt{r^2 - x^2}\,dx = \frac{\pi r^2}{2}.$$

Generally however, computing integrals $\int_a^b f(x)\,dx$ which routinely appear in problems with no apparent notion of area or of average value in sight, can be very difficult. Coming to the rescue in many cases is the Fundamental Theorem of Calculus. With it, many more definite integrals can be computed relatively easily. But this—the most important theorem in all of calculus—gives us a great deal more.

10.1 The Fundamental Theorem

Theorem 10.1. (The Fundamental Theorem of Calculus)

(i) *Let f be continuous on $[a,b]$. Then for each $x \in [a,b]$,*

$$\left(\int_a^x f(t)\,dt\right)' = f(x).$$

P.R. Mercer, *More Calculus of a Single Variable*, Undergraduate Texts in Mathematics, DOI 10.1007/978-1-4939-1926-0_10

(ii) *Let f' be continuous on $[a,b]$. Then for each $x \in [a,b]$,*

$$\int_a^x f'(t)\,dt = f(x) - f(a).$$

Proof. For part (i), set $F(x) = \int_a^x f(t)\,dt$. Then by Lemma 9.22,

$$F(x+h) - F(x) = \int_x^{x+h} f(t)\,dt.$$

So by the Average Value Theorem (Theorem 9.10) there is $c \in [x, x+h]$ such that

$$\frac{1}{(x+h)-x}\int_x^{x+h} f(t)\,dt = \frac{F(x+h)-F(x)}{h} = f(c).$$

Now as $h \to 0$ we must have $c \to x$, and so $f(c) \to f(x)$ because f is continuous. Therefore

$$\lim_{h\to 0} \frac{F(x+h)-F(x)}{h} = f(x), \quad \text{as desired.}$$

For part (ii), we use part (i) to see that $\left(\int_a^x f'(t)\,dt\right)' = f'(x)$. So having the same derivative, the functions $\int_a^x f'(t)\,dt$ and $f(x)$ must differ by a constant, by Lemma 5.6. That is, for some $C \in \mathbf{R}$,

$$f(x) = \int_a^x f'(t)\,dt + C.$$

Therefore $f(x) - f(a) = \left(\int_a^x f'(t)\,dt + C\right) - \left(\int_a^a f'(t)\,dt + C\right) = \int_a^x f'(t)\,dt.$ □

The Fundamental Theorem (Theorem 10.1) contains the astonishing fact that if f' is continuous, then the operations of differentiation and (definite) integration are inverses of one another—that is, except for the "$-f(a)$" which appears in Part (ii).

So virtually any useful or interesting fact about derivatives corresponds to a useful or interesting fact about integrals, and vice-versa. Here is a good example.

Example 10.2. We show how the Mean Value Theorem (Theorem 5.2) yields the Average Value Theorem (Theorem 9.10), and vice versa, by way of the Fundamental Theorem (Theorem 10.1). For f continuous on $[a,b]$, the Fundamental Theorem Part (i) says in particular that $F(x) = \int_a^x f(t)\,dt$ is differentiable. Then the Mean Value Theorem applied to F says that there is $c \in (a,b)$ such that

$$\frac{F(b) - F(a)}{b - a} = F'(c).$$

Again by the Fundamental Theorem Part (i), this reads $\frac{1}{b-a}\int_a^b f(x)\,dx = f(c)$, which is the Average Value Theorem. (In Exercise 10.5 we see how to obtain the Mean Value Theorem for Integrals (Theorem 9.14) this way.) Conversely, if f' is continuous then $\int_a^b f'(x)\,dx = f(b) - f(a)$, by the Fundamental Theorem Part (ii). Then the Average Value Theorem applied to f' implies that there is $c \in (a,b)$ such that

$$f'(c) = \frac{1}{b-a}\int_a^b f'(x)\,dx.$$

This reads

$$f'(c) = \frac{f(b) - f(a)}{b - a},$$

which is the Mean Value Theorem. In this latter analysis, we require that f' is continuous because if f is only differentiable, then $\int_a^b f'(x)\,dx$ may not be defined. And even if it were defined, the Average Value Theorem might not be applicable. A function f for which f' is continuous is often called *continuously differentiable.* ◇

Example 10.3. Here is the integral analogue of Cauchy's Mean Value Theorem (Theorem 5.11). For f and g continuous on $[a,b]$, Cauchy's Mean Value Theorem applied to

$$F(x) = \int_a^x f(t)\,dt \qquad \text{and} \qquad G(x) = \int_a^x g(t)\,dt$$

says that there is $c \in (a,b)$ such that

$$F'(c)\big(G(b) - G(a)\big) = G'(c)\big(F(b) - F(a)\big).$$

That is,

$$f(c)\int_a^b g(t)\,dt = g(c)\int_a^b f(t)\,dt.$$

This is **Cauchy's Mean Value Theorem for Integrals.** ◇

For $f(b) - f(a)$ the notation $f(t)\Big|_a^b$ is commonly used. This way, Part (ii) of the Fundamental Theorem with $x = b$ is written

$$\int_a^b f'(t)\,dt = f(t)\Big|_a^b.$$

As we saw in the proof of Part (ii), an **antiderivative** f of a given f' is not unique: $f + C$ is also an antiderivative of f' for any $C \in \mathbf{R}$. But we also saw in the proof that whichever constant C is chosen does not matter—the C's cancel out. So in practice one usually chooses $C = 0$.

The string of symbols

$$\int f(t)\,dt$$

is used to denote an antiderivative of f, but when written this way it is referred to as an **indefinite integral**. It is the Fundamental Theorem which makes this terminology and notation reasonable; indeed, Part (i) tells how to produce an antiderivative of a continuous function. And again, since any two indefinite integrals of f differ by a constant, the expression

$$\int f(t)\,dt + C$$

is used to denote ***all*** **indefinite integrals** of f, that is, ***all*** **antiderivatives** of f.

The Fundamental Theorem Part (ii) (Theorem 10.1) contains another astonishing fact—that for a continuous function f, evaluating a definite integral $\int_a^b f(x)\,dx$ (which remember, is defined in terms of averages or area) comes down to the apparently very different problem of finding an antiderivative for f.

Example 10.4. In Example 9.7 we showed that

$$\int_a^b \sin(x)\,dx = \cos(a) - \cos(b),$$

by considering an appropriate sequence of Riemann sums. But $(\cos(x))' = -\sin(x)$ and so by the Fundamental Theorem (Theorem 10.1),

$$\int_a^b \sin(x)\,dx = -\cos(x)\Big|_a^b = -(\cos(b) - \cos(a)) = \cos(a) - \cos(b).$$

Likewise, since $(\sin(x))' = \cos(x)$,

$$\int_a^b \cos(x)\,dx = \sin(x)\Big|_a^b = \sin(b) - \sin(a).$$ ◇

Example 10.5. For $r \in \mathbf{R}$ with $r \neq -1$ we have $\left(\dfrac{x^{r+1}}{r+1}\right)' = x^r$, by the Power Rule. So by the Fundamental Theorem,

$$\int_a^b x^r\,dx = \frac{x^{r+1}}{r+1}\Bigg|_a^b = \frac{b^{r+1} - a^{r+1}}{r+1} \qquad (r \neq -1).$$

As an indefinite integral, this reads

$$\int x^r\,dx = \frac{x^{r+1}}{r+1} + C \qquad (r \neq -1).$$

To see what happens when $r = -1$, observe that $\big(\ln(x)\big)' = \dfrac{1}{x}$ for $x > 0$. So by the Fundamental Theorem,

$$\int_a^b \frac{1}{x}\,dx = \ln(x)\Big|_a^b = \ln(b) - \ln(a) = \ln\left(\frac{b}{a}\right) \qquad (a, b > 0).$$

And as an indefinite integral, this reads

$$\int \frac{1}{x}\,dx = \ln(x) + C \qquad (x > 0).$$

Sometimes it is convenient to write $\int_a^b \frac{1}{x}\,dx = \ln(x)\big|_a^b = \ln(b) - \ln(a)$ for $a, b > 0$ in the equivalent form (for $p, q > -1$):

$$\int_p^q \frac{1}{1+x}\,dx = \ln(1+x)\Big|_p^q = \ln(1+q) - \ln(1+p) = \ln\left(\frac{1+q}{1+p}\right).$$ ◇

Example 10.6. Taking $r = -p \neq -1$ in Example 10.5,

$$\lim_{N\to\infty} \int_1^N \frac{1}{x^p}\, dx = \lim_{N\to\infty} \left[\frac{1}{1-p}\left(\frac{N}{N^p} - 1\right)\right].$$

This limit exists if $p > 1$, and it diverges to $+\infty$ if $p < 1$. Also from Example 10.5,

$$\lim_{N\to\infty} \int_1^N \frac{1}{x}\, dx = \lim_{N\to\infty} (\ln(N)),$$

which diverges to $+\infty$. Therefore, by the Integral Test (Theorem 9.24), the **p-series**

$$\sum_{n=1}^{\infty} \frac{1}{n^p} \quad \textit{converges if and only if } p > 1.$$

(We obtained this result differently in Sect. 6.7.) For example, the Harmonic series $\sum_{n=1}^{\infty} \frac{1}{n}$ diverges to $+\infty$ and $\sum_{n=1}^{\infty} \frac{1}{n^2}$ converges. We will see in Theorem 12.7 that in fact,

$$\sum_{n=1}^{\infty} \frac{1}{n^2} = \frac{\pi^2}{6} \cong 1.645.$$

Being the first to find the sum of this series was one of Euler's many triumphs. ◇

Example 10.7. We have met several times, beginning with Example 6.11 (line (6.7)), the useful inequality:

$$\ln(x) \le x - 1 \quad \text{for} \quad x > 0. \tag{10.1}$$

Here is a way to obtain it using integrals. Again by Example 10.5,

$$\int_1^x \frac{1}{t}\, dt = \ln(x) - \ln(1) = \ln(x).$$

Now if $x > 1$, then $\frac{1}{t} \le 1$ on $[1, x]$ and so

$$\ln(x) = \int_1^x \frac{1}{t}\, dt \le \int_1^x 1\, dt = x - 1.$$

If $0 < x < 1$, then $\frac{1}{t} \geq 1$ on $[x, 1]$ and so

$$\int_x^1 \frac{1}{t}\,dt \geq \int_x^1 1\,dt = 1 - x.$$

Therefore

$$\ln(x) = \int_1^x \frac{1}{t}\,dt = -\int_x^1 \frac{1}{t}\,dt \leq -\int_x^1 1\,dt = x - 1.$$

Either way, $\ln(x) \leq x - 1$ for $x > 0$. ◇

Example 10.8. Sometimes a complicated limit can be viewed as the limit of a Riemann sum, so that evaluating the limit comes down to evaluating a definite integral. For example, let us consider the limit (for $k \neq -1$)

$$\lim_{n\to\infty} \left(\frac{1}{n} \frac{1^k + 2^k + \cdots + n^k}{n^k} \right).$$

We write

$$\frac{1}{n} \frac{1^k + 2^k + \cdots + n^k}{n^k} = \frac{1}{n} \left(\left(\frac{1}{n}\right)^k + \left(\frac{2}{n}\right)^k \cdots + \left(\frac{k}{n}\right)^k \right),$$

and notice that this is a Riemann sum for $\int_0^1 x^k\,dx$. Therefore

$$\lim_{n\to\infty} \left(\frac{1}{n} \frac{1^k + 2^k + \cdots + n^k}{n^k} \right) = \int_0^1 x^k\,dx = \frac{1}{k+1}.$$ ◇

Example 10.9. We have seen that for a function u which is differentiable and never zero,

$$(\ln |u(x)|)' = \frac{u'(x)}{u(x)}.$$

So by the Fundamental Theorem,

$$\int \frac{u'(x)}{u(x)}\,dx = \ln |u(x)| + C.$$

Following [46], we use this to show that

$$\int \sec(x)\,dx = \ln\left|\sec(x) + \tan(x)\right| + C.$$

Observe that

$$(\sec(x) + \tan(x))' = \sec(x)\tan(x) + \sec^2(x) = \sec(x)\,(\tan(x) + \sec(x)).$$

Therefore

$$\frac{(\sec(x) + \tan(x))'}{\sec(x) + \tan(x)} = \sec(x),$$

and so indeed

$$\int \sec(x)\,dx = \ln\left|\sec(x) + \tan(x)\right| + C.$$

Similarly (Exercise 10.21),

$$\int \csc(x)\,dx = \ln\left|\csc(x) - \cot(x)\right| + C. \qquad \diamond$$

Example 10.10. Here we show that for $0 < x \le 2$,

$$\begin{aligned}\ln(x) &= (x-1) - \frac{(x-1)^2}{2} + \frac{(x-1)^3}{3} - \frac{(x-1)^4}{4} + \cdots \\ &= \sum_{n=0}^{\infty} \frac{(-1)^n}{n+1}(x-1)^{n+1},\end{aligned}$$

thus extending Example 8.26, in which we obtained this series only for $1/2 \le x \le 2$. And again, setting $x = 2$ here we obtain the sum of the Alternating Harmonic series:

$$\ln(2) = 1 - \frac{1}{2} + \frac{1}{3} - \frac{1}{4} + \cdots = \sum_{n=0}^{\infty} \frac{(-1)^n}{n+1}.$$

For $x \neq -1$, the following identity can be found by doing long division on the left-hand side, or simply verified by cross multiplication:

$$\frac{1 + (-1)^n x^{n+1}}{1 + x} = 1 - x + x^2 - x^3 + \cdots + (-1)^n x^n.$$

Therefore

$$\frac{1}{1+x} = 1 - x + x^2 - x^3 + \cdots + (-1)^n x^n + \frac{(-1)^{n+1}x^{n+1}}{1+x}.$$

Integrating from 0 to $u > -1$ and using Example 10.5, we get

$$\begin{aligned}\ln(1+u) &= \int_0^u \frac{1}{1+x}\,dx \\ &= u - \frac{1}{2}u^2 + \frac{1}{3}u^3 - \cdots + \frac{(-1)^n}{n+1}u^{n+1} + (-1)^{n+1}\int_0^u \frac{x^{n+1}}{1+x}\,dx.\end{aligned}$$

Now for $u \geq 0$ and $x \in [0, u]$, we have $1 \leq 1 + x \leq 1 + u$, so that

$$\frac{x^{n+1}}{1+u} \leq \frac{x^{n+1}}{1+x} \leq x^{n+1}.$$

Integrating with respect to x from 0 to u, we get

$$\frac{1}{1+u}\frac{u^{n+2}}{(n+2)} \leq \int_0^u \frac{x^{n+1}}{1+x}\,dx \leq \frac{u^{n+2}}{(n+2)}.$$

For $-1 < u < 0$ and $t \in [u, 0]$, a similar analysis leads to the same inequalities, but reversed. So either way, if $-1 < u \leq 1$, each of the right-hand and left-hand sides $\to 0$ as $n \to \infty$. Therefore we are justified in writing (for $-1 < u \leq 1$):

$$\ln(1+u) = u - \frac{1}{2}u^2 + \frac{1}{3}u^3 - \frac{1}{4}u^4 + \cdots = \sum_{n=0}^{\infty} \frac{(-1)^n}{n+1}u^{n+1}.$$

Setting $u = x - 1$ here we get, for $0 < x \leq 2$:

$$\ln(x) = (x-1) - \frac{(x-1)^2}{2} + \frac{(x-1)^3}{3} - \frac{(x-1)^4}{4} + \cdots = \sum_{n=0}^{\infty} \frac{(-1)^n}{n+1}(x-1)^{n+1}. \quad \diamond$$

Example 10.11. For $t \in (-\infty, \infty)$, we have $(\arctan(t))' = \frac{1}{1+t^2}$. So by the Fundamental Theorem,

$$\int_0^x \frac{1}{1+t^2}\,dt = \arctan(t)\Big|_0^x = \arctan(x).$$

And in particular,

$$\int_0^1 \frac{1}{1+t^2}\,dt = \frac{\pi}{4}. \qquad \diamond$$

Example 10.12. Here we verify the curious fact that (e.g., [32])

$$\pi = \frac{22}{7} - \int_0^1 \frac{x^4(1-x^4)}{1+x^2}\,dx.$$

Since the integrand here is positive, $\pi < \frac{22}{7}$. In particular $\pi \neq \frac{22}{7}$!! The following identity can be found by doing long division on the left-hand side, or by obtaining a common denominator on the right-hand side:

$$\frac{x^4(1-x^4)}{1+x^2} = x^6 - 4x^5 + 5x^4 - 4x^2 + 4 - \frac{4}{1+x^2}.$$

Then using Examples 10.5 and 10.11,

$$\begin{aligned}\int_0^1 \frac{x^4(1-x^4)}{1+x^2}\,dx &= \int_0^1 (x^6 - 4x^5 + 5x^4 - 4x^2 + 4)\,dx - 4\int_0^1 \frac{1}{1+x^2}\,dx \\ &= \frac{1}{7} - \frac{2}{3} + 1 - \frac{4}{3} + 4 - 4\frac{\pi}{4} \\ &= \frac{22}{7} - \pi.\end{aligned}$$

See Exercises 10.38 and 10.39 for similar results. $\diamond$

Example 10.13. Here (see also [42]) we obtain the well known **Leibniz series**, named for German mathematician Gottfried W. Leibniz (1646–1716):

$$\arctan(u) = u - \frac{1}{3}u^3 + \frac{1}{5}u^5 - \frac{1}{7}u^7 + \cdots = \sum_{n=0}^{\infty} \frac{(-1)^n}{2n+1}u^{2n+1}, \quad \text{for } -1 \le u \le 1.$$

Replacing x with t^2 in the identity (see Example 10.10)

$$\frac{1}{1+x} = 1 - x + x^2 - x^3 + \cdots + (-1)^n x^n + \frac{(-1)^{n+1}x^{n+1}}{1+x},$$

we get

$$\frac{1}{1+t^2} = 1 - t^2 + t^4 - t^6 + \cdots + (-1)^n t^{2n} + \frac{(-1)^{n+1} t^{2n+2}}{1+t^2}.$$

Then integrating from 0 to u and using Examples 10.5 and 10.11, we get

$$\begin{aligned}\arctan(u) &= \int_0^u \frac{1}{1+t^2}\,du \\ &= u - \frac{1}{3}u^3 + \frac{1}{5}u^5 - \cdots + \frac{(-1)^n}{2n+1}u^{2n+1} + (-1)^{n+1}\int_0^u \frac{t^{2n+2}}{1+t^2}\,dt.\end{aligned}$$

Now for $u > 0$ and $t \in [0, u]$, we have $1 \le 1 + t^2 \le 1 + u^2$, so that

$$\frac{t^{2n+2}}{1+u^2} \le \frac{t^{2n+2}}{1+t^2} \le t^{2n+2}.$$

Integrating with respect to t from 0 to u, we get

$$\frac{1}{1+u^2}\frac{u^{2n+3}}{(2n+3)} \le \int_0^u \frac{t^{2n+2}}{1+t^2}\,dt \le \frac{u^{2n+3}}{(2n+3)}.$$

For $u < 0$ and $t \in [u, 0]$, a similar analysis leads to the same inequalities, but reversed. So either way, if $-1 \le u \le 1$, each of the right-hand and left-hand sides $\to 0$ as $n \to \infty$. Therefore we are indeed justified in writing (for $-1 \le u \le 1$):

$$\arctan(u) = u - \frac{1}{3}u^3 + \frac{1}{5}u^5 - \frac{1}{7}u^7 + \cdots = \sum_{n=0}^{\infty} \frac{(-1)^n}{2n+1}u^{2n+1}.$$

Taking $u = 1$ in this, the Leibniz series, we get the **Gregory-Leibniz series** named also for Scottish mathematician James Gregory (1638–1675):

$$\frac{\pi}{4} = 1 - \frac{1}{3} + \frac{1}{5} - \frac{1}{7} + \frac{1}{9} - \frac{1}{11} + \cdots = \sum_{n=0}^{\infty} \frac{(-1)^n}{2n+1}. \qquad \diamond$$

Remark 10.14. A result similar to the Integral Test (Theorem 9.24) is used in [2] to obtain sums of rearrangements of various alternating series. For example, taking three positive terms then two negative terms and so on, in the Gregory-Leibniz series, one has:

$$1+\frac{1}{5}+\frac{1}{9}-\frac{1}{3}-\frac{1}{7}+\frac{1}{13}+\frac{1}{17}+\frac{1}{21}-\frac{1}{11}-\cdots=\frac{\pi}{4}+\frac{1}{4}\ln\left(\frac{3}{2}\right).$$

See also [7]. And taking three positive terms then two negative terms and so on, in the Alternating Harmonic series, one has:

$$1+\frac{1}{3}+\frac{1}{5}-\frac{1}{2}-\frac{1}{4}+\frac{1}{7}+\frac{1}{9}+\frac{1}{11}-\frac{1}{6}-\cdots=\ln(2)+\frac{1}{2}\ln\left(\frac{3}{2}\right).$$

In these formulas, replacing the 3 positive terms and 2 negative terms with m positive terms and n negative terms respectively, we get $\ln\left(\frac{m}{n}\right)$ instead of the $\ln\left(\frac{3}{2}\right)$. ◦

10.2 The Natural Logarithmic and Exponential Functions Again

In Chap. 6 we defined the exponential function then showed that its inverse exists—this is natural logarithmic function. An alternative way is to define the natural logarithmic function then show that its inverse exists—this is the exponential function. Here we outline the latter approach, made possible by the Fundamental Theorem of Calculus (Theorem 10.1).

Let $a, b > 0$. The Power Rule for Rational Powers does not apply to $\int_a^b x^n\,dx$ when $n = -1$, but the integral still makes sense for $n = -1$. Therefore we can *define* the **natural logarithmic function** $\ln(x)$ by

$$\ln(x) = \int_1^x \frac{1}{t}\,dt, \quad \text{for } x > 0.$$

The integrand is positive and so $\ln(x)$ is an increasing function, with $\ln(x) < 0$ for $x \in (0, 1)$, $\ln(1) = 0$, and $\ln(x) > 0$ for $x \in (1, \infty)$. By its very definition, $\ln(x)$ is differentiable by the Fundamental Theorem (Theorem 10.1), and that theorem gives

$$(\ln(x))' = \frac{1}{x} \quad \text{for } x > 0.$$

For $x < 0$ we have $\ln(|x|) = \ln(-x)$, and so the Chain Rule gives

$$(\ln(|x|))' = \frac{1}{x} \quad \text{for } x \neq 0. \quad \text{That is, } \int \frac{1}{x}\,dx = \ln(|x|) + C.$$

Then further by the Chain Rule, for differentiable functions u which are never zero,

$$(\ln|u(x)|)' = \frac{u'(x)}{u(x)}.$$

That is,

$$\int \frac{u'(x)}{u(x)}\, dx = \ln(|u(x)|) + C\,.$$

Lemma 10.15. *The natural logarithmic function* $\ln(x)$ *has the following properties:*

(i) $\ln(ab) = \ln(a) + \ln(b)$ *for* $a, b > 0$,
(ii) $\ln(a/b) = \ln(a) - \ln(b)$ *for* $a, b > 0$,
(iii) $\ln(a^r) = r\ln(a)$ *for* $a > 0$ *and* $r \in \mathbf{Q}$,
(iv) $\ln(x)$ *is a strictly increasing function,*
(v) $\ln(x) \to \infty$ *as* $x \to \infty$,
(vi) $\ln(x) \to -\infty$ *as* $x \to 0^+$.

Proof. For (i), observe that for $a > 0$ we have $(\ln(ax))' = \frac{a}{ax} = \frac{1}{x} = (\ln(x))'$, and so $\ln(ax) = \ln(x) + C$ for some $C \in \mathbf{R}$, by Corollary 5.7. To determine C, set $x = 1$. Now set $x = b$.
For (ii), write $\ln(a) = \ln(\frac{a}{b}b)$ and apply (i).
For (iii), by the Chain Rule and the Power Rule for Rational Powers,

$$(\ln(x^r))' = \frac{1}{x^r}(x^r)' = \frac{1}{x^r}(rx^{r-1}) = \frac{r}{x} = r(\ln(x))' = (r\ln(x))'.$$

Therefore $\ln(x^r) = r\ln(x) + C$ for some $C \in \mathbf{R}$, by Corollary 5.7. To determine C, set $x = 1$. Now set $x = a$.
For (iv) we have observed already that $\frac{1}{x} > 0$ for $x > 0$ and so $\ln(x)$ is in fact *strictly* increasing.
For (v), notice that $\ln(2^n) = n\ln(2)$. Now $\ln(x)$ is increasing, so $\ln(x) > n\ln(2)$ for $x > 2^n$. Therefore, since $\ln(2) > 0$, we can make $\ln(x)$ as large as we please, by taking x large.
For (vi), we simply write $\ln(x) = -\ln(1/x)$ and appeal to (v). □

Since $\ln(x)$ is continuous (it is differentiable) and $\ln(1) = 0$, by Lemma 10.15 part (v) and the Intermediate Value Theorem (Theorem 3.17) we may *define* the number $\mathrm{e} > 1$ as that number for which $\ln(\mathrm{e}) = 1$. That is,

$$\ln(\mathrm{e}) = \int_1^{\mathrm{e}} \frac{1}{t}\, dt = 1.$$

This number is unique, by Rolle's Theorem (Theorem 5.1) and Lemma 10.15 part (iv). Then by Lemma 10.15 parts (iv)–(vi), we see that $f(x) = \ln(x)$ has an inverse function, defined on $(-\infty, \infty)$, with range $(0, \infty)$. It is denoted by $\exp(x)$. Since $\ln(\mathrm{e}) = 1$, we have $\exp(1) = \mathrm{e}$. And since $\ln(1) = 0$, we have $\exp(0) = 1$.

Now any property of $\ln(x)$ gives rise to a property of $\exp(x)$, since the latter is the inverse of the former. We list those, in order, which correspond to the properties listed in Lemma 10.15. We leave their proofs as an exercise.

Lemma 10.16. *The function* $\exp(x)$ *has the following properties:*

(i) $\exp(a+b) = \exp(a)\exp(b)$ *for* $a, b \in \mathbf{R}$,
(ii) $\exp(a-b) = \exp(a)/\exp(b)$ *for* $a, b \in \mathbf{R}$,
(iii) $(\exp(a))^r = \exp(ar)$ *for* $a \in \mathbf{R}$ *and* $r \in \mathbf{Q}$,
(iv) $\exp(x)$ *is a strictly increasing function,*
(v) $\exp(x) \to \infty$ *as* $x \to \infty$,
(vi) $\exp(x) \to 0$ *as* $x \to -\infty$.

Proof. This is Exercise 10.43. □

We now show that $\exp(t)$ is continuous at every $t \in \mathbf{R}$. (Or we could apply Exercise 4.16.) First, using Lemma 10.16 (i) or (ii), observe that

$$\exp(t) - \exp(s) = \exp(t)\,[1 - \exp(s-t)]\,.$$

Therefore, since $\exp(0) = 1$, it suffices to show that $\exp(t)$ is continuous at $t = 0$. We use the inequality (10.1), which we obtained using integrals in Example 10.7:

$$\ln(t) \le t - 1 \quad \text{for} \quad t > 0\,.$$

Replacing t with $\exp(t)$ in this inequality, we get

$$1 + t \le \exp(t) \quad \text{for} \quad t \in \mathbf{R}\,.$$

Notice that $\exp(t)$ is increasing, so

$$1 + t \le \exp(t) \le 1 \quad \text{for} \quad t \le 0.$$

This shows that $\exp(t)$ is continuous from the left at $t = 0$.

Now replacing t with $1/t$ in (10.1) we get

$$1 - \frac{1}{t} < \ln(t) \quad \text{for} \quad t > 0\,.$$

And replacing t with $\exp(t)$ here, we get

$$\exp(t) < 1 + t\exp(t) \quad \text{for} \quad t > 0\,.$$

Again $1 + t \le \exp(t)$ and $\exp(t)$ is increasing, so

$$1 + t \le \exp(t) < 1 + t\exp(t) \le 1 + t\mathrm{e} \quad \text{for} \quad 0 < t \le 1\,.$$

This shows that $\exp(t)$ is continuous from the right at $t = 0$. So $\exp(t)$ is indeed continuous at $t = 0$.

In view of Lemma 10.15 (iii), $a^r = \exp(\ln(a^r)) = \exp(r\ln(a))$ for any $a > 0$ and any rational number r. But since $\exp(x)$ is continuous, for any $a > 0$ and any *real* number x it is reasonable to *define*

$$a^x = \exp(x\ln(a)).$$

(Then Lemma 10.16 (iii) extends to $(\exp(a))^b = \exp(ab)$ for $a, b \in \mathbf{R}$.) Then taking $a = \exp(1) = \mathrm{e}$, we get

$$\mathrm{e}^x = \exp(x\ln(\mathrm{e})) = \exp(x).$$

This being the case, it is most customary to denote the inverse of $\ln(x)$ by e^x instead of $\exp(x)$.

This function is called the **exponential function**. It satisfies

$$\mathrm{e}^{\ln(x)} = x \quad \text{for } x > 0 \qquad \text{and} \qquad \ln(\mathrm{e}^x) = x \quad \text{for } x \in \mathbf{R}.$$

The graphs of $\ln(x)$ and e^x are shown in **Fig. 10.1**.

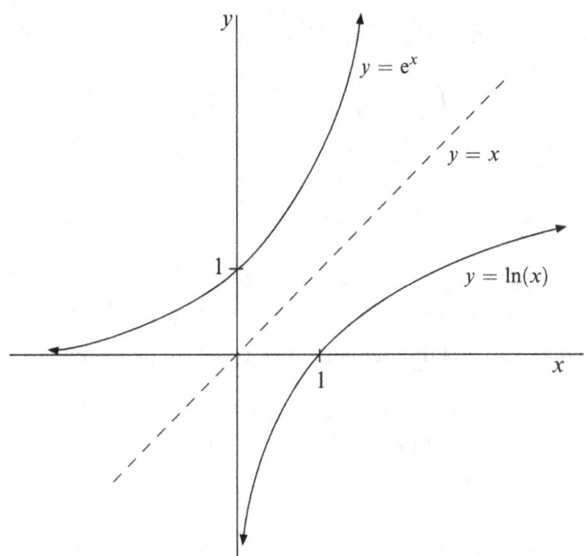

Fig. 10.1 The graphs of $y = \ln(x)$ and its inverse, $y = \mathrm{e}^x$. Each is the graph of the other, reflected in the line $y = x$

The relationship $\ln(\mathrm{e}^x) = x$ seems to imply, by the Chain Rule, that $\frac{1}{\mathrm{e}^x}(\mathrm{e}^x)' = 1$ and so $(\mathrm{e}^x)' = \mathrm{e}^x$. But this only shows that *if* $(\mathrm{e}^x)'$ exists, then it equals e^x. One really must show first that $(e^x)'$ exists. This can be done using Exercise 4.16, but here we do so more directly. We show that e^x is differentiable at every $x \in \mathbf{R}$, and we find its derivative while doing so.

In (10.1), we replace x with each of e^{x-y} and e^{y-x}, to get

$$1 + (x - y) \le e^{x-y}$$

and

$$1 + (y - x) \le e^{y-x}.$$

Taken together, these give (for $x \le y$, say):

$$e^x < \frac{e^y - e^x}{y - x} \le e^y.$$

Therefore, since e^t is continuous, letting $y \to x$ (or $x \to y$), we get $(e^x)' = e^x$ for $x \in \mathbf{R}$.

Then for differentiable functions u the Chain Rule gives

$$(e^{u(x)})' = e^{u(x)}\, u'(x).$$

That is, by the Fundamental Theorem (Theorem 10.1),

$$\int e^{u(x)} u'(x)\, dx = e^{u(x)} + C.$$

Exercises

10.1. Let f be continuous on $[a, b]$, and let $A = \frac{1}{b-a} \int_a^b f(x)\, dx$. Apply Rolle's Theorem (Theorem 5.1) and the Fundamental Theorem (Theorem 10.1) to

$$h(x) = \int_a^x (f(t) - A)\, dt$$

to prove the Average Value Theorem (Theorem 9.10).

10.2. Let f and g be continuous on $[a, b]$ with $\int_a^b f(x)\, dx = \int_a^b g(x)\, dx$. Show that there is $c \in [a, b]$ such that $f(c) = g(c)$.

10.3. Show that $\int_0^x (x - u) f(u)\, du = \int_0^x \int_0^u f(t)\, dt\, du$, for f continuous on $\mathbf{R}$. Hint: Differentiate each side with respect to x.

10.4. Assume the hypotheses which yield Cauchy's Mean Value Theorem for Integrals (Example 10.3). Show how, if we assume further that one of the two functions is never zero, we can obtain the Mean Value Theorem for Integrals (Theorem 9.14).

10.5. [41] Let f and g be continuous on $[a,b]$, with g nonnegative and not identically zero. Set

$$F(x) = \int_a^x f(t)g(t)\,dt \quad \text{and} \quad G(x) = \int_a^x g(t)\,dt.$$

Apply Rolle's Theorem (Theorem 5.1) and the Fundamental Theorem (Theorem 10.1) to

$$H(x) = F(x)G(b) - F(b)G(x)$$

to obtain the Mean Value Theorem for Integrals (Theorem 9.14). (The function $H(x)$ is $\pm$ twice the area of the triangle determined by the points $(0,0)$, $(F(x), F(b))$, and $(G(x), G(b))$. See Exercise 1.12.)

10.6. [51] Let f be continuous on $[1,2]$, with $\int_1^2 f(x)\,dx = 0$. Show that there is $c \in (1,2)$ such that

$$\int_c^2 f(t)\,dt = cf(c).$$

Hint: Consider

$$F(x) = x\int_1^2 f(t)\,dt.$$

10.7. Here is a slightly different proof of Part (i) of the Fundamental Theorem (Theorem 10.1). **(a)** Again set $F(x) = \int_a^x f(t)\,dt$, and show that

$$\left|\frac{F(x+h) - F(x)}{h} - f(x)\right| = \frac{1}{|h|}\left|\int_x^{x+h} (f(t) - f(x))\,dt\right|.$$

(b) Now use the fact that f is continuous.

10.8. Here is another proof of Part (i) of Fundamental Theorem (Theorem 10.1).

(a) Again set $F(x) = \int_a^x f(t)\,dt$. Use the Extreme Value Theorem (Theorem 3.23) to show that there are t_m and t_M (depending on x and h) such that $f(t_m) \le f(t) \le f(t_M)$ for all $x \in [x, x+h]$.
(b) Show that $hf(t_m) \le F(x+h) - F(x) \le hf(t_M)$, then use the continuity of f.

10.9. [6, 9] Fill in the details of the following proof of Part (ii) of the Fundamental Theorem (Theorem 10.1), which does not rely on Part (i). With f' continuous on $[a,b]$, let $a = x_0 < x_1 < \cdots < x_{n-1} < x_N = b$, with $x_{j+1} - x_j = \frac{(b-a)}{N}$ for each j.

(a) Verify that

$$f(b) - f(a) = \sum_{j=1}^{N} \left(f(x_j) - f(x_{j-1})\right).$$

(b) Apply the Mean Value Theorem (Theorem 5.2) to each term in the sum.
(c) Let $N \to \infty$ in a suitable way.

10.10. Let $f(t) = 0$ for $t < 0$, and $f(t) = 1$ for $t \ge 0$.
(a) For $a < 0$, compute $F(x) = \int_a^x f(t)\,dt$.
(b) Is F differentiable?
(c) How, if at all, does this fit into the Fundamental Theorem (Theorem 10.1)?

10.11. [16] Use the Fundamental Theorem (Theorem 10.1) to show that if f is continuous on **R** and periodic with period T, then $\int_a^{a+T} f(x)\,dx$ is independent of a. Conclude that $\int_a^{a+T} f(x)\,dx = \int_0^T f(x)\,dx$.

10.12. Let f be a function defined (for simplicity) on all of **R**. Then f is an **odd function** if $f(-x) = -f(x)$ for all x and f is an **even function** if $f(-x) = f(x)$ for all x.

(a) Show that x, x^3 and $\sin(x)$ are odd, while $1, x^2$, and $\cos(x)$ are even.
(b) Use the Fundamental Theorem (Theorem 10.1) to show that if f is odd and continuous then $\int_{-a}^{a} f(x)\,dx = 0$ for all a.
(c) Use the Fundamental Theorem (Theorem 10.1) to show that if f is even and continuous then $\int_{-a}^{a} f(x)\,dx = 2\int_0^a f(x)\,dx$ for all a.
(d) Show that any function defined on **R** can be written as the sum of an odd function and an even function. (For example, $e^x = \sinh(x) + \cosh(x)$.)

10.13. [18] Show that

$$\lim_{n\to\infty} \left(\int_0^1 (1+x^n)^n\,dx \right)^{1/n} = 2.$$

Hint: First show that $\int_0^1 (1+x^n)^n\,dx \le 2^n$. Then use the AGM Inequality (Theorem 2.10) to show that $\int_0^1 (1+x^n)^n\,dx \ge \int_0^1 2^n x^{n^2/2}\,dx \ge 2^n \frac{2}{n^2+2}$.

10.14. [5] We have seen that the Harmonic series diverges, so for each $n = 1, 2, 3, \ldots$ there is a least positive integer a_n such that

$$\frac{1}{n} + \frac{1}{n+1} + \frac{1}{n+2} + \cdots + \frac{1}{a_n} > 1.$$

Show that $\frac{a_n}{n} \to \mathrm{e}$ as $n \to \infty$.

10.15. [10, 39]

(a) Let $a > 0$. Show that

$$\lim_{n\to\infty}\left(\frac{1}{n} + \frac{1}{n+a} + \frac{1}{n+2a} + \cdots + \frac{1}{n+(n-1)a}\right) = \frac{1}{a}\ln(1+a).$$

(b) Find

$$\lim_{n\to\infty}\left(\frac{1}{n} + \frac{2}{n} + \frac{3}{n} + \cdots + \frac{1}{2n-1}\right).$$

(c) Let $a > 1$. Find

$$\lim_{n\to\infty}\left(\frac{1}{n} + \frac{1}{n+1} + \frac{1}{n+2} + \cdots + \frac{1}{n+na}\right).$$

10.16. [14] Let $a > 0$. Find

$$\lim_{n\to\infty}\prod_{j=1}^{n}\left(a + \frac{j-1}{n}\right)^{1/n}.$$

Hint: For $x > 0$, $\left(x\ln(x) - x\right)' = \ln(x)$, and so $\int \ln(x)\,dx = x\ln(x) - x + C$.

10.17. [55] Consider the function

$$w_r(x) = \int_0^x t^{r-1}\,dt = \begin{cases} \dfrac{x^r - 1}{r} & \text{if } r \ne 0 \\ \ln(x) & \text{if } r = 0. \end{cases}$$

(a) Show that $w_r(x)$ increases with r.
(b) Use (a) to show that for $\alpha > 1$,

$$(1+x)^\alpha \geq 1+\alpha x \quad \text{for } x > -1,$$

and that this inequality persists for $\alpha < 0$, and is reversed for $0 < \alpha < 1$. Hint: Substitute $x = t+1$ and use (a) for the two cases $r \leq 1$ and $1 \leq r$.

This is the improved Bernoulli's Inequality from Example 6.12, extended as in Exercise 6.5.

10.18. [55] (cf. Exercises 8.48 and 9.51.) For f continuous on $[0,1]$ and $r > 0$, define the Power Mean M_r by

$$M_r = \left(\int_0^1 f(t)^r \, dt \right)^{1/r}.$$

Here we show that $M_r < M_s$ for $r < s$. Consider the function

$$w_r(x) = \int_0^x t^{r-1} \, dt = \begin{cases} \dfrac{x^r - 1}{r} & \text{if } r \neq 0 \\ \ln(x) & \text{if } r = 0. \end{cases}$$

(a) Show that $w_r(x)$ increases with r.
(b) Show that $\int_0^1 w_r(\frac{f(t)}{M_r}) \, dt = 0$.
(c) Conclude that

$$0 = \int_0^1 w_r\left(\frac{f(t)}{M_r}\right) dt < \int_0^1 w_s\left(\frac{f(t)}{M_r}\right) dt \quad \text{for } r < s.$$

(d) Deduce the desired result.
(e) Bonus: What is the natural thing to take as the definition for M_0 ?

10.19. Here is another way to obtain inequality (6.7): $\ln(x) \leq x-1$, for $x > 0$.

(a) Verify that

$$\big(t - \ln(t)\big)' = \frac{t-1}{t},$$

and so

$$\int_1^x \frac{t-1}{t}\,dt = x - \ln(x) - 1.$$

(b) Consider $x \geq 1$ and $0 < x < 1$ and show that either way, $x - \ln(x) - 1 \geq 0$.

10.20. **(a)** Find the length of the curve $y = \frac{x^3}{6} + \frac{1}{2x}$, from $x = 1$ to $x = 2$.

(b) Find the length of the curve $y = \frac{x^2}{8} - \ln x$, from $x = 1$ to $x = 4$.

10.21. [15]

(a) Show that

$$\int \csc(x)\,dx = \ln|\csc(x) - \cot(x)| + C.$$

(b) Use (a) to find the length of the curve $y = \frac{1}{2}\ln(\sin(2x))$ from $x = \pi/6$ to $x = \pi/3$.

10.22. [13] Let f be defined on $[a, b]$ with f' continuous there, and let P and Q be points on the graph of f. Denote by $L(P, Q)$ the length of the curve $y = f(x)$ from P to Q and denote by $D(P, Q)$ the length of the chord from P to Q. Show that

$$\lim_{Q\to P} \frac{L(P, Q)}{D(P, Q)} = 1.$$

Hint: First take $[a, b] = [0, 1]$, $f(0) = 0$, and $P = (0, 0)$.

10.23. [53] Let $0 < a < \mathrm{e}$. **(a)** Show that $a < \mathrm{e}^x$ on the interval $(\ln(a), 1)$.
(b) Show that

$$x - \ln(a) < \mathrm{e} - x \quad \text{for} \quad 0 < a < \mathrm{e}.$$

10.24. [49] Here is another way to show that

$$\mathrm{e} = \sum_{n=0}^{\infty} \frac{1}{n!}.$$

Let $0 < t \leq 1$.

(a) Show that there is $c \in (0, t)$ such that

$$\frac{\mathrm{e}^t - 1}{t} = \mathrm{e}^c.$$

(b) Conclude that $1 + et > e^t > 1 + t$.

(c) Integrate from 0 to x to show that

$$1 + x + \frac{ex^2}{2} > e^x > 1 + x + \frac{x^2}{2}.$$

(d) Continue, showing that

$$1 + x + \frac{x^2}{2} + \frac{x^3}{3!} + \cdots + \frac{ex^{n+1}}{(n+1)!} > e^x > 1 + x + \frac{x^2}{2} + \frac{x^3}{3!} + \cdots + \frac{x^{n+1}}{(n+1)!}.$$

(e) Now set $x = 1$, and let $n \to \infty$.

10.25. Let $x > 0$. Since $e^x - 1 = \int_0^x e^t\,dt \geq \int_0^x 1\,dt = x$ we have (again) $e^x \geq 1 + x$.

(a) Continue this, showing that

$$e^x \geq 1 + x + \frac{x^2}{2!} + \frac{x^3}{3!} + \cdots + \frac{x^n}{n!} \quad \text{for } x > 0.$$

(b) Show that if n is even this inequality persists for $x \geq 0$, but is reversed for $x \leq 0$.

10.26. [4, 30] Let $x > 0$. Since $\cos(t) \leq 1$, we have $\int_0^x \cos(t)\,dt \leq \int_0^x dt$. That is, $\sin(x) \leq x$. Integrating again we get $1 - \cos(x) \leq x^2/2 \leq 1$.

(a) Continue this, showing that

$$x - \frac{x^3}{3!} \leq \sin(x) \leq x$$

$$1 - \frac{x^2}{2!} + \frac{x^4}{4!} - \frac{x^6}{6!} \leq \cos(x) \leq 1 - \frac{x^2}{2!} + \frac{x^4}{4!}$$

$$x - \frac{x^3}{3!} + \frac{x^5}{5!} - \frac{x^7}{7!} \leq \sin(x) \leq x - \frac{x^3}{3!} + \frac{x^5}{5!}, \quad \text{etc.}$$

(b) Show that for $x < 0$ these inequalities persist for $\cos(x)$ and are reversed for $\sin(x)$. (We remark that sine is an odd function: $\sin(-u) = -\sin(u)$, and cosine is an even function: $\cos(-u) = \cos(u)$.)

10.27. [22] Observe that $\frac{1}{t^2} \leq \frac{1}{t} \leq 1$ for $t \geq 1$, and $1 \leq \frac{1}{t} \leq \frac{1}{t^2}$ for $0 < t \leq 1$.

(a) Integrate these from 1 to $1 + x$, (for $x > -1$) to obtain

$$\frac{x}{1+x} \leq \ln(1+x) \leq x.$$

(b) Integrate this from 0 to y (for $y > 0$) to obtain, after some rearrangement,

$$y - \ln(1+y) \leq \ln(1+y) + y\ln(y) - y \leq \frac{y^2}{2}.$$

(c) Show that the inequalities in (b) can be rearranged to improve the inequality $\frac{y}{1+\frac{1}{2}y} \leq \ln(1+y)$ for $y \geq 0$. (This was in Exercise 6.23.)

10.28. [20, 28] Let $x \in \mathbf{R}$ and $\varepsilon > 0$. Suppose that f is defined on $[x-\varepsilon, x+\varepsilon]$, with $f^{(3)}$ continuous there. Show that

$$f'(x) = \lim_{h\to 0}\left(\frac{3}{2h^3}\int_{-h}^{h} tf(x+t)\,dt\right).$$

10.29. [45] Suppose that f has two continuous derivatives on $[a,b]$, with f' and f'' each positive there. Set

$$M_f(a,b) = \int_a^b \frac{xf'(x)}{f(b)-f(a)}\,dx.$$

(a) Show that $M_f(a,b)$ is a *mean*: $a \leq b$ implies $a \leq M_f(a,b) \leq b$.
(b) Show that

$$\frac{1}{b-a}\int_a^b f(x)\,dx < f(M_f(a,b)).$$

(c) Use Chebyshev's Integral Inequality (Exercise 9.44) to show that $A = \frac{a+b}{2} < M_f$.

10.30. [11] Let f be defined on $(0,\infty)$ with f'' continuous and positive. For $0 < a < b$, set

$$S_f(a,b) = \frac{\int_a^b xf''(x)\,dx}{\int_a^b f''(x)\,dx}.$$

(a) Show that $S_f(a, b)$ is a *mean*: $a < S_f(a, b) < b$.
(b) Now let $f(x) = x^r$. Compute $S_f(a, b)$ for $r = -1,\ 1/2$, and 2. Do these look familiar?
(c) Compute $S_f(a, b)$ for $f(x) = x \ln(x)$. Does this look familiar?

Hint: For $x > 0$, $(x \ln(x) - x)' = \ln(x)$, and so $\int \ln(x)\, dx = x \ln(x) - x + C$.

10.31. [37] In Example 8.14 we met **Jordan's Inequality**:

$$\sin(x) \geq \frac{2}{\pi} x \quad \text{for } x \in [0, \pi/2].$$

Fill in the details of the following way to obtain this inequality using integrals.

(a) Show that

$$\sec^2(t) \geq 1 \quad \text{for } t \in [0, \pi/2).$$

(b) Conclude that

$$\tan(x) \geq x \quad \text{for } x \in [0, \pi/2).$$

(c) Show that (b) implies that the function

$$g(x) = \begin{cases} \dfrac{\sin x}{x} & \text{if } x \neq 0 \\ 1 & \text{if } x = 0 \end{cases}$$

is decreasing on $(0, \pi/2)$.
(d) Use the fact that g is continuous everywhere to deduce Jordan's Inequality.

10.32. [50] Let f be continuous on $[0, 1]$, with $\int_0^1 f(x)\, dx = 0$. Show, as follows, that there is $c \in (0, 1)$ such that

$$\int_0^c (x + x^2) f(x)\, dx = c^2 f(c).$$

Let $F(x) = x^2 f(x) - \int_0^x (t + t^2) f(t)\, dt$. Show that F is continuous and that F changes signs on $[0, 1]$, then use Bolzano's Theorem (Theorem 3.7). (See Exercise 10.33 for another approach.)

10.33. [50] Let f be continuous on $[0, 1]$, with $\int_0^1 f(x)\,dx = 0$. Show, as follows, that there is $c \in (0, 1)$ such that

$$\int_0^c (x + x^2) f(x)\,dx = c^2 f(c).$$

(a) Show that there is $a \in (0, 1)$ such that $\int_0^a x f(x)\,dx = 0$.

Hint: Consider $F(x) = x \int_0^x f(t)\,dt - \int_0^x t f(t)\,dt$ and use Flett's Mean Value Theorem (Theorem 7.4).

(b) Show that there is $b \in (0, a)$ such that $\int_0^b x f(x)\,dx = b f(b)$.

Hint: Consider $G(x) = \mathrm{e}^{-x} \int_0^x t f(t)\,dt$.

(c) Now consider $H(x) = x \int_0^x t f(t)\,dt - \int_0^x (t + t^2) f(t)\,dt$, and apply Flett's Mean Value Theorem (Theorem 7.4). (See Exercise 10.32 for another approach.)

10.34. [27] Show, as follows, that

$$\tan(x) < \frac{\pi x}{\pi - 2x} \quad \text{for } x \in (0, \pi/2).$$

(a) Show that for $t \in (0, \pi/2)$, we have $\cos(t) > 1 - 2t/\pi$.
(b) Conclude that for such for t we have $(\sec(t))^2 < \frac{1}{(1-2t/\pi)^2}$.
(c) Integrate both sides of (b) from 0 to x to obtain the desired result.

10.35. Show that

$$\lim_{n\to\infty} \left(\sum_{j=1}^{n} \frac{n}{n^2 + j^2} \right) = \frac{\pi}{4}.$$

10.36. [24, 25] We saw in Example 10.10 that for $-1 < x \le 1$, we have

$$\ln(1 + x) = x - \frac{1}{2}x^2 + \frac{1}{3}x^3 - \frac{1}{4}x^4 + \cdots.$$

(a) Show that for $-1 \le x < 1$, we have

$$\ln(1 - x) = -x - \frac{1}{2}x^2 - \frac{1}{3}x^3 - \frac{1}{4}x^4 - \cdots.$$

(b) Use these to find a series for $\ln(\frac{1+x}{1-x})$, for $|x| < 1$.
(c) Set $x = 1/(2n)$ in the series in (b), then replace n with $n + 1/2$ to show that

$$e < \left(1 + \frac{1}{n}\right)^{n+1/2},$$

which we obtained in Sect. 6.3.
(d) Solve $t = \frac{1+x}{1-x}$ for x and use the series in (b) to show that for $t > 1$,

$$\ln(t) = 2\left(\left(\frac{t-1}{t+1}\right) + \frac{1}{3}\left(\frac{t-1}{t+1}\right)^3 + \frac{1}{5}\left(\frac{t-1}{t+1}\right)^5 + \cdots\right).$$

10.37. In summing the Alternating Harmonic series and in obtaining the Leibniz series we showed, respectively, that as $n \to \infty$,

$$\int_0^1 \frac{x^{n+1}}{1+x}\,dx \to 0 \qquad \text{and} \qquad \int_0^x \frac{t^{2n+2}}{1+t^2}\,dt \to 0\,.$$

Show these by instead using the Mean Value Theorem for Integrals (Theorem 9.14).

10.38. [32] We saw in Example 10.12 that

$$\int_0^1 \frac{x^4(1-x^4)}{1+x^2}\,dx = \frac{22}{7} - \pi\,, \quad \text{and so} \quad \pi < \frac{22}{7}.$$

Verify that $\int_0^1 x^4(1-x^4)\,dx = 1/630$ and use this to obtain the improvement

$$\frac{22}{7} - \frac{1}{630} < \pi < \frac{22}{7} - \frac{1}{1{,}260}.$$

10.39. [31] It is a fact that

$$\begin{aligned}\frac{x^8(1-x)^8(25+816x^2)}{1+x^2} &= 12{,}656x^4 - 12{,}656x^2 - 12{,}656x^6 + 12{,}681x^8 - 200x^9 \\ &- 11{,}165x^{10} - 7{,}728x^{11} + 35{,}763x^{12} - 39{,}368x^{13} + 22{,}057x^{14} \\ &- 6{,}528x^{15} + 816x^{16} - \frac{12{,}656}{x^2+1} + 12{,}656,\end{aligned}$$

which makes

$$\int_0^1 \frac{x^8(1-x)^8(25+816x^2)}{1+x^2}\, dx$$

easy (but still horribly tiresome) to evaluate. In fact,

$$\frac{1}{3{,}164}\int_0^1 \frac{x^8(1-x)^8(25+816x^2)}{1+x^2}\, dx = \frac{355}{113} - \pi.$$

Therefore $\pi < 355/113$. Can you get better estimates, as in Exercise 10.38 ?

10.40. [8]

(a) Justify the following:

$$\begin{aligned}\ln(2) &= 1 - \frac{1}{2} + \frac{1}{3} - \frac{1}{4} + \frac{1}{5} - \frac{1}{6} + \cdots \\ &= \left(1 - \frac{1}{2}\right) + \left(\frac{1}{3} - \frac{1}{4}\right) + \left(\frac{1}{5} - \frac{1}{6}\right) + \cdots \\ &= \frac{1}{1\cdot 2} + \frac{1}{3\cdot 4} + \frac{1}{5\cdot 6} + \cdots .\end{aligned}$$

(b) Justify the following:

$$\begin{aligned}\frac{\pi}{4} &= 1 - \frac{1}{3} + \frac{1}{5} - \frac{1}{7} + \frac{1}{9} - \frac{1}{11} + \cdots \\ &= \left(1 - \frac{1}{3}\right) + \left(\frac{1}{5} - \frac{1}{7}\right) + \left(\frac{1}{9} - \frac{1}{11}\right) + \cdots \\ &= 2\left(\frac{1}{1\cdot 3} + \frac{1}{5\cdot 7} + \frac{1}{9\cdot 11} + \cdots\right).\end{aligned}$$

10.41. [17]

(a) Verify that for $k \neq 0$, $\displaystyle\int_1^x t^{k-1}\, dt = \frac{x^k - 1}{k}$.

(b) Evaluate $\lim_{k\to 0} \frac{x^k-1}{k}$. (The reader may find this comforting.)

10.42. **(a)** Set $x = 1/\sqrt{3}$ in the Leibniz series (Example 10.13) to obtain a series for $\pi/6$.

(b) Set $\alpha = 1/2$ and $\beta = 1/3$ in the trigonometric identity

$$\tan(\alpha + \beta) = \frac{\tan\alpha + \tan\beta}{1 - \tan\alpha\tan\beta}$$

to obtain **Euler's formula**

$$\frac{\pi}{4} = \arctan\left(\frac{1}{2}\right) + \arctan\left(\frac{1}{3}\right).$$

(c) Use Euler's formula to obtain another series for $\pi/4$, which converges much more quickly than does the Gregory-Leibniz series (Example 10.13).

Note: **Machin's formula**, obtained in 1706 by John Machin,

$$\frac{\pi}{4} = 4\arctan\left(\frac{1}{5}\right) - \arctan\left(\frac{1}{239}\right)$$

yields a series which converges even more quickly. William Shanks, around 1873, used Machin's formula to compute π to 707 decimal places. It took him 15 years. It was discovered in the 1950s with the aid of contemporary computers that his computation was incorrect at the 528th decimal place.

10.43. Prove Lemma 10.16.

10.44. [12, 33]

(a) Show that $f(x) = c\ln(x)$, for $x > 0$ and arbitrary $c \in \mathbf{R}$, is the *only* continuous function which satisfies

$$f(xy) = f(x) + f(y) \quad \text{for} \quad x, y > 0.$$

(b) Find all continuous functions f which satisfy

$$f(xy) = yf(x) + xf(y) \quad \text{for} \quad x, y > 0.$$

10.45. [34] Let $x > 0$. **(a)** Use $\int_x^{x+1} \frac{1}{t}\,dx < \frac{1}{x}$ and $\int_{x+1}^{x+2} \frac{1}{t}\,dx > \frac{1}{x+2}$ to show that

$$\ln\left(\frac{x+1}{x}\right)\frac{1}{x+2} < \ln\left(\frac{x+2}{x+1}\right)\frac{1}{x}.$$

(b) Show that $f(x) = \frac{\ln\left(\frac{x+2}{x+1}\right)}{\ln\left(\frac{x+1}{x}\right)}$ is an increasing function on $(0, \infty)$.

10.46. [3] Here's another way to obtain inequalities (6.1):

$$\left(1+\frac{1}{n}\right)^n < \mathrm{e} < \left(1+\frac{1}{n}\right)^{n+1} \quad \text{for } n = 1, 2, 3, \ldots.$$

Let $r = 1 + \frac{1}{n}$. **(a)** For the left side, show that $\sum_{j=1}^{n} \frac{1}{r^{j-1}}(r^j - r^{j-1}) = 1$ and that

$$\sum_{j=1}^{n} \frac{1}{r^{j-1}}(r^j - r^{j-1}) > \int_1^{r^n} \frac{1}{x}\,dx.$$

(b) For the right side, show that $\sum_{j=1}^{n+1} \frac{1}{r^j}(r^j - r^{j-1}) = 1$ and that

$$\sum_{j=1}^{n+1} \frac{1}{r^j}(r^j - r^{j-1}) < \int_1^{r^{n+1}} \frac{1}{x}\,dx.$$

10.47. [38]

(a) Use $(x\ln(x) - x)' = \ln(x)$ to show that $\lim_{a\to 0+} \int_a^1 \ln(x)\,dx = -1$.

(b) Prove that

$$\lim_{n\to\infty} \frac{\sqrt[n]{n!}}{n} = \frac{1}{\mathrm{e}}.$$

Hint: Show that $\ln\left(\frac{\sqrt[n]{n!}}{n}\right) = \sum_{k=1}^{n} \frac{1}{n}\ln\left(\frac{k}{n}\right)$, and recognize this as a Riemann sum related to (a).

(c) Denote by A_n the Arithmetic Mean and by G_n the Geometric Mean, of the first n natural numbers. Show that the result in (b) is the same as

$$\lim_{n\to\infty} \frac{G_n}{A_n} = \frac{2}{\mathrm{e}}.$$

Other methods can be found in [29,47,54], and are generalized in [26] and [48]. (Also, cf. Exercise 6.12.)

10.48. Fill in the details of the following argument, which shows how to obtain Jensen's Inequality (Theorem 8.17) from Steffensen's Inequality (Exercise 9.52): *Let* f *and* g *be continuous on* $[a, b]$, *with* f *increasing*, $0 \le g \le 1$, *and* $\lambda = \int_a^b g(x)\,dx$. *Then*

$$\int_a^{a+\lambda} f(x)\,dx \le \int_a^b f(x)g(x)\,dx.$$

(a) Let $a = x_0 \le x_1 \le x_2 \le \cdots \le x_n$ and let $w_1, \ldots, w_n$ be positive, with $\sum_{j=1}^{n} w_j = 1$. Define g on $[a, x_n]$ via

$$g = \sum_{j=1}^{k} w_j \quad \text{on } [x_{k-1}, x_k], \quad \text{for } k = 1, 2, \cdots, n.$$

Verify that $0 \le g \le 1$ and that

$$\lambda = \int_0^{x_n} g(t)\,dt = \sum_{j=1}^{n} x_j w_j.$$

(b) If f is convex then $f'' \ge 0$, so that f' is increasing. Apply Steffensen's Inequality to f' and g as above, and use the Fundamental Theorem (Theorem 10.1).

10.49. [43,44,52]

(a) Prove the following **integral analogue of Flett's Mean Value Theorem** (Theorem 7.4). Let f be continuous on $[a, b]$ with $f(a) = f(b)$. Show that there is $c \in (a, b)$ such that

$$f(c) = \frac{1}{c-a} \int_a^c f(x)\,dx.$$

Hint: Consider $F(t) = (t-a)f(t) - \int_a^t f(x)\,dx.$

(b) Draw a picture which shows what this result says geometrically.

(c) Apply the result to the function $g(x) = f(x) - \frac{f(b)-f(a)}{b-a}(x-a)$ to obtain a version which does not require that $f(a) = f(b)$.

10.50. [21] Let f be continuously differentiable on $[a, b]$ with $f'(a) \ne 0$. Let $c \in [a, b]$ be as given by the Average Value Theorem (Theorem 9.10):

$$f(c) = \frac{1}{b-a} \int_a^b f(x)\,dx.$$

(a) Show that if b is near enough to a, then c is unique.
(b) Evaluate

$$\lim_{b\to a}\frac{\int_a^b f(x)\,dx-(b-a)f(a)}{(b-a)^2}$$

two different ways, to show that

$$\lim_{b\to a}\frac{c-a}{b-a}=\frac{1}{2}.$$

(This result is generalized to the case where $f'(a)=0$ in [1, 19, 40], and in other ways in [23, 36].)
(c) Draw a picture which shows that the result in (b) is really not very surprising—remember that $f'(a)\neq 0$.

10.51. [44] Let f be continuously differentiable on $[a,b]$ with $f''(a)\neq 0$. Let $c\in[a,b]$ be as given by the Mean Value Theorem (Theorem 5.2):

$$\frac{f(b)-f(a)}{b-a}=f'(c).$$

(a) Show that if b is near enough to a, then c is unique.
(b) Use the Fundamental Theorem (Theorem 10.1) along with the result of Exercise 10.50 to show that if f'' is continuous and $f''(a)\neq 0$ then

$$\lim_{c\to a}\frac{c-a}{b-a}=\frac{1}{2}.$$

(c) Prove the result in (b) above directly—that is, without the Fundamental Theorem. (All of this is generalized considerably, in [35].)

References

1. Bao-lin, Z.: A note on the mean value theorem for integrals. Am. Math. Mon. **104**, 561–562 (1997)
2. Beigel, R.: Rearranging terms in alternating series. Math. Mag. **54**, 244–246 (1981)
3. Bird, M.T.: Approximations of e. Am. Math. Mon. **75**, 286–288 (1968)
4. Bo, D.: A simple derivation of the Maclaurin series for sine and cosine. Am. Math. Mon. **97**, 836 (1990)
5. Bobo, E.R.: A sequence related to the harmonic series. Coll. Math. J. **26**, 308–310 (1995)
6. Bond, R.A.B.: An alternative proof of the fundamental theorem of calculus. Math. Gaz. **65**, 288–289 (1981)
7. Brown, F., Cannon, L.O., Elich, J., Wright, D.G.: On rearrangements of the alternating harmonic series. Coll. Math. J. **16**, 135–138 (1985)

8. Burk, F.: Summing series via integrals. Coll. Math. J. **31**, 178–181 (2000)
9. Cunningham, F., Jr.: The two fundamental theorems of calculus. Am. Math. Mon. **72**, 406–407 (1965)
10. Cusumano, A., Diminnie, C., Havlak, K.: Problem 895. Coll. Math. J. **33**, 58–59 (2002)
11. Dietel, B.C., Gordon, R.A.: Using tangent lines to define means. Math. Mag. **76**, 52–61 (2003)
12. Ebanks, B.: Looking for a few good means. Am. Math. Mon. **119**, 658–669 (2012)
13. Eenigenburg, P.: A note on the ratio of arc length to chordal length. Coll. Math. J. **28**, 391–393 (1997)
14. Euler, R., Klein, B.G.: Problem 1178. Math. Mag. **57**, 302 (1984)
15. Ferdinands, J.: Finding curves with computable arc length. Coll. Math. J. **38**, 221–223 (2007)
16. Fettis, H.E.: Problem Q678. Math. Mag. **55**, 300, 307 (1982)
17. Finlayson, H.C.: The place of $\ln(x)$ among the powers of x. Am. Math. Mon. **94**, 450 (1987)
18. Furdui, O.: Problem Q1014. Math. Mag. **84**, 297, 303 (2011)
19. Gluskin, E.: An observation related to the integral average value theorem. Int. J. Math. Educ. Sci. Techol. **28**, 132–134 (1997)
20. Groetsch, C.W.: Lanczos' generalized derivative. Am. Math. Mon. **105**, 320–326 (1998)
21. Jacobson, B.: On the mean value theorem for integrals. Am. Math. Mon. **89**, 300–301 (1982)
22. Jahnke, T.: Those logarithmic inequalities again. Math. Gaz. **65**, 130–132 (1981)
23. Khalili, P., Vasiliu, D.: An extension of the mean value theorem for integrals. Int. J. Math. Educ. Sci. Technol. **41**, 707–710 (2010)
24. Khattri, S.K.: Three proofs of the inequality $\mathrm{e} < \left(1 + \frac{1}{x}\right)^{x+1/2}$. Am. Math. Mon. **117**, 273–277 (2010)
25. Körner, T.W.: A Companion to Analysis. American Mathematical Society, Providence (2004)
26. Kubelka, R.: Means to an end. Math. Mag. **74**, 141–142 (2001)
27. Laforgia, A., Wong, Y.: Problem 1215. Math. Mag. **59**, 119 (1986)
28. Lanczos, C.: Applied Analysis. Dover, Mineola (1988) (originally published 1956)
29. Lang, S.: A First Course in Calculus, Addison Wesley, Reading (1964)
30. Leonard, I.E., Duemmel, J.: More – and Moore – power series without Taylor's theorem. Am. Math. Mon. **92**, 588–589 (1985)
31. Lucas, S.K.: Integral proofs that $355/113 > \pi$. Aust. Math. Soc. Gaz. **32**, 263–266 (2005)
32. Lucas, S.K.: Approximations to π derived from integrals with nonnegative integrands. Am. Math. Mon. **116**, 166–172 (2009)
33. Luthar, R.S., Chernoff, P.: Problem E2329. Am. Math. Mon. **79**, 1139–1140 (1972)
34. Margolis, B., Hejhal, D.A.: Problem E2086. Am. Math. Mon. **76**, 421–422 (1969)
35. Mera, R.: On the determination of the intermediate point in Taylor's theorem. Am. Math. Mon. **99**, 56–58 (1992)
36. Mercer, P.R.: The number c in Cauchy's average value theorem. Int. J. Math. Educ. Sci. Technol. **35**, 118–122 (2004)
37. Mitrinovic, D.S.: Analytic Inequalities. Springer, Berlin/New York (1970)
38. Mumma II, C.C.: $N!$ and the root test. Am. Math. Mon. **93**, 561 (1986)
39. Penney, D.E., Vanden Eynden, T.L.: Problem 608. Coll. Math. J. **20**, 334–335 (1998)
40. Polezzi, M.: On the weighted mean value theorem for integrals. Int. J. Math. Educ. Sci. Technol. **37**, 868–870 (2006)
41. Putney, T.: Proof of the first mean value theorem of the integral calculus. Am. Math. Mon. **60**, 113–114 (1953)
42. Roy, R.: The discovery of the series formula for π by Leibniz, Gregory and Nilakantha. Math. Mag. **63**, 291–306 (1990)
43. Sahoo, P.K.: Some results related to the integral mean value theorem. Int. J. Math. Educ. Sci. Technol. **38**, 818–822 (2007)
44. Sahoo, P.K., Riedel, T.: Mean Value Theorems and Functional Equations. World Scientific, Singapore/River Edge (1998)
45. Sandor, J.: On means generated by derivatives of functions. Int. J. Math. Educ. Sci. Technol. **28**, 146–148 (1997)

46. Schaumberger, N.: $\int \sec(\theta)d\theta$. Am. Math. Mon. **68**, 565 (1961)
47. Schaumberger, N.: Alternate approaches to two familiar results. Coll. Math. J. **15**, 422–423 (1984)
48. Schaumberger, N.: A generalization of $\lim \sqrt[n]{n!}/n = \mathrm{e}^{-1}$. Coll. Math. J. **20**, 416–418 (1989)
49. Schaumberger, N.: Another proof of the formula $\mathrm{e} = \sum_{0}^{\infty} \frac{1}{n!}$. Coll. Math. J. **25**, 38–39 (1994)
50. Thong, D.V., Botsko, W.W., Chen, H.: Problem 11555. Am. Math. Mon. **119**, 704–706 (2012)
51. Thong, D.V., Plaza, A., Falcón, S.: Problem 1837. Math. Mag. **84**, 65 (2011)
52. Wayment, S.G.: An integral mean value theorem. Math. Gaz. **54**, 300–301 (1970)
53. Wilanski, A.: Problem Q676. Math. Mag. **55**, 237, 244 (1982)
54. Wilker, J.B.: Stirling ideas for freshman calculus. Math. Mag. **57**, 209–214 (1984)
55. Witkowski, A.: A new proof of the monotonicity of power means. J. Inequal. Pure Appl. Math. **5**, article 6 (2004)

Chapter 11
Techniques of Integration

Let us be resolute in prosecuting our ends, and mild in our methods of doing so.

—Claudio Aquaviva

By way of the Fundamental Theorem of Calculus (Theorem 10.1), many properties of integrals come from properties of derivatives and vice-versa. For example, the most basic *technique of integration* is to recognize the integrand as the derivative of some particular function. We saw a few examples of this sort of thing in the previous chapter. Here we focus on arguably the next two most important techniques of integration: *u-Substitution* which comes from the Chain Rule for derivatives, and *Integration by Parts* which comes from the Product Rule for derivatives.

11.1 Integration by u-Substitution

Let F be an antiderivative of f, that is $F' = f$. Then for differentiable functions u, an application of the Chain Rule gives $(F(u(x)))' = f(u(x))u'(x)$. Integrating through with respect to x (and using the Fundamental Theorem (Theorem 10.1)) we get the following technique of integration.

u-Substitution: $\displaystyle\int f(u(x))u'(x)\,dx = F(u(x)) + C, \quad \text{where} \quad F' = f.$

For definite integrals this translates to the statement

$$\int_a^b f(u(x))u'(x)\,dx = F(u(x))\Big|_a^b .$$

The general strategy for employing u-Substitution is as follows: Faced with something which resembles the left-hand side above, we look for a part of the integrand (which we will call u) whose derivative is very much like another part of the integrand—this will be the u'. At the same time we try to recognize the f, and find an antiderivative (which we have denoted by F).

P.R. Mercer, *More Calculus of a Single Variable*, Undergraduate Texts in Mathematics, DOI 10.1007/978-1-4939-1926-0_11

Example 11.1. Beginning simply, we evaluate $\int \cos(x^2+1)2x\,dx$. Observe first that $(x^2+1)' = 2x$, so we set $u = x^2+1$. Then the integral reads

$$\int \cos(x^2+1)2x\,dx = \int \cos(u(x))u'(x)\,dx.$$

Here the f is $f(u) = \cos(u)$, and so $F(u) = \int \cos(u)\,du = \sin(u) + C$. Finally then,

$$\int \cos(x^2+1)2x\,dx = \sin(x^2+1) + C. \qquad \diamond$$

Let us look again at Example 11.1, to illustrate an informal scheme for carrying out u-Substitution which is typically used in practice. We set

$$u = x^2 + 1,$$

so that

$$\frac{du}{dx} = 2x.$$

Now (this is the key part) we write this as

$$du = 2x\,dx.$$

This last step, treating $\frac{du}{dx}$ as a fraction and clearing its denominator, can be made entirely rigorous in a number of ways; see [9], for example. Here, du is called the *differential* of the function $u = x^2 + 1$. Substituting these into the integral, we get

$$\begin{aligned}\int \cos(x^2+1)2x\,dx &= \int \cos(u)\,du \\ &= \sin(u) + C \\ &= \sin(x^2+1) + C.\end{aligned}$$

We shall employ this scheme again in the next few examples.

Example 11.2. We evaluate $\int \tan(x)\,dx = \int \frac{\sin(x)}{\cos(x)}\,dx$. First we make the observation that $(\cos(x))' = -\sin(x)$, and so we set $u = \cos(x)$. Then $du = -\sin(x)dx$ and the integral reads

$$\int \tan(x)\,dx = -\int \frac{1}{u}\,du \;=\; -\ln|u| + C$$
$$= -\ln\big|\cos(x)\big| + C \;=\; \ln\big|\sec(x)\big| + C.$$

Then, for example,

$$\int_0^{\pi/4} \tan(x)\,dx = \ln\big|\sec(x)\big|\Big|_0^{\pi/4} = \ln(\sqrt{2}) = \frac{1}{2}\ln(2).$$ ◇

Example 11.3. To evaluate $\int \frac{\ln(x)}{x}\,dx$, we notice that $\big(\ln(x)\big)' = \frac{1}{x}$, and so we set $u = \ln(x)$. Then $du = \frac{1}{x}dx$ and the integral reads

$$\int \frac{\ln(x)}{x}\,dx = \int \ln(x)\frac{1}{x}\,dx = \int u\,du$$
$$= \frac{1}{2}u^2 + C = \frac{1}{2}\big(\ln(x)\big)^2 + C.$$

Then, for example,

$$\int_1^{e} \frac{\ln(x)}{x}\,dx = \frac{1}{2}\big(\ln(x)\big)^2\Big|_1^{e} = \frac{1}{2}.$$ ◇

Example 11.4. Let $n \geq 1$ be an integer. Using a trigonometric identity,

$$\int \tan^{n+2}(x)\,dx = \int \tan^n(x)\tan^2(x)\,dx = \int \tan^n(x)(\sec^2(x) - 1)\,dx$$
$$= \int \tan^n(x)\sec^2(x)\,dx - \int \tan^n(x)\,dx.$$

Then since $\big(\tan(x)\big)' = \sec^2(x)$, we set $u = \tan(x)$ to get $du = \sec^2(x)dx$. Doing this, we obtain the *reduction formula*

$$\int \tan^{n+2}(x)\,dx = \int u^n\,du - \int \tan^n(x)\,dx$$
$$= \frac{1}{n+1}u^{n+1} - \int \tan^n(x)\,dx.$$
$$= \frac{1}{n+1}\tan^{n+1}(x) - \int \tan^n(x)\,dx.$$

Then, for example, setting $I_n = \int_0^{\pi/4} \tan^n(x)\,dx$, we obtain $I_{n+2} = \frac{1}{n+1} - I_n$. And taking $n = 3$ here, for example, we get (using also the last part of Example 11.2):

$$\int_0^{\pi/4} \tan^5(x)\,dx = \frac{1}{4} - \int_0^{\pi/4} \tan^3(x)\,dx = \frac{1}{4} - \left(\frac{1}{2} - \int_0^{\pi/4} \tan(x)\,dx\right) = -\frac{1}{4} + \frac{1}{2}\ln(2).$$

◇

After a walk in the mountains, a hiker must have achieved some particular average elevation. Whether the independent variable is time or distance or perhaps something else, the hiker's average elevation should remain the same. This simple fact is the essence of the following, which describes how a definite integral transforms under a u-Substitution.

Theorem 11.5. (Change of Variables) *Let u have a continuous derivative on $[a, b]$ and let f be continuous on an interval I which contains the range of u. Then*

$$\int_a^b f(u(x))u'(x)\,dx = \int_{u(a)}^{u(b)} f(u)\,du.$$

Proof. For $x \in I$, let $F(x) = \int_{u(a)}^{x} f(t)\,dt$, and $H(x) = F(u(x))$. Then by the Chain Rule and the Fundamental Theorem (Theorem 10.1),

$$H'(x) = F'(u(x))u'(x) = f(u(x))u'(x).$$

Therefore

$$\int_{u(a)}^{u(b)} f(u)\,du = F(u(b)) - F(u(a)) = H(b) - H(a)$$

$$= \int_a^b H'(x)\,dx = \int_a^b f(u(x))u'(x)\,dx,$$

as we wanted to show. □

Example 11.6. For an integral of the form $\int \frac{g'(x)}{g(x)}\,dx$, where g' is continuous, we set $u = g(x)$, so that $du = g'(x)dx$. Therefore

$$\int \frac{g'(x)}{g(x)}\,dx = \int \frac{1}{u}\,du.$$

Then by Theorem 11.5, as a definite integral this translates to

$$\int_a^b \frac{g'(x)}{g(x)}\,dx = \int_{g(a)}^{g(b)} \frac{1}{u}\,du,$$

as long as $g(a)$ and $g(b)$ are on the same side of zero, that is, $g(a)g(b) > 0$. So for example, with $u = 1 + x^2$, we have

$$\int_0^1 \frac{2x}{1+x^2}\,dx = \int_1^2 \frac{1}{u}\,du = \ln(2).$$

Or, looking again at Example 11.2, wherein $u = \cos(x)$,

$$\int_0^{\pi/4} \tan(x)\,dx = -\int_1^{1/\sqrt{2}} \frac{1}{u}\,du = \ln\left(\sqrt{2}\right) = \frac{1}{2}\ln(2). \qquad \diamond$$

11.2 Integration by Parts

If F is an antiderivative of f, that is $F' = f$, then by the Product Rule we have $(Fg)' = F'g + Fg' = fg + Fg'$. That is, $fg = (Fg)' - Fg'$. Then integrating through, using the Fundamental Theorem (Theorem 10.1) and taking the $+C = 0$, we get the following technique of integration.

Integration by Parts:

$$\int f(x)g(x)\,dx = F(x)g(x) - \int F(x)g'(x)\,dx, \quad \text{where } F' = f.$$

For definite integrals this translates to the statement

$$\int_a^b f(x)g(x)\,dx = F(x)g(x)\Big|_a^b - \int_a^b F(x)g'(x)\,dx. \qquad (F' = f.)$$

The strategy for employing Integration by Parts is (vaguely) as follows: Part of the integrand (the f) will be integrated and the rest of it will be differentiated (the g). If some particular choice of these two parts appears to make things simpler, that is, $\int F(x)g'(x)\,dx$ is simpler than $\int f(x)g(x)\,dx$, then it is probably worth pursuing.

Example 11.7. To evaluate $\int e^x x\,dx$, we notice that differentiating e^x or integrating e^x really makes no difference as far as making things simpler goes (particularly if we choose the $+C = 0$). But differentiating x, to get 1, makes things much simpler than integrating x, which gives $(1/2)x^2$. So setting $f(x) = e^x$ and $g(x) = x$, we get

$$\int e^x x\,dx = \int f(x)g(x)\,dx = F(x)g(x) - \int F(x)g'(x)\,dx.$$

That is,

$$\int e^x x\,dx = e^x x - \int e^x 1\,dx = xe^x - e^x + C. \qquad \diamond$$

Remark 11.8. Taking the $+C = 0$ as we did in Example 11.7 is not always advantageous. See Exercise 11.30, and the proof of Theorem 11.17 below. ∘

Example 11.9. Here we evaluate $\int t\sin(t)\,dt$. Again, differentiating or integrating $\sin(t)$ makes little difference, but differentiating t makes things much simpler than integrating t. Therefore, setting $f(t) = \sin(t)$ and $g(t) = t$, we obtain

$$\int \sin(t)(t)\,dt = \int f(t)g(t)\,dt = F(t)g(t) - \int F(t)g'(t)\,dt.$$

That is,

$$\int t\sin(t)\,dt = -\cos(t)(t) + \int \cos(t)1\,dt = -t\cos(t) + \sin(t) + C. \qquad \diamond$$

Example 11.10. In Example 8.14 we met **Jordan's Inequality**:

$$\sin(x) \geq \frac{2}{\pi}x \quad \text{for } x \in [0, \pi/2].$$

In Exercise 10.31 we saw a way to obtain this using integrals. Here is another way to obtain Jordan's Inequality using integrals, which begins with Example 11.9. (This method lends itself nicely to a refinement; see Exercise 11.24.) Observe that

$$t\sin(t) \geq 0 \quad \text{for } t \in [0, \pi/2],$$

and so

$$\int_0^u t\sin(t)\,dt = \sin(u) - u\cos(u) \geq 0 \quad \text{for } u \in [0, \pi/2].$$

Therefore

$$\left(\frac{\sin(u)}{u}\right)' = \frac{u\cos(u) - \sin(u)}{u^2}$$

$$= -\frac{1}{u^2}\int_0^u t\sin(t)\,dt \le 0 \quad \text{for } u \in (0, \pi/2].$$

Then by the Fundamental Theorem (Theorem 10.1),

$$\int_x^{\pi/2} \left(\frac{\sin(u)}{u}\right)' du = \frac{\sin(u)}{u}\Big|_x^{\pi/2}$$

$$= \frac{2}{\pi} - \frac{\sin(x)}{x} \le 0 \quad \text{for } x \in (0, \pi/2],$$

which yields Jordan's Inequality. ◇

Example 11.11. To evaluate the integral $\int \ln(x)\,dx$, we have little choice: we cannot set $f(x) = \ln(x)$ because the problem itself asks for an antiderivative of $\ln(x)$. So we set $f(x) = 1$ and $g(x) = \ln(x)$, to obtain

$$\begin{aligned}\int \ln(x)\,dx &= \int 1\ln(x)\,dx = \int f(x)g(x)\,dx \\ &= F(x)g(x) - \int F(x)g'(x)x\,dx \\ &= x\ln(x) - \int x\frac{1}{x}\,dx \\ &= x\ln(x) - x + C.\end{aligned}$$

◇

Remark 11.12. Example 11.11 is an important one. And it has an equally important partner which is evaluated in a very similar way:

$$\int \tan^{-1}(x)\,dx = \frac{1}{1+x^2} + C.$$

See Exercise 11.25. ○

Example 11.13. In Sects. 6.5, 8.5, and 10.1 we summed the Alternating Harmonic series:

$$\ln(2) = 1 - \frac{1}{2} + \frac{1}{3} - \frac{1}{5} + \cdots.$$

Here is a series of *positive* terms for ln(2), which converges much more quickly. Integrating by parts repeatedly (the first step has $f(x) \equiv 1$ and $g(x) = 1/(1+x)$), we get

$$\begin{aligned}
\ln(2) &= \int_0^1 \frac{1\,dx}{1+x} = x\frac{1}{1+x}\Big|_0^1 + \int_0^1 \frac{x\,dx}{(1+x)^2} \\
&= \frac{1}{2} + \frac{1}{2}x^2\frac{1}{(1+x)^2}\Big|_0^1 + \int_0^1 \frac{1}{2}x^2\frac{2\,dx}{(1+x)^3} \\
&= \frac{1}{2} + \frac{1}{2}\frac{1}{2^2} + \frac{1}{3}x^3\frac{1}{(1+x)^3}\Big|_0^1 + \int_0^1 \frac{1}{3}x^3\frac{3\,dx}{(1+x)^4} \\
&= \frac{1}{2} + \frac{1}{2}\frac{1}{2^2} + \frac{1}{3}\frac{1}{2^3} + \cdots + \frac{1}{n2^n} + \int_0^1 \frac{x^n}{(1+x)^{n+1}}\,dx.
\end{aligned}$$

Now $0 \le \dfrac{x^n}{(1+x)^{n+1}} \le x^n$ for $x \in [0,1]$ and so

$$0 \le \int_0^1 \frac{x^n}{(1+x)^{n+1}}\,dx \le \int_0^1 x^n\,dx = \frac{1}{n+1}.$$

Therefore, letting $n \to \infty$, we may write

$$\ln(2) = \frac{1}{2} + \frac{1}{2}\frac{1}{2^2} + \frac{1}{3}\frac{1}{2^3} + \cdots = \sum_{n=1}^{\infty} \frac{1}{n2^n}.$$

This series was obtained first by—you guessed it—Euler. A similar method can be applied with an eye on the Gregory-Leibniz series for $\pi/4$ which we obtained in Sect. 10.1. See Exercise 11.35. ◇

11.3 Two Consequences

Integration by Parts yields the following classical result.

Theorem 11.14. (Second Mean Value Theorem for Integrals) *Let f and g be continuous on $[a,b]$, with f' continuous, and $f' > 0$. Then there is $c \in [a,b]$ such that*

$$\int_a^b f(x)g(x)\,dx = f(a)\int_a^c g(x)\,dx + f(b)\int_c^b g(x)\,dx.$$

Proof. On $[a,b]$, define $G(x) = \int_a^x g(t)\,dt$. Integrating by parts, and using the Fundamental Theorem (Theorem 10.1), we get

$$\int_a^b f(x)g(x)\,dx = f(x)G(x)\Big|_a^b - \int_a^b f'(x)G(x)\,dx = f(b)G(b) - \int_a^b f'(x)G(x)\,dx.$$

Now since f' is continuous and $f' > 0$, we may apply the Mean Value Theorem for Integrals (Theorem 9.14) to see that there is $c \in [a,b]$ such that this

$$= f(b)G(b) - G(c)\int_a^b f'(x)\,dx = f(b)G(b) - G(c)\big(f(b) - f(a)\big),$$

again by the Fundamental Theorem (Theorem 10.1). Therefore

$$\int_a^b f(x)g(x)\,dx = f(b)\int_a^b g(x)\,dx - f(b)\int_a^c g(x)\,dx + f(a)\int_a^c g(x)\,dx$$

$$= f(b)\int_c^b g(x)\,dx + f(a)\int_a^c g(x)\,dx,$$

which is what we wanted to show. □

See Exercise 11.40 for another proof of this theorem. As well as having intrinsic appeal, the Second Mean Value Theorem for Integrals has applications in what is known as *series of functions*, and *trigonometric series* in particular. If $g(x) \equiv 1$, its conclusion reads

$$\int_a^b f(x)\,dx = f(a)[c-a] + f(b)[b-c].$$

That is to say, in **Fig. 11.1** the two shaded regions have equal area.

Integration by Parts together with u-Substitution yields the following classical inequality. Its geometric interpretation is shown in **Fig. 11.2.**

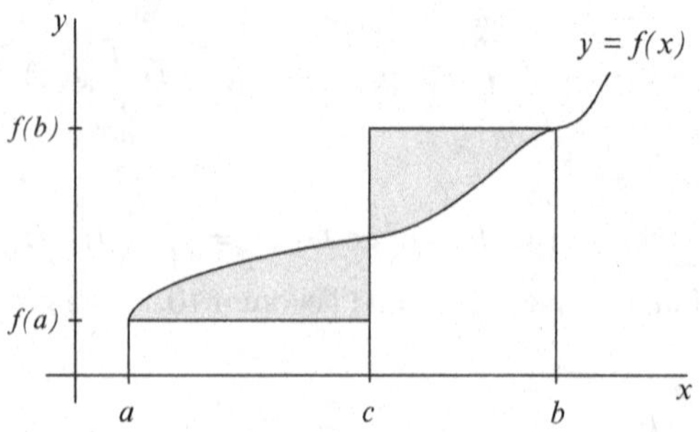

Fig. 11.1 The Second Mean Value Theorem for Integrals (Theorem 11.14) with $g(x) \equiv 1$: the two shaded regions have equal area

Theorem 11.15. (Young's Integral Inequality) *Let f be continuous on $[0, A]$ and strictly increasing there (so that f^{-1} exists and is continuous) and let $f(0) = 0$. Let $a \in [0, A]$ and $b \in [0, f(A)]$. Then*

$$\int_0^a f(x)\,dx + \int_0^b f^{-1}(x)\,dx \geq ab.$$

Further, there is equality if and only if $b = f(a)$. And the inequality is reversed for f strictly decreasing.

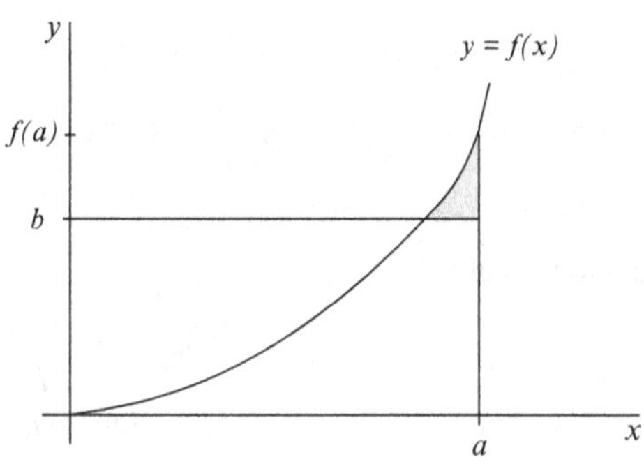

Fig. 11.2 For Young's Inequality (Theorem 11.15): The difference between the left-hand side and the right-hand side is the area of the shaded region

Proof. Using the u-Substitution $u = f^{-1}(x)$, so that $x = f(u)$ and $dx = f'(u)du$,

$$\int_0^b f^{-1}(x)\,dx = \int_0^{f^{-1}(b)} u f'(u)\,du.$$

Integrating by Parts we obtain

$$\int_0^b f^{-1}(x)\,dx = \int_0^{f^{-1}(b)} uf'(u)\,du = uf(u)\Big|_0^{f^{-1}(b)} - \int_0^{f^{-1}(b)} f(u)\,du$$

$$= f^{-1}(b)b - \int_0^{f^{-1}(b)} f(u)\,du.$$

Then adding $\int_0^a f(x)\,dx$ to each side we get

$$\int_0^a f(x)\,dx + \int_0^b f^{-1}(x)\,dx = bf^{-1}(b) + \int_0^a f(x)\,dx - \int_0^{f^{-1}(b)} f(x)\,dx$$

$$= bf^{-1}(b) + \int_{f^{-1}(b)}^a f(x)\,dx$$

$$= ab + \int_{f^{-1}(b)}^a \big(f(x) - b\big)\,dx.$$

Now on the right-hand side, if $f^{-1}(b) < a$ then the integrand is positive and so the integral is positive. If $f^{-1}(b) > a$, then the integrand is negative and so the integral is positive. In either case we have $\int_0^a f(x)\,dx + \int_0^b f^{-1}(x)\,dx > ab$, as desired. The condition for equality is clear, as well as the last statement in the theorem. □

Remark 11.16. The proof above follows [6]; the method there is refined somewhat in [19]. See also [2, 34], and [43]. ○

If we apply Young's Integral Inequality (Theorem 11.15) to $f(x) = x^{p-1}$ $(p > 1)$, we obtain Young's Inequality (Corollary 6.19):

$$\frac{1}{p}a^p + \frac{1}{q}b^q \geq ab \qquad \Big(\text{for } \frac{1}{p} + \frac{1}{q} = 1\Big).$$

And (as we have seen) replacing a with $a^{1/p}$ and b with $b^{1/p}$ yields the weighted AGM Inequality with $n = 2$ (Corollary 6.16). And taking $p = q = 2$ gives Lemma 2.7.

11.4 Taylor's Theorem Again

We employed Integration by Parts repeatedly in Example 11.13. Here is another result that is obtained by repeatedly employing Integration by Parts.

Theorem 11.17. (Integration by Parts Identity) *Let $f,\ f',\ f'',\ldots,\ f^{(n+1)}$ each be continuous on some open interval I containing x_0. Then for $x \in I$,*

$$f(x) = f(x_0) + (x-x_0)f'(x_0) + \frac{(x-x_0)^2}{2!}f''(x_0)$$

$$+\cdots+\frac{(x-x_0)^n}{n!}f^{(n)}(x_0) + \int_{x_0}^{x}\frac{(x-t)^n}{n!}f^{(n+1)}(t)\,dt.$$

Proof. The Fundamental Theorem (Theorem 10.1) gives, for $x \in I$,

$$f(x) = f(x_0) + \int_{x_0}^{x} f'(t)\,dt.$$

Using Integration by Parts while choosing the $+C = -x$, we get

$$f(x) = f(x_0) - (x-t)f'(t)\Big|_{x_0}^{x} + \int_{x_0}^{x}(x-t)f''(t)\,dt.$$

The desired result is now obtained by employing Integration by Parts repeatedly in the usual way. We show just the first few steps, up to $n = 3$.

$$f(x) = f(x_0) - (x-t)f'(t)\Big|_{x_0}^{x} + \int_{x_0}^{x}(x-t)f''(t)\,dt$$

$$= f(x_0) + (x-x_0)f'(x_0) - \tfrac{(x-t)^2}{2}f''(t)\Big|_{x_0}^{x} + \int_{x_0}^{x}\tfrac{(x-t)^2}{2}f^{(3)}(t)\,dt.$$

$$= f(x_0) + (x-x_0)f'(x_0) + \tfrac{(x-x_0)^2 f''(x_0)}{2} + \int_{x_0}^{x}\tfrac{(x-t)^2}{2}f^{(3)}(t)\,dt.$$

$$= f(x_0)+(x-x_0)f'(x_0)+\tfrac{(x-x_0)^2 f''(x_0)}{2} - \tfrac{(x-t)^3}{6}f^{(3)}(t)\Big|_{x_0}^{x} + \int_{x_0}^{x}\tfrac{(x-t)^3}{6}f^{(4)}(t)dt.$$

□

Example 11.18. Suppose that each of f, f', f'', ..., $f^{(n+1)}$ is continuous on an open interval I containing x_0, with

$$0 = f'(x_0) = f''(x_0) = \cdots = f^{(n)}(x_0), \quad \text{but} \quad f^{(n+1)}(x_0) \neq 0.$$

Since $f^{(n+1)}$ is continuous, $f^{(n+1)}(x) \neq 0$ for x sufficiently close to x_0, by Lemma 3.4. So if $f^{(n+1)}(x_0) > 0$ and n is even say, then for such x the Integration by Parts Identity (Theorem 11.17) reads:

$$f(x) = f(x_0) + 0 + 0 + \cdots + 0 + \int_{x_0}^{x} \frac{(x-t)^n}{n!} f^{(n+1)}(t)\,dt > f(x_0).$$

Therefore x_0 yields a local minimum for f. The reader may wish to consider other possibilities, for example $f^{(n+1)}(x_0) < 0$ and n odd, etc. ◊

For $n = 1$ Example 11.18 is the Second Derivative Test (Exercise 8.18), so this example extends the Second Derivative Test. Example 11.18 also follows from the following consequence of the Integration by Parts Identity (Theorem 11.17); that was the content of Exercise 8.54.

Corollary 11.19. (Taylor's Theorem (Theorem 8.20)) *Let f, f', f'', ..., $f^{(n+1)}$ each be continuous on some open interval I containing x_0. For each $x \in I$, there is c between x and x_0 such that*

$$f(x) = \sum_{k=0}^{n} \frac{f^{(k)}(x_0)}{k!}(x-x_0)^k + \frac{f^{(n+1)}(c)}{(n+1)!}(x-x_0)^{n+1}.$$

Proof. For $x \in I$, the Integration by Parts Identity (Theorem 11.17) reads

$$f(x) = f(x_0) + (x-x_0)f'(x_0) + \frac{(x-x_0)^2}{2!} f''(x_0)$$
$$+ \cdots + \frac{(x-x_0)^n}{n!} f^{(n)}(x_0) + \int_{x_0}^{x} \frac{(x-t)^n}{n!} f^{(n+1)}(t)\,dt.$$

Now whether $x_0 \leq x$ or $x < x_0$, the function $(x-t)^n$ does not change sign over the interval of integration. So by the Mean Value Theorem for Integrals (Theorem 9.14), there is c between x and x_0 such that

$$\int_{x_0}^{x} \frac{(x-t)^n}{n!} f^{(n+1)}(t)\,dt = f^{(n+1)}(c) \int_{x_0}^{x} \frac{(x-t)^n}{n!}\,dt = \frac{f^{(n+1)}(c)}{(n+1)!}(x-x_0)^{n+1}.$$

□

Remark 11.20. Corollary 11.19 is in fact a little weaker than Taylor's Theorem (Theorem 8.20), because we require that $f^{(n+1)}$ is continuous. In Taylor's Theorem we only require that $f^{(n+1)}$ exists. ○

Exercises

11.1. Let f be a function defined (for simplicity) on $\mathbf{R}$. Then f is an **odd function** if $f(-x) = -f(x)$ for all x and f is an **even function** if $f(-x) = f(x)$ for all x.

(a) Show that 1, x^2, and $\cos(x)$ are even functions.
(b) Show that x, x^3, $\sin(x)$, and $\arctan(x)$ are odd functions.
(c) Use a u-Substitution to show that if f is odd and continuous, then for all $a \in \mathbf{R}$,

$$\int_{-a}^{a} f(x)\,dx = 0$$

Interpret this geometrically.
(d) Use a u-Substitution to show that if f is even and continuous, then for all $a \in \mathbf{R}$,

$$\int_{-a}^{a} f(x)\,dx = 2\int_{0}^{a} f(x)\,dx$$

Interpret this geometrically.
(e) Show that any function defined on $\mathbf{R}$ can be written as the sum of an odd function and an even function. (For example, $e^x = \sinh(x) + \cosh(x)$.)

11.2. [21]

(a) Show that if f is an **even function** on $\mathbf{R}$ (i.e., $f(-x) = f(x)$ for all x) then for $a \in \mathbf{R}$ and $r > 0$,

$$\int_{-r}^{r} \frac{f(x)}{e^{ax}+1}\,dx = \int_{0}^{r} f(x)\,dx.$$

Hint: Verify that for $u \neq -1$, $\frac{1}{1+u} + \frac{1}{1+1/u} = 1$.
(b) Evaluate $\int_{-r}^{r} \frac{1}{e^{ax}+1}\,dx$, $\int_{-\pi/2}^{\pi/2} \frac{\cos(x)}{e^{ax}+1}\,dx$, and $\int_{-r}^{r} \frac{1}{(x^2+1)(e^{ax}+1)}\,dx$.

11.3. [32]

(a) Let f be continuous on $[a,b]$ with $f(x)+f(a+b-x)$ constant for all $x \in [a,b]$. Show that

$$\int_a^b f(x)\,dx = (b-a)\,f(\tfrac{a+b}{2}) = \frac{b-a}{2}\,[f(a)+f(b)]\,.$$

(b) Can you interpret this geometrically?

(c) Evaluate $\int_0^{\pi/2} (\sin(x))^2\,dx$.

11.4. Evaluate

$$\int_0^\infty e^{-x^2} x\,dx = \lim_{A\to\infty} \int_0^A e^{-x^2} x\,dx.$$

11.5. Show that

$$\int_0^1 \frac{1}{\sqrt{1-x^2}}\,dx = \lim_{a\to 1+} \int_0^a \frac{1}{\sqrt{1-x^2}}\,dx = \frac{\pi}{2}.$$

11.6. [39] Use the Cauchy–Schwarz Integral Inequality (Corollary 9.28) and a u-Substitution to show that

$$\int_1^2 \frac{x\,dx}{\sqrt{x^3+8}} \le \sqrt{\tfrac{2}{3}\ln\tfrac{4}{3}} \cong 0.4379\,.$$

11.7. Show that $\ln(ab) = \ln(a) + \ln(b)$ for $a, b > 0$ as follows.

(a) Verify that

$$\ln(ab) = \ln(a) + \int_a^{ab} \frac{1}{x}\,dx.$$

(b) In the integral, make the substitution $u = x/a$.

11.8. **(a)** Let $a > 0$ and $r \in \mathbf{Q}$. Make the substitution $u = x^{1/r}$ in the integral

$$\ln(a^r) = \int_1^{a^r} \frac{1}{x}\,dx\,,$$

to show that $\ln(a^r) = r\ln(a)$.

(b) Combine (a) with Exercise 11.7 to show that $\ln(a/b) = \ln(a) - \ln(b)$.

11.9. **(a)** Make the substitution $x = \tan(u)$ to verify that

$$\int \frac{1}{1+x^2}\,dx = \arctan(x) + C.$$

(b) Make the substitution $x = \sin(u)$ to find

$$\int \frac{1}{\sqrt{1-x^2}}\,dx.$$

(c) Make the substitution $x = \sec(u)$ to find

$$\int \sqrt{x^2-1}\,dx.$$

Theses are called *trigonometric substitutions.*

11.10. [5] Substituting $u = 1/x$ to evaluate $I = \int_{-1}^{1} \frac{1}{1+x^2}\,dx$ yields $I = 0$, which clearly cannot be correct. Verify this, then explain what is wrong here.

11.11. [1, 4] Make the substitution $x = \pi/2 - u$ to show that for $\alpha \in \mathbf{R}$,

$$\int_0^{\pi/2} \frac{dx}{1+(\tan(x))^{\alpha}} = \frac{\pi}{4}.$$

11.12. [16]

(a) Make the substitution $x = \pi/2 - u$ to show that

$$I = \int_0^{\pi/2} \frac{\sqrt{\sin x}}{\sqrt{\sin x} + \sqrt{\cos x}}\,dx = \frac{\pi}{4}.$$

(b) Make the substitution $x = \pi - u$ to show that

$$I = \int_0^{\pi} \frac{x \sin x}{1+\cos^2 x}\,dx = \frac{\pi^2}{4}.$$

11.13. Evaluate:

(a) $\int \frac{dx}{x\ln(x)}$ **(b)** $\int_t^{t^2} \frac{dx}{x\ln(x)}$ **(c)** $\int \frac{\ln(\ln(x))}{x\ln(x)}\,dx$ **(d)** $\int \frac{\ln(\ln(\ln(x)))}{x\ln(x)}\,dx.$

11.14. Let $x, y > 0$ with $x \neq y$.

(a) [12] Show that their Logarithmic Mean $L = L(x, y) = \frac{x-y}{\ln(x)-\ln(y)}$ satisfies

$$L = \left(\int_0^1 \left(tx + (1-t)y \right)^{-1} dt \right)^{-1}.$$

For $r \neq 0$, set

$$Q_r = \int_0^1 \left(tx^r + (1-t)y^r \right)^{1/r} dt.$$

(b) Verify that $Q_1 = \frac{x+y}{2} = A$, the Arithmetic Mean of x and y and use (a) to see that $Q_{-1} = G^2/L$, where $G = \sqrt{xy}$, the Geometric Mean of x and y.

(c) Use Jensen's Inequality to show that $[(1-t)x^r + ty^r]^{1/r}$ decreases as r decreases, so that Q_r decreases as r decreases.

(d) To make Q_r continuous on $(-\infty, \infty)$, show that we should we define

$$Q_0 = \int_0^1 x^t y^{(1-t)} dt = L.$$

(e) Conclude that $G \leq L \leq A$.

(f) Show that $Q_0 < Q_{1/2}$ implies that $L < \frac{1}{3}G + \frac{2}{3}A$. Note: We obtained the sharper inequality $L < \frac{1}{3}A + \frac{2}{3}G$ in Exercise 6.43. We shall meet this again in Sect. 14.5.

11.15. [15] Show that

$$\lim_{n \to \infty} \frac{1}{n} \int_0^n \frac{x \ln(1 + x/n)}{1 + x} dx = 2\ln(2) - 1.$$

11.16. [8] Let f be nonnegative, continuous, and concave on $[0, 1]$, with $f(0) = 1$.

(a) Show that

$$2 \int_0^1 x^2 f(x)\, dx + \frac{1}{12} \leq \left(\int_0^1 f(x)\, dx \right)^2.$$

(b) Show that equality holds if and only if $f(x) = 1 - x$.

11.17. [33]

(a) Let G be increasing and differentiable on $[a,b]$ and let f be continuous and decreasing on some interval which contains $a, b, G(a)$, and $G(b)$. Prove that if $G(t) \geq t$, then

$$\int_a^b f(t)G'(t)\,dt \geq \int_{G(a)}^{G(b)} f(t)\,dt.$$

(b) Show that if $G(t) \leq t$, the inequality is reversed.

(c) Let $G(t) = b - \int_t^b g(x)\,dx$, where g is continuous and $0 \leq g(x) \leq 1$, to obtain the left side of **Steffensen's Inequalities** (see also Exercises 9.52 and 10.48): *Let f and g be continuous on $[a,b]$, with f decreasing and $0 \leq g \leq 1$. Then for $\lambda = \int_a^b g(x)\,dx$,*

$$\int_{b-\lambda}^{b} f(x)\,dx \leq \int_a^b f(x)g(x)\,dx \leq \int_0^{\lambda} f(x)\,dx.$$

(d) Prove the right side of Steffensen's Inequalities.

11.18. **(a)** Evaluate $\int \sin^4(x)\cos(x)\,dx$ by u-Substitution.

(b) Evaluate $\int \sin^4(x)\cos(x)\,dx$ by Integration by Parts.

11.19. By employing Integration by Parts on the right-hand side, show that for f continuous on $\mathbf{R}$, $\int_0^x (x-u)f(u)\,du = \int_0^x \int_0^u f(t)\,dt\,du$, (See also Exercise 10.3.)

11.20. [18] Show that

$$\int_0^1 t^{n-1}\frac{\ln\left((1+t^{n^2})(1+t)\right)}{1+t^n}\,dt = \frac{(\ln(2))^2}{n}$$

two ways: using u-Substitution and using Integration by Parts.

11.21. Evaluate $\int x\ln(x)\,dx$ by Integration by Parts, two ways:

(a) Integrate x and differentiate $\ln(x)$. **(b)** Integrate $\ln(x)$ and differentiate x.

11.22. Show that

$$\int \sec^3(x)\,dx = \frac{1}{2}\Big(\sec(x)\tan(x) + \ln\big|\sec(x)+\tan(x)\big| + C\Big).$$

Notice that the right-hand side here is the average (Arithmetic Mean) of the derivative of $\sec(x)$ and the (most general) antiderivative of $\sec(x)$. The interesting paper [17] investigates this further.

11.23. For $n \geq 2$ an integer, verify the reduction formula

$$\int \sec^n(x)\,dx = \frac{1}{n-1}\sec^{n-1}(x)\sin(x) + \frac{n-2}{n-1}\int \sec^{n-2}(x)\,dx.$$

11.24. [28] Here is a refinement of Jordan's Inequality

$$\sin(t) \geq \frac{2}{\pi}t \quad \text{for } t \in [0, \pi/2].$$

(a) Agree that in Example 11.10 we showed that

$$\frac{\sin x}{x} = \frac{2}{\pi} + \int_x^{\pi/2} \frac{1}{u^2}\int_0^u t\sin(t)\,dt\,du.$$

(b) Substitute Jordan's Inequality into the integrand to show that, in fact,

$$\sin(x) \geq \left(\frac{2}{\pi} + \frac{1}{12\pi}\left(\pi^2 - 4x^2\right)\right)x \quad \text{for } x \in [0, \pi/2].$$

11.25. **(a)** Evaluate $\int \tan^{-1}(x)\,dx$ by Integration by Parts.

(b) Show that $\int_0^1 \tan^{-1}(x)\,dx = \frac{\pi}{4} - \frac{1}{2}\ln 2.$

(c) Find $\int x\tan^{-1}(x)\,dx.$

11.26. **(a)** Let f'' and g'' be continuous on $[a, b]$. Show that

$$\int \left[f''(x)g(x) - g''(x)f(x)\right]dx = f'(x)g(x) - g'(x)f(x) + C.$$

(b) Use this to evaluate, for example, $\int e^{ax}\sin(bx)\,dx.$

11.27. [14, 37]

(a) Show that

$$\int \frac{f'(x)}{g(x)}\,dx = \frac{f(x)}{g(x)} - \int \frac{f(x)g(x)}{g(x)^2}\,dx.$$

(b) Show that

$$\int \frac{F(x)g'(x)}{g(x)^2}\,dx = \int \frac{f(x)}{g(x)}\,dx - \frac{F(x)}{g(x)}, \quad \text{where } F' = f.$$

(c) Evaluate $\int \frac{\ln(x)}{x^2}\,dx$.

11.28. [26] Use Integration by Parts to show that

$$\int_{3\pi/4}^{\pi} \left[(\tan(x))^2 - \tan(x)\right]\mathrm{e}^{-x}\,dx = \mathrm{e}^{-\pi}.$$

11.29. [23] Show that for $x > 0$,

$$\left|\int_{x}^{x+1} \sin(t^2)\,dt\right| < \frac{1}{x}.$$

Hint: Write $\int_x^{x+1} \sin(t^2)\,dt = \int_x^{x+1} t\frac{\sin(t^2)}{t}\,dt$ and use Integration by Parts.

11.30. [7, 36, 40] In the Integration by Parts formula

$$\int f(x)g(x)\,dx = F(x)g(x) - \int F(x)g'(x)\,dx$$

we are implicitly taking $C = 0$, when any other C might also do.

(a) Show that evaluating $\int x\tan^{-1}(x)\,dx$ with $f(x) = x$ and $g(x) = \tan^{-1}(x)$ is considerably easier if we take $C = 1/2$ than if we take $C = 0$.

(b) Evaluate $\int \frac{1}{x}\,dx$ by parts, setting $f(x) \equiv 1$ and $g(x) = \frac{1}{x}$, to show that choosing $C = 0$ can be spurious.

(c) Show that the Integration by Parts formula, even with the arbitrary C, is of no use if Fg is constant. Can you find another instance for which the Integration by Parts formula is of no use?

11.31. For $n = 0, 1, 2, \ldots$, the **Legendre polynomials** are given (among other ways) by Rodrigues's formula

$$P_n(x) = \frac{1}{2^n n!} \frac{d^n}{dx^n} \left[(1 - x^2)^n\right].$$

(a) Verify that $P_0(x) = 1$, $P_1(x) = x$, $P_2(x) = \frac{3x^2-1}{2}$, and $P_3(x) = \frac{5x^3-3x}{2}$.
(b) Verify that for $n = 0, 1, 2$ and 3, the P_n's in (a) satisfy Legendre's Differential Equation

$$\frac{d}{dx}\left[(1 - x^2)\frac{d}{dx}P_n(x)\right] + n(n + 1)P_n(x) = 0.$$

(c) [13] Use Rodrigues's formula and Integration by Parts repeatedly to prove the following result, which is evocative of the Mean Value Theorem for Integrals (Theorem 9.14): *Let f be continuous on $[-1, 1]$. Then there is $c \in [-1, 1]$ such that*

$$\int_{-1}^{1} f(x)P_n(x)\,dx = \frac{f^{(n)}(c)}{n!} \int_{-1}^{1} x^n P_n(x)\,dx.$$

11.32. [35] Let f be such that $f'' > 0$ on $[a, b]$.

(a) Show that if $f(a) = f(a) = 0$ then

$$\int_a^b f(t)\,dt > \frac{(b - a)^2}{2} \frac{f'(a)f'(b)}{f'(b) - f'(a)}.$$

Hint: Apply the Cauchy–Schwarz Integral Inequality (Corollary 9.28) to $t\sqrt{f''(t)}$ and $\sqrt{f''(t)}$, then use Integration by Parts.
(b) Let $c = \frac{f(b)-f(a)}{b-a}$. Show, not assuming $f(a) = f(a) = 0$, that

$$\int_a^b f(t)\,dt > (b - a)\frac{f(a) + f(b)}{2} - \frac{(b - a)^2}{2} \frac{(c - f'(a))(f'(b) - c)}{f'(b) - f'(a)}.$$

Hint: Transform the situation here to that of (a), using a certain auxiliary function—just as we transformed the situation of the Mean Value Theorem (Theorem 5.2) to that of Rolle's Theorem (Theorem 5.1). See also Exercise 5.9.

11.33. [11, 22, 31, 42] Here is a unified way to obtain the Gregory-Leibniz series and the Alternating Harmonic series. Consider again the reduction formula that we obtained in Example 11.4:

$$I_n = \int_0^{\pi/4} \tan^n(x)\,dx = \frac{1}{n-1} - \int_0^{\pi/4} \tan^{n-2}(x)\,dx = \frac{1}{n-1} - I_{n-2}.$$

(a) Let $n = 2m$ and apply the formula m times to obtain

$$\int_0^{\pi/4} \tan^{2m}(x)\,dx = (-1)^m \left[\frac{\pi}{4} - \left(1 - \frac{1}{3} + \frac{1}{5} - \cdots + \frac{(-1)^{m+1}}{2m-1}\right)\right].$$

(b) Let $n = 2m+1$ and apply the formula m times to obtain

$$\int_0^{\pi/4} \tan^{2m+1}(x)\,dx = \frac{(-1)^m}{2}\left[\ln(2) - \left(1 - \frac{1}{2} + \frac{1}{3} - \cdots + \frac{(-1)^{m+1}}{m}\right)\right].$$

(c) Obviously, $I_n + I_{n-2} = \frac{1}{n-1}$. Verify that $I_n + I_{n+2} = \frac{1}{n+1}$.

(d) Verify that $\int_0^{\pi/4} \tan^n x\,dx$ decreases as n increases.

(e) Show that $\frac{1}{2(n+1)} \le I_n \le \frac{1}{2(n-1)}$, and therefore $I_n \to 0$.

(f) Conclude that $\frac{\pi}{4} = \sum_{n=0}^{\infty} \frac{(-1)^n}{2n+1}$ and $\ln(2) = \sum_{n=1}^{\infty} \frac{(-1)^{n+1}}{n}$.

11.34. [11, 42] Exercise 11.33 contained the fact that $\int_0^{\pi/4} \tan^n x\,dx \to 0$ as $n \to \infty$.

(a) Fill in the details of another way to show this: Use the fact that $\tan(x)$ is convex on $[0, \frac{\pi}{4}]$ to show that

$$x \le \tan x \le \frac{4}{\pi}x.$$

(b) Fill in the details of yet another way to show this: In the integral, make the substitution $u = \tan(x)$ then find bounds for the integrand.

11.35. Write $\frac{\pi}{4} = \tan^{-1}(1) = \int_0^1 \frac{dx}{1+x^2}$, then proceed similarly to how we did in Example 11.13 which dealt with $\ln(2)$, to show that

$$\frac{\pi}{4} = \frac{1}{2}\left(1 + \frac{1}{3} + \frac{1\cdot 2}{3\cdot 5} + \frac{1\cdot 2\cdot 3}{3\cdot 5\cdot 7} + \frac{1\cdot 2\cdot 3\cdot 4}{3\cdot 5\cdot 7\cdot 9} + \cdots\right).$$

11.36. **(a)** Show that if α' is continuous on $[a, b]$, then

$$\int_a^b \alpha(x)\alpha'(x)\,dx = \frac{1}{2}[\alpha(b)^2 - \alpha(a)^2].$$

(b) Let g be continuous on $[0, 1]$, with $0 \le g \le 1$. Apply the formula in (a), with $\alpha(x) = \int_0^x g(u)\,du$, to show that

$$\int_0^{\lambda} x\,dx \le \int_0^1 xg(x)\,dx, \quad \text{where } \lambda = \int_0^1 g(u)\,du.$$

Note: This is a special case of **Steffensen's Inequalities**—see Exercises 9.52, 10.48, and 11.17.

(c) [38] Let f be continuous and nonconstant on $[0, 1]$ with $m \le f \le M$ there, and $\int_0^1 f(x)\,dx = 0$. Show that

$$\left|\int_0^1 xf(x)\,dx\right| \le \frac{-mM}{2(M-m)}.$$

11.37. [10] Let f be continuous on $[0, 1]$, with f' continuous there also. Apply Integration by Parts to $\int_0^1 (1-x)f'(x)\,dx$, then the Cauchy–Schwarz Integral Inequality (Corollary 9.28), to show that

$$\left(\int_0^1 f(x)\,dx\right)^2 \le \frac{1}{3}\int_0^1 |f'(x)|^2\,dx.$$

11.38. For $x > 0$, the **Gamma function** is defined by

$$\Gamma(x) = \int_0^{\infty} t^{x-1}e^{-t}\,dt = \lim_{A\to\infty}\int_0^{A} t^{x-1}e^{-t}\,dt.$$

(a) Show that $\Gamma(1) = 1$, $\Gamma(2) = 1$, and $\Gamma(3) = 2$.

(b) Use Integration by Parts to show that if n is a positive integer then $\Gamma(n+1) = n!$.

(c) Use Hölder's Integral Inequality to show that $\Gamma(x)$ is *logarithmically convex* (see also Exercise 8.35):

$$\ln\left(\Gamma((1-t)x + ty)\right) \le (1-t)\ln(\Gamma(x)) + t\ln(\Gamma(y)).$$

($\Gamma(n+1) = n!$ says that Γ essentially interpolates the factorial function at the positive integers. Of course there are many functions which interpolate the factorial function at the positive integers; it is the property in (c) that makes Γ so special. But see also [24].)

11.39. [25] Let f be continuous on $[a,b]$ with f' continuous also, and $|f'(x)| \le M$ for $x \in [a,b]$. Show that

$$\left| \int_a^{(a+b)/2} f(x)\,dx - \int_{(a+b)/2}^{b} f(x)\,dx \right| \le M\left(\frac{b-a}{2}\right)^2 .$$

Hint: Let $g(x) = \begin{cases} 1 & \text{if } x \in [a, \frac{a+b}{2}] \\ -1 & \text{if } x \in (\frac{a+b}{2}, b] \end{cases}$ then integrate $\int_a^b f(x)g(x)\,dx$ by parts.

11.40. Fill in the details, as follows, of another proof of the Second Mean Value Theorem for Integrals (Theorem 11.14). **(a)** Setting

$$h(x) = f(a)\int_a^x g(t)\,dt + f(b)\int_x^b g(t)\,dt,$$

show that

$$h(b) \le \int_a^b f(t)g(t)\,dt \le h(a).$$

(b) Apply the Intermediate Value Theorem (Theorem 3.17) to h.

11.41. [3]

(a) Use the Second Mean Value Theorem for Integrals (Theorem 11.14) to prove the following. *Let f'' be continuous and nonzero on $[a,b]$, with $f'(x) \ge m > 0$ for all x in $[a,b]$. Then*

$$\left| \int_a^b \sin(f(x))\,dx \right| \le \frac{4}{m}.$$

(b) Let $a > 0$. Show that for all $b > a$,

$$\left| \int_a^b \sin(x^2)\,dx \right| \le \frac{2}{a}.$$

11.42. [29] Let f and g be continuous on $[a,b]$, with f' continuous also, and $f' > 0$. (These are the hypotheses for the Second Mean Value Theorem for

Integrals (Theorem 11.14)). Suppose further that $\int_a^t g(x)\,dx$ and $\int_t^b g(x)\,dx$ are each nonnegative for $t \in [a,b]$. Prove that there is $c \in [a,b]$ such that

$$\int_a^b f(x)g(x)\,dx = f(c)\int_a^b g(x)\,dx.$$

(b) This resembles the Mean Value Theorem for Integrals (Theorem 9.14), but without requiring that $g \geq 0$. Find a g that changes sign, but satisfies these hypotheses.

11.43. Bonnet's form of the Second Mean Value Theorem for Integrals is: *Let f and g be continuous on $[a,b]$, with f nonnegative, f' continuous and $f' > 0$. Then there is $c \in [a,b]$ such that*

$$\int_a^b f(x)g(x)\,dx = f(b)\int_c^b g(x)\,dx.$$

(a) Show that this is a consequence of the Second Mean Value Theorem for Integrals (Theorem 11.14). Hint: Set $F(x) = f(x)$ on $(a,b]$ and $F(a) = 0$, then apply the Second Mean Value Theorem for Integrals to F. (The Second Mean Value Theorem for Integrals is sometimes referred to as **Weierstrass's form of Bonnet's Theorem**.)
(b) Prove Bonnet's form directly—that is, without any appeal to the Second Mean Value Theorem for Integrals.
(c) Formulate an equivalent version of Bonnet's form for $f' < 0$.
(d) Prove that for $0 < a < b$,

$$\left|\int_a^b \frac{\sin x}{x}\,dx\right| < \frac{2}{a}.$$

11.44. [30] Let f and g be positive, with f' and g' nonnegative and continuous on $[0,b]$. Let $f(0) = 0$. Show that for $0 < a \leq b$,

$$f(a)g(b) \leq \int_0^a g(x)f'(x)\,dx + \int_0^b f(x)g'(x)\,dx,$$

with equality if and only if either $a = b$, or $a < b$ and g is constant.

11.45. [41] Fill in the details of another proof of Young's Integral Inequality (Theorem 11.15). Observe that if f is strictly increasing, then its antiderivative is strictly convex. Therefore, for any $0 < c \neq a < A$,

$$\int_0^a f(t)\,dt > \int_0^c f(t)\,dt + f(c)(a-c).$$

Now set $c = f^{-1}(a)$ and use (which we also used in the proof in the text):

$$\int_0^b f^{-1}(x)\,dx = bf^{-1}(b) - \int_0^{f^{-1}(b)} f(u)\,du.$$

11.46. [19] Apply Young's Integral Inequality (Theorem 11.15) to

$$f(x) = \sqrt[4]{x^4+1} - 1$$

to show that

$$\int_0^3 \sqrt[4]{x^4+1}\,dx + \int_1^3 \sqrt[4]{x^4-1}\,dx \geq 9.$$

11.47. Use the substitution $t = (1-u)a + ub$ for $u \in [0,1]$ to show that (c.f. the Integration by Parts Identity (Theorem 11.17)):

$$\int_a^b \frac{(b-t)^n}{n!} f^{(n+1)}(t)\,dt = \frac{(b-a)^n}{n!} \int_0^1 (1-u)^n f^{(n+1)}((1-u)a+ub)\,du.$$

11.48. [27] Here's a proof of the Integration by Parts Identity that doesn't use Integration by Parts! Set

$$F_t(x) = f(t) + (x-t)f'(t) + \frac{(x-t)^2}{2} f''(t) + \frac{(x-t)^3}{3!} f^{(3)}(t) + \cdots + \frac{(x-t)^n}{n!} f^{(n)}(t).$$

(a) Treat x as fixed and show that

$$\frac{d}{dt} F_t(x) = \frac{(x-t)^n}{n!} f^{(n+1)}(t).$$

(b) Integrate from $t = a$ to $t = x$, and use the Fundamental Theorem (Theorem 10.1).

11.49. [20] Show that for $x < 1$ and $n = 0, 1, 2, \ldots$,

$$(n+1)\int_0^x \frac{(x-t)^n}{(1-t)^{n+2}}\,dt = \frac{x^{n+1}}{1-x}.$$

11.50. Show that the conclusion of Taylor's Theorem (Theorem 8.20 or Corollary 11.19) can be written (this form for the remainder is called *Cauchy's* form):

$$f(x) = \sum_{k=0}^{n} \frac{f^{(k)}(x_0)}{k!}(x - x_0)^k + \frac{f^{(n+1)}(c)(x - c)^n}{n!}(x - x_0).$$

References

1. Ailawanda, S., Oltikar, B.C., Spiegel, M.R.: Problem 260. Coll. Math. J. **16**, 305–306 (1985)
2. Anderson, N.: Integration of inverse functions. Math. Gaz. **54**, 52–53 (1970)
3. Apostol, T.: Calculus, vol. 1, p. 362. Blaisdell, New York (1961)
4. Arora, A.K., Sudhir, K.G., Rodriguez, D.M.: Special integration techniques for trigonometric integrals. Am. Math. Mon. **95**, 126–130 (1988)
5. Beetham, R.: An integral. Math. Gaz. **47**, 60 (1963)
6. Boas, R.P., Jr., Marcus, M.B.: Inverse functions and integration by parts. Am. Math. Mon. **81**, 760–761 (1974)
7. Borman, J.L.: A remark on integration by parts. Am. Math. Mon. **51**, 32–33 (1944)
8. Bracken, P., Seiffert, H.J.: Problem 11133. Am. Math. Mon. **114**, 360–361 (2007)
9. Bressoud, D.M.: Second Year Calculus, from Celestial Mechanics to Special Relativity. Springer, New York (1991)
10. Burk, F.: Numerical integration via integration by parts. Coll. Math. J. **17**, 418–422 (1986)
11. Burk, F.: $\pi/4$ and $\ln 2$ recursively. Coll. Math. J. **18**, 51 (1987)
12. Carlson, B.C.: The logarithmic mean. Am. Math. Mon. **79**, 615–618 (1972)
13. Cioranescu, N.: La généralisation de la première formule de la moyenne. L'Enseignement Mathématique **37**, 292–302 (1938)
14. Deveau, M., Hennigar, R.: Quotient rule integration by parts. Coll. Math J. **43**, 254–256 (2012)
15. Díaz-Barrero, J.L., Herman, E.: Problem 11225. Am. Math. Mon. **114**, 750 (2007)
16. Fitt, A.D.: What they don't teach you about integration at school. Math. Gaz. **72**, 11–15 (1988)
17. Frohliger, J., Poss, R.: Just an average integral. Math. Mag. **62**, 260–261 (1989)
18. Glasser, M.L.: Problem 580. Coll. Math. J. **28**, 235–236 (1997)
19. Hoorfar, A., Qi, F.: A new refinement of Young's inequality. Math. Inequal. Appl. **11**, 689–692 (2008)
20. Howard, J., Schlosser, J.: Problem Q700. Math. Mag. **58**, 238, 245 (1985)
21. Jameson, G.J.O., Jameson, T.P.: Some remarkable integrals derived from a simple algebraic identity. Math. Gaz. **97**, 205–209 (2013)
22. Kazarinoff, D.K.: A simple derivation of the Leibnitz-Gregory series. Am. Math. Mon. **62**, 726–727 (1955)
23. Landau, M.D., Gillis, J., Shimshoni, M.: Problem E2211. Am. Math. Mon. **77**, 1107–1108 (1970)
24. Laugwitz, D., Rodewald, B.: A simple characterization of the Gamma function. Am. Math. Mon. **94**, 534–536 (1987)
25. Lupu, C., Lupu, T.: Problem 927 (unpublished solution by Mercer, P.R.) Coll. Math. J. **41**, 242 (2010)
26. Luthar, R.S., Lindstom, P.A.: Problem 288. Coll. Math. J. **17**, 361–362 (1986)
27. Mazzone, E.F., Piper, B.R.: Animating nested Taylor polynomials to approximate a function. Coll. Math. J. **41**, 405–408 (2010)
28. Mercer, A.McD.: Unpublished solution (by proposer) for Problem E2952. Amer. Math. Mon. **93**, 568–569 (1986)
29. Mercer, A.McD.: A new mean value theorem for integrals. Math. Gaz. **97**, 510–512 (2013)

30. Mitrinovic, D.S.: Analytic Inequalities, p. 49. Springer, Berlin/New York (1970)
31. Nash, C.: Infinite series by reduction formulae. Math. Gaz. **74**, 140–143 (1990)
32. Nelson, R.B.: Symmetry and integration. Coll. Math. J. **26**, 39–41 (1995)
33. Pecaric, J.E.: Connection between some inequalities of Gauss, Steffensen and Ostrowski. Southeast Asian Bull. Math. **13**, 89–91 (1989)
34. Schnell, S., Mendoza, C.: A formula for integrating inverse functions. Math. Gaz. **84**, 103–104 (2000)
35. Sieffert, H.J., Chico Problem Group. Cal. State – Chico: Problem 291. Coll. Math. J. **17**, 444–445 (1986)
36. Smith, C.D.: On the problem of Integration by Parts. Math. News Lett. **3**, 7–8 (1928)
37. Switkes, J.: A quotient rule integration by parts formula. Coll. Math. J. **36**, 58–60 (2005)
38. Thong, D.V.: Problem 11581 (unpublished solution by Mercer, P.R.). Am. Math. Mon. **118**, 557 (2011)
39. Underhill, W.V.: Finding bounds for definite integrals. Coll. Math. J. **15**, 426–429 (1984)
40. Watson, H.: A fallacy by parts. Math. Gaz. **69**, 122 (1985)
41. Witkowski, A.: On Young's inequality. J. Inequal. Pure Appl. Math. **7**, 164 (2006)
42. Zheng, L.: An elementary proof for two basic alternating series. Am. Math. Mon. **109**, 187–188 (2002)
43. Zhu, L.: On Young's inequality. Int. J. Math. Educ. Sci. Technol. **35**, 601–603 (2004)

Chapter 12
Classic Examples

It can be of no practical use to know that π is irrational, but if we can know, it surely would be intolerable not to know.

—E.C. Titchmarsh

In this chapter we allow ourselves another brief diversion. Except for Wallis's product for π (Lemma 12.1), this chapter is independent of all subsequent chapters.

After obtaining Wallis's product, we show that π is irrational. Then we show that e^r is irrational whenever $r \neq 0$ is rational. This implies in particular that the sum of the Alternating Harmonic series, namely $\ln(2)$, is irrational.

We also show that Euler's sum $\sum_{n=1}^{\infty} 1/n^2 = \pi^2/6$, and that the sum $\sum 1/p$, of the reciprocals of all of the prime numbers, diverges. Among other things, the exercises contain Vieta's formula, an evaluation of the Probability integral, and a tiny glimpse of transcendental numbers.

12.1 Wallis's Product

We begin by deriving another reduction formula, to be used presently and in the next section.

Let $n \geq 2$ be an integer. Applying Integration by Parts (differentiating the $\cos^{n-1}(x)$ and integrating the $\cos(x)$) to

$$\int \cos^n(x)\,dx = \int \cos^{n-1}(x)\cos(x)\,dx,$$

we get

$$\begin{aligned}\int \cos^n(x)\,dx &= \cos^{n-1}(x)\sin(x) + \int (n-1)\cos^{n-2}(x)\sin^2(x)\,dx \\ &= \cos^{n-1}(x)\sin(x) + (n-1)\int \cos^{n-2}(x)(1-\cos^2(x))\,dx.\end{aligned}$$

P.R. Mercer, *More Calculus of a Single Variable*, Undergraduate Texts in Mathematics, DOI 10.1007/978-1-4939-1926-0_12

Then solving for $\int \cos^n(x)\,dx$, we obtain the reduction formula:

$$\int \cos^n(x)\,dx = \frac{1}{n}\cos^{n-1}(x)\sin(x) + \frac{n-1}{n}\int \cos^{n-2}(x)\,dx.$$

In particular,

$$\int_0^{\pi/2} \cos^n(x)\,dx = \frac{n-1}{n}\int_0^{\pi/2} \cos^{n-2}(x)\,dx.$$

In an entirely similar way, which we leave for Exercise 12.1, we have

$$\int \sin^n(x)\,dx = -\frac{1}{n}\sin^{n-1}(x)\cos(x) + \frac{n-1}{n}\int \sin^{n-2}(x)\,dx,$$

and in particular,

$$\int_0^{\pi/2} \sin^n(x)\,dx = \frac{n-1}{n}\int_0^{\pi/2} \sin^{n-2}(x)\,dx.$$

The following beautiful infinite product for π was obtained in 1658 by English mathematician John Wallis (1616–1703). Our proof uses the above reduction formula involving $\cos^n(x)$.

Lemma 12.1. (Wallis's Product)

$$\frac{\pi}{2} = \lim_{m\to\infty}\left(\frac{2}{1}\frac{2}{3}\frac{4}{3}\frac{4}{5}\frac{6}{5}\frac{6}{7}\cdots\frac{2m}{2m-1}\frac{2m}{2m+1}\right) = \prod_{m=1}^{\infty}\frac{2m}{2m-1}\frac{2m}{2m+1}.$$

Proof. Setting $I_n = \int_0^{\pi/2} \cos^n(x)\,dx$, we obtained the reduction formula

$$I_n = \frac{n-1}{n} I_{n-2}. \tag{12.1}$$

If we set $n = 2m$ and apply (12.1) m times, we get (using $I_0 = \pi/2$):

$$I_{2m} = \frac{2m-1}{2m} I_{2m-2} = \frac{2m-1}{2m}\frac{2m-3}{2m-2} I_{2m-4} = \cdots$$

$$= \frac{2m-1}{2m}\frac{2m-3}{2m-2}\cdots\frac{3}{4}\frac{1}{2}\frac{\pi}{2}.$$

On the other hand, if we set $n = 2m + 1$ and apply (12.1) m times, we get (using $I_1 = 1$):

$$I_{2m+1} = \frac{2m}{2m+1} I_{2m-1} = \frac{2m}{2m+1}\frac{2m-2}{2m-1} I_{2m-3} = \cdots$$

$$= \frac{2m}{2m+1}\frac{2m-2}{2m-1} \cdots \frac{4}{5}\frac{2}{3}(1).$$

Therefore,

$$\frac{I_{2m}}{I_{2m+1}} = \frac{\frac{2m-1}{2m}\frac{2m-3}{2m-2}\cdots\frac{3}{4}\frac{1}{2}\frac{\pi}{2}}{\frac{2m}{2m+1}\frac{2m-2}{2m-1}\cdots\frac{4}{5}\frac{2}{3}}$$

$$= \frac{2m+1}{2m}\frac{2m-1}{2m}\frac{2m-1}{2m-2}\cdots\frac{7}{6}\frac{5}{6}\frac{5}{4}\frac{3}{4}\frac{3}{2}\frac{1}{2}\frac{\pi}{2}.$$

We claim that

$$\frac{I_{2m}}{I_{2m+1}} \to 1 \quad \text{as } m \to \infty,$$

which would yield Wallis's product:

$$\frac{\pi}{2} = \lim_{m\to\infty}\left(\frac{2}{1}\frac{2}{3}\frac{4}{3}\frac{4}{5}\frac{6}{5}\frac{6}{7}\cdots\frac{2m}{2m-1}\frac{2m}{2m+1}\right) = \prod_{m=1}^{\infty}\frac{2m}{2m-1}\frac{2m}{2m+1}.$$

Now to verify the claim. Again by (12.1) with $m = 2m + 1$,

$$I_{2m+1} = \frac{2m}{2m+1} I_{2m-1},$$

and so

$$\frac{I_{2m-1}}{I_{2m+1}} = 1 + \frac{1}{2m}.$$

Then since $0 \le \cos(x) \le 1$ on $[0, \pi/2]$,

$$1 \le \frac{I_{2m}}{I_{2m+1}} \le \frac{I_{2m-1}}{I_{2m+1}} = 1 + \frac{1}{2m},$$

and so the claim is verified and the proof is complete. □

The above treatment is essentially from [4, 12], but see also [19]. For an elementary (but still tricky) proof, see [28]. For an interesting geometric interpretation of

Wallis's product, see [25]. For an analogous infinite product for e, see [24], and for other related infinite products, see [26].

We remark that each factor in Wallis's product is >1. Therefore for any $N \in \mathbf{N}$,

$$\prod_{m=1}^{N} \frac{2m}{2m-1}\frac{2m}{2m+1} < \frac{\pi}{2}.$$

See Exercise 12.2 for a related upper estimate for $\pi/2$.

12.2 π Is Irrational

In this section we provide the wonderful 1947 proof [20] by the American (Canadian born) mathematician Ivan Niven (1915–1999), that π is irrational. (See also [8,31].) This fact was first proved in 1761 by the Swiss mathematician J.H. Lambert (1728–1777).

Theorem 12.2. *π is irrational.*

Proof. Seeking a contradiction we assume that $\pi = p/q$ where p and q are positive integers. (And $p > q$, since $\pi > 1$.) For the polynomial of degree $2n$ given by

$$g(x) = \frac{q^n}{n!}x^n(\pi - x)^n,$$

we have $g^{(2n)}(x) \equiv (-1)^n q^n (2n)(2n-1)\cdots(n+1)$, and for $k \geq 2n+1$ we have $g^{(k)}(x) \equiv 0$. We make the following additional observations about g:

(i) $g^{(k)}(0) = 0$ for $k = 0, 1, 2, \ldots, n-1$;
(ii) $g^{(k)}(0)$ is an integer for $n \leq k \leq 2n$.

Also $g(x) = g(\pi - x)$, and so $g^{(k)}(0) = (-1)^k g^{(k)}(\pi)$. Therefore,

(iii) $g^{(k)}(\pi) = 0$ for $k = 0, 1, 2, \ldots, n-1$, and
(iv) $g^{(k)}(\pi)$ is an integer for $n \leq k \leq 2n$.

In summary then: $g^{(2n)}(x)$ is an integer, and $g^{(k)}(0)$ and $g^{(k)}(\pi)$ are integers for $k = 0, 1, 2, 3, \ldots$.

Now let us consider the integral $\displaystyle\int_0^{\pi} \sin(x)g(x)\,dx$. Since g is a polynomial of degree $2n$, to evaluate this integral, we shall employ Integration by Parts $2n$ times. In doing so, we shall integrate the $\sin(x)$ part and all of its *descendants*, while differentiating the $g(x)$ part and all of its *descendants*. The first four steps look like this:

$$\int_0^\pi \sin(x)g(x)\,dx = -\cos(x)g(x)\Big|_0^\pi + \int_0^\pi \cos(x)g'(x)\,dx$$

$$= \text{integer } + \sin(x)g'(x)\Big|_0^\pi - \int_0^\pi \sin(x)g''(x)\,dx$$

$$= \text{integer } + 0 + \cos(x)g''(x)\Big|_0^\pi - \int_0^\pi \cos(x)g^{(3)}(x)\,dx$$

$$= \text{integer } + 0 + \text{ integer } + \sin(x)g^{(3)}(x)\Big|_0^\pi + \int_0^\pi \sin(x)g^{(4)}(x)\,dx\,.$$

Then after $2n$ steps, we obtain

$$\int_0^\pi \sin(x)g(x)\,dx = \text{integer } + (-1)^n \int_0^\pi g^{(2n)}(x)\sin(x)\,dx$$

$$= \text{integer } + (\text{integer}) \cdot \int_0^\pi \sin(x)\,dx,$$

which is an integer. Observe now that

$$0 < \int_0^\pi \sin(x)g(x)\,dx < \int_0^\pi g(x)\,dx < \pi\frac{q^n}{n!}\pi^n\pi^n = \pi\frac{p^n}{n!}\pi^n = \pi\frac{(p\pi)^n}{n!}.$$

In Example 8.25 we showed that for any $u \in \mathbf{R}$,

$$\frac{u^n}{n!} \to 0 \quad \text{as } n \to \infty.$$

So if we choose n very large, then $0 < \int_0^\pi \sin(x)g(x)\,dx < 1$ and so $\int_0^\pi \sin(x)g(x)\,dx$ *cannot* be an integer. This is a contradiction, and so π is irrational. □

12.3 More Irrational Numbers

Niven's proof above that π is irrational (Theorem 12.2) is a simplification of a more general approach which goes back to the French mathematician Charles Hermite (1822–1901). See [7, 23], also [2]. Here is another instance of Hermite's approach.

Theorem 12.3. e^u *is irrational for any nonzero integer* u.

Proof. For the polynomial of degree $2n$ given by

$$g(x) = \frac{x^n(1-x)^n}{n!},$$

we have $g^{(2n)}(x) \equiv (-1)^n(2n)(2n-1)\cdots(n+1)$, and for $k \geq 2n+1$ we have $g^{(k)}(x) \equiv 0$. We make the following additional observations about g:

(i) $g^{(k)}(0) = 0$ for $k = 0, 1, 2, \ldots, n-1$;
(ii) $g^{(k)}(0)$ is an integer for $n \leq k \leq 2n$.

Also $g(x) = g(1-x)$, and so $g^{(k)}(0) = (-1)^k g^{(k)}(1)$. Therefore

(iii) $g^{(k)}(1) = 0$ for $k = 0, 1, 2, \ldots, n-1$, and
(iv) $g^{(k)}(1)$ is an integer for $n \leq k \leq 2n$.

In summary then: $g^{(2n)}(x)$ is an integer, and $g^{(k)}(0)$ and $g^{(k)}(1)$ are integers for $k = 0, 1, 2, 3, \ldots$.

We need only show that e^u is irrational for any positive integer u. (If e^u is irrational then so is e^{-u}.) Seeking a contradiction let us assume that $e^u = p/q$, where p and q are positive integers. Consider now

$$qu^{2n+1}\int_0^1 e^{ux}g(x)\,dx\,.$$

Since g is a polynomial of degree $2n$, to evaluate the integral here, we shall employ Integration by Parts $2n$ times. In doing so, we shall we integrate the e^{ux} part and all of its *descendants*, while differentiating the $g(x)$ part and all of its *descendants*. With our assumption that $e^u = p/q$, the first three integrations by parts yield:

$$\begin{aligned}
& qu^{2n+1}\int_0^1 e^{ux}g(x)\,dx \\
&= qu^{2n+1}\left(\frac{1}{u}e^{ux}g(x)\Big|_0^1 - \int_0^1 \frac{1}{u}e^{ux}g'(x)\,dx\right) \\
&= \text{integer} + qu^{2n+1}\left(-\int_0^1 \frac{1}{u}e^{ux}g'(x)\,dx\right) \\
&= \text{integer} + qu^{2n+1}\left(-\frac{1}{u^2}e^{ux}g'(x)\Big|_0^1 + \int_0^1 \frac{1}{u^2}e^{ux}g''(x)\,dx\right)
\end{aligned}$$

$$= \text{integer} + \text{integer} + qu^{2n+1}\left(\int_0^1 \frac{1}{u^2}\mathrm{e}^{ux}g''(x)\,dx\right)$$

$$= \text{integer} + \text{integer} + qu^{2n+1}\left(\frac{1}{u^3}\mathrm{e}^{ux}g''(x)\Big|_0^1 - \int_0^1 \frac{1}{u^3}\mathrm{e}^{ux}g^{(3)}(x)\,dx\right)$$

$$= \text{integer} + \text{integer} + \text{integer} + qu^{2n+1}\left(-\int_0^1 \frac{1}{u^3}\mathrm{e}^{ux}g^{(3)}(x)\,dx\right).$$

Then after $2n$ integrations by parts, we get

$$qu^{2n+1}\int_0^1 \mathrm{e}^{ux}g(x)\,dx = \text{integer } + qu^{2n+1}\left(\int_0^1 \frac{1}{u^{2n}}\mathrm{e}^{ux}g^{(2n)}(x)\,dx\right)$$

$$= \text{integer } + qu^{2n+1}(\text{integer})\left(\int_0^1 \frac{1}{u^{2n}}\mathrm{e}^{ux}\,dx\right)$$

$$= \text{integer} + qu^{2n+1}(\text{integer})\cdot \frac{1}{u^{2n+1}}\mathrm{e}^{ux}\Big|_0^1,$$

which is an integer. Observe now that

$$0 < qu^{2n+1}\int_0^1 \mathrm{e}^{ux}g(x)\,dx < qu^{2n+1}\frac{\mathrm{e}^u}{n!} = p\frac{u^{2n+1}}{n!}.$$

In Example 8.25 we showed that for any $u \in \mathbf{R}$,

$$\frac{u^n}{n!} \to 0 \quad \text{as } n \to \infty.$$

This is easily extended (see Exercise 12.7) to show that for any $u \in \mathbf{R}$,

$$\frac{u^{2n+1}}{n!} \to 0 \quad \text{as } n \to \infty.$$

So if we choose n very large, then $0 < qu^{2n+1}\int_0^1 \mathrm{e}^{ux}g(x)\,dx < 1$ and therefore $qu^{2n+1}\int_0^1 \mathrm{e}^{ux}g(x)\,dx$ *cannot* be an integer. This is a contradiction, and so e^u is irrational. □

Corollary 12.4. e^r *is irrational for any nonzero rational number r.*

Proof. Let $r = p/q$ for integers p and q, with $q \neq 0$. If $e^r = x$ is rational then $e^p = x^q$ is also rational. But by Theorem 12.3 this cannot be the case, unless $p = 0$. □

Remark 12.5. The German mathematician F. Lindemann (1852–1939) proved in 1882 that e^r is in fact *transcendental* if r is a nonzero rational number. That is, it is not the root of any polynomial of any degree, with integer coefficients. Lindemann actually proved much more: e^r is transcendental if r is nonzero and not transcendental. Readers who have studied complex variables will know that $e^{i\pi} = -1$. (This formula is attributed to, yes, L. Euler.) Therefore π is transcendental. The Russian mathematician A.O. Gelfond (1906–1968) proved in 1929 that e^{π} is transcendental. It is currently not known if any of the numbers π^e, πe, $\pi + e$, or $\ln(\pi)$ are even irrational, much less transcendental. ∘

Corollary 12.6. $\ln(t)$ *is irrational for any positive rational number* $t \neq 1$.

Proof. If $\ln(t) = \ln(p/q) = u$ is rational, then $e^u = p/q$ is rational. But this contradicts Corollary 12.4 unless $u = 0$, in which case $t = 1$. □

In Sect. 6.5, and again in Examples 8.26 and 10.10, we showed that the Alternating Harmonic series

$$\sum_{n=1}^{\infty}(-1)^{n+1}\frac{1}{n} = 1 - \frac{1}{2} + \frac{1}{3} - \frac{1}{4} + \cdots = \ln(2) \cong 0.693147.$$

Corollary 12.6 says in particular that this number is irrational.

12.4 Euler's Sum $\sum 1/n^2 = \pi^2/6$

In Sect. 6.5 we saw that the series $\sum_{n=1}^{\infty} \frac{1}{n^2}$ converges, and that its sum is < 2. Showing that a particular series converges is one thing, but finding its sum is often much more difficult.

The following monumental result was discovered by L. Euler, in 1741. Our proof mainly follows the modifications given in [5], of the argument in [18]. A good number of other proofs are known (see for example [3, 9]), but most of them extend beyond the scope of this book.

Theorem 12.7. $\displaystyle\sum_{n=1}^{\infty} \frac{1}{n^2} = \frac{\pi^2}{6}.$

Proof. Consider the integral $I_n = \int_0^{\pi/2} \cos^{2n}(x)\,dx$. The reduction formula (12.1) gives

$$I_n = \frac{2n-1}{2n} I_{n-1},$$

which is easily rewritten as

$$\frac{2n-1}{nI_n} = \frac{2}{I_{n-1}}. \tag{12.2}$$

Or, by replacing n with $n+1$ in (12.2) and then doing a little algebra we get

$$I_n - I_{n+1} = \frac{I_n}{2(n+1)}. \tag{12.3}$$

Each of (12.2) and (12.3) will be of use in what follows. Now for $n \geq 1$, Integration by Parts employed twice gives

$$\begin{aligned}
I_n &= \int_0^{\pi/2} 1 \cos^{2n}(x)\,dx = x\cos^{2n}(x)\Big|_0^{\pi/2} + \int_0^{\pi/2} x 2n \cos^{2n-1}(x)\sin(x)\,dx \\
&= n \int_0^{\pi/2} 2x \cos^{2n-1}(x)\sin(x)\,dx \\
&= nx^2 \cos^{2n-1}(x)\sin(x)\Big|_0^{\pi/2} - \int_0^{\pi/2} nx^2 \Big[\cos^{2n}(x) - (2n-1)\cos^{2n-2}(x)\sin^2(x)\Big]\,dx \\
&= -\int_0^{\pi/2} nx^2 \Big[\cos^{2n}(x) - (2n-1)\cos^{2n-2}(x)(1-\cos^2(x))\Big]\,dx \\
&= \int_0^{\pi/2} nx^2 \Big[(2n-1)\cos^{2n-2}(x) - 2n\cos^{2n}(x)\Big]\,dx.
\end{aligned}$$

So setting

$$J_n = \int_0^{\pi/2} x^2 \cos^{2n}(x)\,dx,$$

this reads

$$I_n = n(2n-1)J_{n-1} - 2n^2 J_n.$$

Therefore,

$$\frac{1}{n^2} = \frac{2n-1}{n}\frac{J_{n-1}}{I_n} - 2\frac{J_n}{I_n}.$$

And using (12.2), this reads

$$\frac{1}{n^2} = 2\frac{J_{n-1}}{I_{n-1}} - 2\frac{J_n}{I_n}.$$

Then summing from 1 to N we get lots of cancellation. More precisely, the sum is *telescoping*:

$$\sum_{n=1}^{N}\frac{1}{n^2} = 2\sum_{n=1}^{N}\left(\frac{J_{n-1}}{I_{n-1}} - \frac{J_n}{I_n}\right) = 2\left(\frac{J_0}{I_0} - \frac{J_N}{I_N}\right).$$

Now

$$I_0 = \int_0^{\pi/2} 1\,dx = \frac{\pi}{2} \quad \text{and} \quad J_0 = \int_0^{\pi/2} x^2\,dx = \frac{\pi^3}{24},$$

and so

$$2\frac{J_0}{I_0} = \frac{\pi^2}{6}.$$

We claim now that $\frac{J_N}{I_N} \to 0$ as $N \to \infty$, which will complete the proof. Jordan's Inequality from Example 8.14 (also Exercise 10.31 and Example 11.10) reads:

$$\frac{2}{\pi}x \le \sin(x) \quad \text{for } x \in [0, \pi/2].$$

Therefore

$$\begin{aligned} J_N = \int_0^{\pi/2} x^2\cos^{2N}(x)\,dx &\le \frac{\pi^2}{4}\int_0^{\pi/2} \sin^2(x)\cos^{2N}(x)\,dx \\ &= \frac{\pi^2}{4}\int_0^{\pi/2} \left(1-\cos^2(x)\right)\cos^{2N}(x)\,dx \\ &= \frac{\pi^2}{4}\left(I_N - I_{N+1}\right). \end{aligned}$$

Then using (12.3),

$$J_N \le \frac{\pi^2}{4}\left(I_N - I_{N+1}\right) = \frac{\pi^2}{4}\frac{I_N}{2(N+1)}.$$

Finally then,

$$0 \le \frac{J_N}{I_N} \le \frac{\pi^2}{4}\frac{1}{2(N+1)}.$$

So the claim is verified and the proof is complete. □

Remark 12.8. It is the case that π^2 is irrational (Exercise 12.6), and so $\sum_{n=1}^{\infty} \frac{1}{n^2}$ is irrational. Euler found exactly all the sums $\sum_{n=1}^{\infty} \frac{1}{n^k}$, for k even. For example, $\sum_{n=1}^{\infty} \frac{1}{n^4} = \frac{\pi^4}{90}$ and $\sum_{n=1}^{\infty} \frac{1}{n^6} = \frac{\pi^6}{945}$. Very little is known about these sums for $k \ge 3$ odd. It was proved only in 1979, by the French mathematician R. Apéry (1916–1994), that $\sum_{n=1}^{\infty} \frac{1}{n^3}$ is irrational. (Very much less if known if $k > 1$ is not an integer!) ∘

12.5 The Sum $\sum 1/p$ of the Reciprocals of the Primes Diverges

Below we prove, following [13], the interesting fact that the sum of the reciprocals of all of the prime numbers diverges to infinity. Another proof is outlined in Exercise 12.15.

This fact implies immediately that there are infinitely many prime numbers. But we have also seen that

$$\sum_{n=1}^{\infty} \frac{1}{n^2} < +\infty,$$

(this sum $= \pi^2/6$, by Theorem 12.7) so this fact also suggests—at least in a vague way—that there are many more prime numbers than there are perfect squares.

Theorem 12.9. *Denote by* $\Pi = \{2, 3, 5, 7, 11, 13, 17, \dots\}$ *the set of prime numbers. Then*

$$\sum_{p \in \Pi} \frac{1}{p} = +\infty.$$

Proof. For $m \geq 2$, let $M = \{p\ prime : 2 \leq p \leq m\}$. Observe that

$$\prod_{p \in M}\left(1+\frac{1}{p-1}\right) = \prod_{p \in M}\left(\frac{1}{1-1/p}\right) = \prod_{p \in M}\left(1+\frac{1}{p}+\frac{1}{p^2}+\cdots\right),$$

wherein each $\left(1+\frac{1}{p}+\frac{1}{p^2}+\cdots\right)$ is a convergent geometric series (Example 2.3), since $p \geq 2$. Now any integer k, with $2 \leq k \leq m$, is a product of powers of primes and each of these primes is obviously $\leq m$. So any $1/k$ arising from such a k must appear somewhere in the product

$$\prod_{p \in M}\left(1+\frac{1}{p}+\frac{1}{p^2}+\cdots\right).$$

(For example,

$$\frac{1}{2{,}793} = \frac{1}{3 \cdot 7^2 \cdot 19} = 1 \cdot \frac{1}{3} \cdot 1 \cdot \frac{1}{7^2} \cdot 1 \cdot 1 \cdot 1 \cdot \frac{1}{19} \cdot 1 \cdot \cdots \cdot 1\text{.)}$$

Therefore,

$$\sum_{k=1}^{m} \frac{1}{k} < \prod_{p \in M}\left(1+\frac{1}{p-1}\right).$$

We saw in Example 6.11 that $\ln(1+x) \leq x$ for $x > -1$, and so

$$\ln\left(\sum_{k=1}^{m} \frac{1}{k}\right) < \ln\left(\prod_{p \in M}\left(1+\frac{1}{p-1}\right)\right) = \sum_{p \in M} \ln\left(1+\frac{1}{p-1}\right) \leq \sum_{p \in M} \frac{1}{p-1}.$$

Now since $p \geq 2$,

$$\sum_{p \in M} \frac{1}{p-1} \leq \sum_{p \in M} \frac{2}{p}.$$

Therefore

$$\frac{1}{2} \ln\left(\sum_{k=1}^{m} \frac{1}{k}\right) < \sum_{p \in M} \frac{1}{p}.$$

And finally, since the Harmonic series diverges to infinity (Sect. 6.5), the proof is complete upon letting $m \to \infty$. □

Remark 12.10. *Twin primes* are two prime numbers p and q such that $p + 2 = q$. For example, 3 & 5 are twin primes, as are 5 & 7, 11 & 13, and 101 & 103. It was proved in 1919 by Norwegian mathematician V. Brun (1885–1978) that $\sum 1/p$ converges, where the sum is taken over all twin primes. It is not known whether there are infinitely many twin primes. If there are not, then Brun's result is trivial. Either way, it is known that this sum (called *Brun's constant*) is between 1.9021 and 1.9022. It was proved in 2013 by Tom Zhang [30] of the University of New Hampshire that there does exist some integer N such that there infinitely many primes p and q satisfying $p + N = q$. He also showed that $N \leq 70{,}000{,}000$. (N is certainly even!) Since then, mathematicians have been continually working to improve Zhang's estimate for N. It is currently know that $N \leq 246$. See the website *Polymath8* for updates. ∘

Exercises

12.1. **(a)** Show that

$$\int \sin^n(x)\,dx = -\frac{1}{n}\sin^{n-1}(x)\cos(x) + \frac{n-1}{n}\int \sin^{n-2}(x)\,dx,$$

and consequently

$$\int_0^{\pi/2} \sin^n(x)\,dx = \frac{n-1}{n}\int_0^{\pi/2} \sin^{n-2}(x)\,dx.$$

(b) Derive Wallis's product (Lemma 12.1) using *this* reduction formula.

12.2. [19] We observed at the end of Sect. 12.1 that in Wallis's product (Lemma 12.1), each factor in the product is >1 and so for any $N \in \mathbf{N}$,

$$\prod_{m=1}^{N} \frac{2}{1}\frac{2}{3}\frac{4}{3}\frac{4}{5}\frac{6}{5}\frac{6}{7}\cdots\frac{2m}{2m-1}\frac{2m}{2m+1} < \frac{\pi}{2}.$$

Fill in the details, as follows, for obtaining an upper estimate for $\pi/2$.

(a) Write

$$\begin{aligned}\frac{\pi}{2} &= \prod_{m=1}^{\infty}\frac{2m}{2m-1}\frac{2m}{2m+1} = \prod_{m=1}^{\infty}\frac{(2m)^2}{(2m)^2-1}\\ &= \prod_{m=1}^{N}\frac{(2m)^2}{(2m)^2-1}\prod_{m=N+1}^{\infty}\frac{(2m)^2}{(2m)^2-1},\end{aligned}$$

and then

$$\prod_{m=N+1}^{\infty} \frac{(2m)^2}{(2m)^2-1} = \exp\left(\ln \prod_{m=N+1}^{\infty} \frac{(2m)^2}{(2m)^2-1}\right)$$
$$= \exp\left(\sum_{m=N+1}^{\infty} \ln\left(\frac{(2m)^2}{(2m)^2-1}\right)\right)$$
$$= \exp\left(\sum_{m=N+1}^{\infty} \ln\left(1+\frac{1}{(2m)^2-1}\right)\right).$$

(b) Now use $\ln(1+x) < x$ (obtained in Example 6.11) and notice that the resulting sum *telescopes* to $\frac{1}{2(2N+1)} = \frac{1}{4N+2}$.

(c) Conclude that

$$\frac{\pi}{2} < \prod_{m=1}^{N} \frac{(2m)^2}{(2m)^2-1} e^{\frac{1}{4N+2}}.$$

(d) Show that this upper estimate is closer to $\frac{\pi}{2}$ than is the lower estimate $\prod_{m=1}^{N} \frac{(2m)^2}{(2m)^2-1}$.

12.3. [27] Show that

$$1+\sum_{n=1}^{\infty} \frac{1}{n+1}\left(\frac{1\cdot 3\cdots(2n-1)}{2\cdot 4\cdots(2n)}\right)^2 = \frac{4}{\pi}.$$

12.4. Derive, as follows, F. Vieta's 1593 formula

$$\frac{2}{\pi} = \sqrt{\frac{1}{2}}\sqrt{\frac{1}{2}+\frac{1}{2}\sqrt{\frac{1}{2}}}\sqrt{\frac{1}{2}+\frac{1}{2}\sqrt{\frac{1}{2}+\frac{1}{2}\sqrt{\frac{1}{2}}}}\cdots.$$

(a) Show that

$$\frac{\sin(\theta)}{\theta} = \cos(\theta/2)\frac{\sin(\theta/2)}{\theta/2} = \cos(\theta/2)\cos(\theta/4)\frac{\sin(\theta/4)}{\theta/4} = \cdots$$
$$= \cos(\theta/2)\cos(\theta/4)\cdots\cos(\theta/2^n)\frac{\sin(\theta/2^n)}{\theta/2^n}.$$

(b) Show that $\lim_{n\to\infty} \frac{\sin(\theta/2^n)}{\theta/2^n} = 1$, and so we may write

$$\frac{\sin(\theta)}{\theta} = \cos(\theta/2)\cos(\theta/4)\cdots\cos(\theta/2^n)\cdots.$$

(c) Write $\cos(\theta/2) = \sqrt{\frac{1}{2}(1+\cos(\theta))}$, and set $\theta = \pi/2$.

12.5. [15] (See also [16].) Here we evaluate the **Probability integral**

$$\int_0^\infty e^{-x^2}\,dx = \lim_{n\to\infty}\int_0^n e^{-x^2}\,dx = \frac{\sqrt{\pi}}{2}.$$

The idea is to replace the e^{-x^2} in the integrand with $\left(1-\frac{x^2}{n}\right)^n$, then let $n\to\infty$.

(a) Show that for $x \in [0,n]$,

$$0 \le e^{-x} - \left(1-\frac{x}{n}\right)^n \le \frac{e^{-1}}{n}.$$

(The right-hand inequality is the harder one.)

(b) Show that

$$0 \le \int_0^{\sqrt{n}} e^{-x^2}\,dx - \int_0^{\sqrt{n}} \left(1-\frac{x^2}{n}\right)^n dx \le \frac{e^{-1}}{\sqrt{n}}.$$

(c) Conclude that

$$\int_0^\infty e^{-x^2}\,dx = \lim_{n\to\infty}\int_0^{\sqrt{n}} \left(1-\frac{x^2}{n}\right)^n dx.$$

(d) Set $x = \sqrt{n}\sin(t)$ and use Integration by Parts to show that

$$\int_0^{\sqrt{n}} \left(1-\frac{x^2}{n}\right)^n dx = \sqrt{n}\int_0^{\pi/2} \cos^{2n+1}(t)\,dt.$$

(e) Use Wallis's product (Lemma 12.1) to show that this last integral $\to \sqrt{\pi}/2$ as $n\to\infty$.

12.6. [22] Look carefully at the proof that π is irrational. Now prove the stronger statement that π^2 is irrational. (This is not easy.)

12.7. Show that for any $u \in \mathbf{R}$,

$$\frac{u^{2n+1}}{n!} \to 0 \quad \text{as } n \to \infty.$$

(Example 8.25 might be helpful.)

12.8. Use $\sum_{n=1}^{\infty} \frac{1}{n^2} = \frac{\pi^2}{6}$ to show that $\sum_{n=1}^{\infty} \frac{1}{(2n-1)^2} = \frac{\pi^2}{8}$.

12.9. [5] Show that

$$0 \leq \frac{\pi^2}{6} - \sum_{n=1}^{\infty} \frac{1}{n^2} \leq \frac{\pi^2}{4(n+1)}.$$

12.10. [10] For natural numbers a and b, denote by $\text{lcm}(a, b)$ and $\gcd(a, b)$ their least common multiple and greatest common divisor respectively.

(a) Show that $\text{lcm}(a, b)\gcd(a, b) = ab$.
(b) Let $\{a_n\}$ be a strictly increasing sequence of natural numbers. Show that

$$\sum_{n=1}^{\infty} \frac{1}{\text{lcm}(a_n, a_{n+1})} \quad \text{converges.}$$

Hint: Use (a) to write this as a *telescoping* series.

12.11. (e.g., [11, 29]) Consider the **Liouville number**

$$x_0 = \sum_{j=1}^{\infty} \frac{1}{10^{j!}} = 0.1100010 \ldots,$$

which has a 1 in the $(j!)$th decimal place, and zeros elsewhere. Liouville showed in 1844 that x_0 is *transcendental*—that is, x_0 is not the root of any polynomial of any degree, with integer coefficients. Fill in the details of the following proof that x_0 is transcendental. Denote by a_j/b_j the fraction obtained by truncating x_0 after the $(j!)$th decimal place (e.g., $a_3/b_3 = 0.110001$).

(a) Show that $b_j = 10^{j!}$.
(b) Show that

$$\left|\frac{a_j}{10^{j!}} - x_0\right| < \frac{2}{10^{(j+1)!}} = \frac{2}{10^{(j+1)j!}} \leq \frac{2}{(10^{j!})^{n+1}} \quad \text{for } j \geq n.$$

(c) Looking for a contradiction, suppose that x_0 is a solution to $P(x) = 0$, where P is a polynomial of degree n with integer coefficients. Apply the Mean Value Theorem (Theorem 5.2) to $P(x)$ on $[\frac{a_j}{10^{j!}}, x_0]$ to show that there is $M > 0$ such that

$$\left|P\left(\frac{a_j}{10^{j!}}\right)\right| \leq \frac{2M}{(10^{j!})^{n+1}}.$$

(d) Conclude that $\left|P(\frac{a_j}{10^{j!}})(10^{j!})^n\right| \leq \frac{2M}{10^{j!}}$, and the left side is necessarily an integer. Now take j large to obtain the desired contradiction.

Remark 12.11. Truly, this exercise is just the tip of the iceberg—see Exercise 12.12 below. By the late 1830s, mathematicians believed that numbers like π and e were transcendental, yet it hadn't even been shown that transcendental numbers exist. This is the significance of Liouville's 1844 result [11].

12.12. [21] Much of this exercise appeared already as Exercise 1.44. A set A is *countable* if there is a one-to-one onto function $\sigma : \mathbf{N} \to A$. (So its elements can be listed off: $\sigma(1)$, $\sigma(2)$, $\sigma(3)$, ...) Fill in the details of the following proof that the set of algebraic numbers—that is, all roots of all polynomials of any degree, with integer coefficients—is countable. This amazing fact was discovered in 1871 by the German mathematician Georg Cantor (1845–1918).

(a) Let $a_0, a_1, \ldots, a_n$ be integers and consider the polynomial equation

$$p(x) = a_n x^n + a_{n-1} x^{n-1} + \cdots + a_1 x + a_0 = 0,$$

which has at most n roots. We may assume that $a_n \geq 1$. Why?

(b) Define the *index* of any such polynomial as

$$\mathit{index}(p) = |a_n| + |a_{n-1}| + \cdots + |a_1| + |a_0|.$$

Show, for example, that there is only one such polynomial with index 2. There are 4 such polynomials with index 3. There are 11 such polynomials with index 4. Show that there are only finitely many polynomials with a given index.

(c) Now use the index to show how the algebraic numbers can be put in one-to-one onto correspondence with the natural numbers. (This proof actually shows that the set of *all* algebraic numbers, not just the real ones, is countable.)

(d) An immediate consequence is that the set of rational numbers is countable—explain.

(e) Show that **R** is not countable. Conclude that the set of transcendental numbers is not countable—not intending to take anything away from Liouville's excellent result from Exercise 12.11!

12.13. [1, 9, 14] Denote by $\gcd(a, b)$ the greatest common divisor of integers a, b and let p be the probability that $\gcd(a, b) = 1$.

(a) For integers a,b, show that the probability that n divides both a and b is $1/n^2$.

(b) Show that the probability that $\gcd(a, b) = n$ is p/n^2.

(c) Show that $\sum_{n=1}^{\infty} p/n^2 = 1$ and conclude that $p = 6/\pi^2 \cong 0.608$.

(The probability that three integers, a, b, c have gcd $= 1$ is $\sum_{n=1}^{\infty} 1/n^3$, etc.)

12.14. [17] Show that

$$\sum_{n=2}^{\infty}\sum_{p\in\Pi}\frac{1}{p^n} < \frac{3}{2} - \ln(2) \cong 0.8069.$$

12.15. [6] Fill in the details, as follows, of another proof that the sum of the reciprocals $\sum 1/p$ of all of the prime numbers p diverges.

(a) Verify that

$$\left(1+\frac{1}{p}\right)\left(1+\frac{1}{p^2}+\frac{1}{p^4}+\frac{1}{p^6}+\cdots+\frac{1}{p^{2k}}\right)$$

$$=\left(1+\frac{1}{p}+\frac{1}{p^2}+\frac{1}{p^3}+\cdots+\frac{1}{p^{2k+1}}\right).$$

(b) Let $M = \{p \text{ prime: } 2 \le p \le m\}$. Show that

$$\prod_{M}\left(1+\frac{1}{p}\right)\sum_{\substack{n \text{ integer with} \\ \text{all divisors in } M}}\frac{1}{n^2} = \sum_{\substack{n \text{ integer with} \\ \text{all divisors in } M}}\frac{1}{n}.$$

(c) Conclude that $\prod_{M}\left(1+\frac{1}{p}\right) \to \infty$ as $m \to \infty$.

(d) Finally, use $1 + x \le e^x$ (there's that inequality again) to show that $\sum 1/p$ diverges.

References

1. Abrams, A.D., Paris, M.J.: The probability that $(a, b) = 1$. Coll. Math. J. **23**, 47 (1992)
2. Beatty, T., Jones, T.W.: A simple proof that $e^{p/q}$ is irrational. Math. Mag. **87**, 50–51 (2014)
3. Chapman, R.: Evaluating $\zeta(2)$ (2014). On the world wide web at http://www.uam.es/personal_pdi/ciencias/cillerue/Curso/zeta2.pdf
4. Courant, R.: Differential and Integral Calculus, 2nd edn. Wiley Classics Library, Hoboken (1988)
5. Daners, D.: A short elementary proof of $\sum 1/k^2 = \pi^2/6$. Math. Mag. **85**, 361–364 (2012)
6. Eynden, C.V.: Proofs that $\sum 1/p$ diverges. Am. Math. Mon. **87**, 394–397 (1980)
7. Hardy, G.H., Wright, E.M.: Introduction to the Theory of Numbers, 4th edn. Oxford University Press, Oxford (1959)
8. Jones, T.W.: Discovering and proving that π is irrational. Am. Math. Mon. **117**, 553–557 (2010)
9. Kalman, D.: Six ways to sum a series. Coll. Math. J. **24**, 402–421 (1993)
10. Khare, C.B.: Problem Q741. Math. Mag. **61**, 315, 325 (1988)
11. Körner, T.W.: Fourier Analysis. Cambridge University Press, Cambridge (1989)
12. Lang, S.: A First Course in Calculus. Addison Wesley, Reading (1964)
13. Leavitt, W.G.: The sum of the reciprocals of the primes. Coll. Math. J. **10**, 198–199 (1979)

14. Leonard, B., Shultz, H.S.: A computer verification of a pretty mathematical result. Math. Gaz. **72**, 7–10 (1988)
15. Levrie, P., Daems, W.: Evaluating the probability integral using Wallis's product formula for π. Am. Math. Mon. **116**, 538–541 (2009)
16. Lord, N.: An elementary single-variable proof of $\int_{-\infty}^{\infty} e^{-x^2} dx = \sqrt{\pi}$. Math. Gaz. **87**, 308–311 (2003)
17. Luthar, R.S., Lossers, O.P.: Problem E2192. Am. Math. Mon. **77**, 769–770 (1970)
18. Matsuoka, Y.: An elementary proof of the formula $\sum_{n=1}^{\infty} 1/n^2 = \pi^2/6$. Am. Math. Mon. **68**, 485–487 (1961)
19. Nash, C.: Infinite series by reduction formulae. Math. Gaz. **74**, 140–143 (1990)
20. Niven, I.M.: A simple proof that π is irrational. Bull. Am. Math. Soc. **53**, 509 (1947)
21. Niven, I.M.: Numbers Rational and Irrational. Mathematical Association of America, Washington DC (1961)
22. Niven, I.M.: Irrational Numbers. Mathematical Association of America, Washington DC (2005)
23. Parks, A.E.: π, e, and other irrational numbers. Am. Math. Mon. **93**, 722–723 (1986)
24. Pippenger, N.: An infinite product for e. Am. Math. Mon. **87**, 391 (1980)
25. Rummler, H.: Squaring the circle with holes. Am. Math. Mon. **100**, 858–860 (1993)
26. Sondow, J.: New Wallis- and Catalan- type infinite products for π, e, and $\sqrt{2+\sqrt{2}}$. Am. Math. Mon. **117**, 912–917 (2010)
27. Van Hamme, L.: Problem Q745. Math. Mag. **62**, 137, 143 (1989)
28. Wästlund, J.: An elementary proof of the Wallis product formula for π. Am. Math. Mon. **114**, 914–917 (2007)
29. Wilker, J.B.: Transcendentals galore. Math. Gaz. **66**, 258–261 (1982)
30. Zhang, Y.: Bounded gaps between primes. Ann. Math. **179**, 1121–1174 (2014)
31. Zhou, L., Markov, L.: Recurrent proofs of the irrationality of certain trigonometric values. Am. Math. Mon. **117**, 360–362 (2010)

Chapter 13
Simple Quadrature Rules

It is the mark of an educated mind to rest satisfied with the degree of precision which the nature of the subject admits.

—Aristotle

In practice most definite integrals cannot be evaluated exactly. In such cases one must resort to various approximation methods, which can be quite complicated. Any method used to approximate a definite integral is called a *quadrature rule.* (*Quadrature* is any process used to construct a square equal in area to that of some given figure.) But in this chapter we see that even the simplest of quadrature rules can be useful, even when the exact value of the integral *is* known.

13.1 The Rectangle Rules

For a function f defined on $[a, b]$, the **Left Rectangle Rule** is the approximation

$$\int_a^b f(x)\,dx \cong f(a)\big[b-a\big].$$

The quantity $f(a)[b-a]$ is the signed area of the rectangle with base $[a, b]$, and height $f(a)$. See **Fig. 13.1**.

Likewise, the **Right Rectangle Rule** is the approximation

$$\int_a^b f(x)\,dx \cong f(b)\big[b-a\big].$$

The quantity $f(b)[b-a]$ is the signed area of the rectangle with base $[a, b]$, and height $f(b)$. See **Fig. 13.2**.

The following result is pretty well obvious, upon drawing a picture. We have already seen it (essentially) in our proof of the Integral Test (Theorem 9.24).

P.R. Mercer, *More Calculus of a Single Variable*, Undergraduate Texts in Mathematics, DOI 10.1007/978-1-4939-1926-0_13

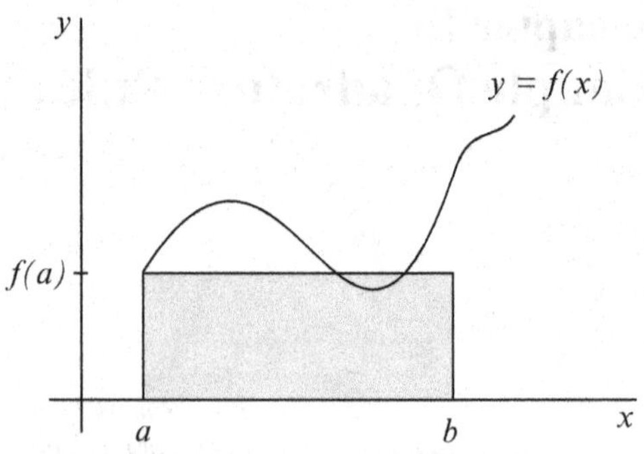

Fig. 13.1 Left Rectangle Rule: $\int_a^b f(x)\,dx$ is approximated by the area of the shaded rectangle

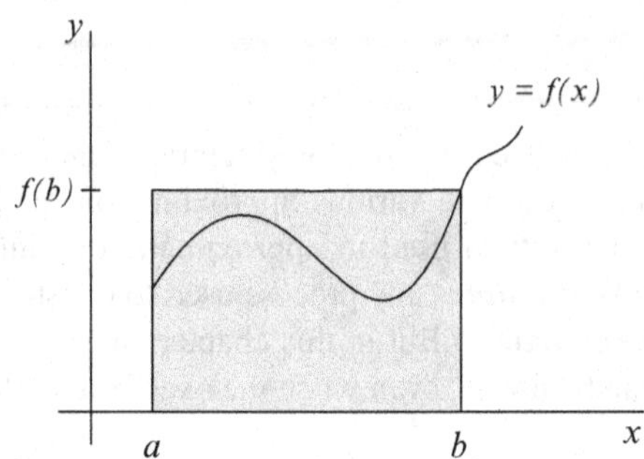

Fig. 13.2 Right Rectangle Rule: $\int_a^b f(x)\,dx$ is approximated by the area of the shaded rectangle

Lemma 13.1. *Let f be continuous and increasing on $[a, b]$. Then*

$$f(a)\big[b-a\big] \;\leq\; \int_a^b f(t)\,dt \;\leq\; f(b)\big[b-a\big],$$

and these inequalities are reversed for f continuous and decreasing.

Proof. This is Exercise 13.1. (See the proof of the Integral Test (Theorem 9.24).) □

Example 13.2. In (6.8) we obtained the estimates

$$\frac{1}{x+1} < \ln\left(\frac{x+1}{x}\right) < \frac{1}{x} \qquad \text{for } x > 0\,. \tag{13.1}$$

These follow quickly from Lemma 13.1. Indeed, for $x > 0$,

$$\int_x^{x+1} \frac{1}{t}\,dt \;=\; \ln\left(\frac{x+1}{x}\right).$$

Then since $1/t$ is strictly decreasing we have, by Lemma 13.1,

$$\frac{1}{x+1}[(x+1)-x] < \int_x^{x+1} \frac{1}{t}\,dt < \frac{1}{x}[(x+1)-x] \qquad \text{for } x > 0,$$

from which (13.1) follows. The inequalities in (13.1) are equivalent to

$$\left(1+\frac{1}{x}\right)^x < \mathrm{e} < \left(1+\frac{1}{x}\right)^{x+1} \qquad \text{for } x \in \mathbf{R},$$

which we have seen and used many times. ◇

Example 13.3. Let $0 \le a < b$ and $n \in \mathbf{N}$. Observe that

$$\int_a^b x^n\,dx = \frac{b^{n+1}-a^{n+1}}{n+1}.$$

Then since x^n is strictly increasing, Lemma 13.1 gives

$$a^n(b-a) < \frac{b^{n+1}-a^{n+1}}{n+1} < b^n(b-a).$$

We obtained these inequalities differently, and used them, in Example 1.32, and in Exercises 1.39 and 6.59. ◇

13.2 The Trapezoid and Midpoint Rules

For a function f defined on $[a,b]$, the **Trapezoid Rule** is the approximation

$$\int_a^b f(x)\,dx \cong \frac{f(a)+f(b)}{2}[b-a].$$

The quantity $\frac{f(a)+f(b)}{2}[b-a]$ is the signed area of the trapezoid with base $[a,b]$, and heights $f(a)$ and $f(b)$. See **Fig. 13.3**.

The **Midpoint Rule** is the approximation

$$\int_a^b f(x)\,dx \cong f\left(\frac{a+b}{2}\right)[b-a].$$

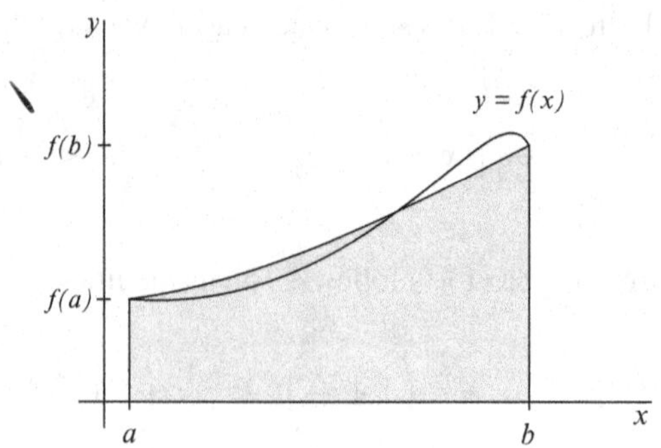

Fig. 13.3 Trapezoid Rule: $\int_a^b f(x)\,dx$ is approximated by the area of the shaded trapezoid

The quantity $f(\frac{a+b}{2})[b-a]$ is the signed area of the rectangle with base $[a,b]$, and height $f(\frac{a+b}{2})$. See **Fig. 13.4**.

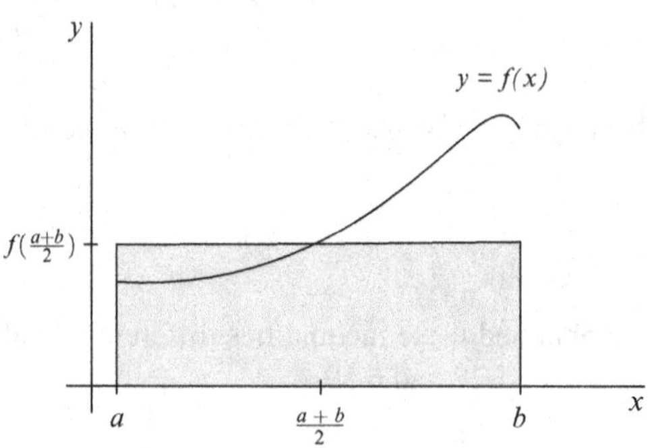

Fig. 13.4 Midpoint Rule: $\int_a^b f(x)\,dx$ is approximated by the area of the shaded rectangle

The following inequalities are named for French mathematicians Charles Hermite (1822–1901) and Jacques Hadamard (1865–1963). We shall refer to these as the **HH Inequalities** for short [36]. These are similar in spirit to those of Lemma 13.1 and are also fairly obvious, upon drawing a picture: See **Fig. 13.5**—the Midpoint Rule is sometimes called the *Tangent Rule*.

Lemma 13.4. (the HH Inequalities) *Let f be defined on $[a,b]$ with $f'' \geq 0$, so that f is convex. Then*

$$f(\frac{a+b}{2})[b-a] \leq \int_a^b f(x)\,dx \leq \frac{f(a)+f(b)}{2}[b-a].$$

And these inequalities are reversed for $f'' \leq 0$.

Proof. For the right-hand inequality we let $x = (1-t)a + tb$, so that $dx = (b-a)\,dt$. Then since f is convex,

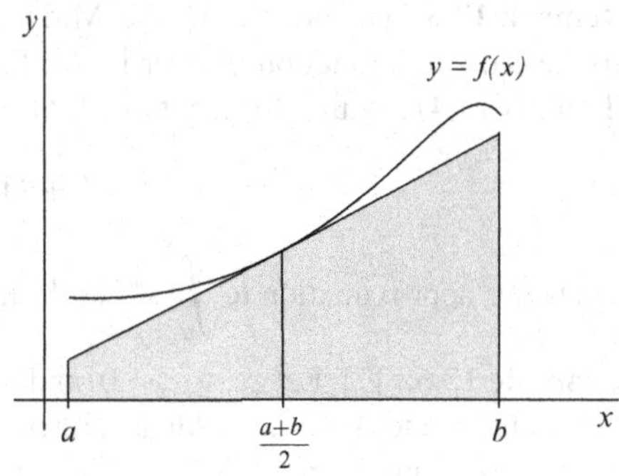

Fig. 13.5 The Midpoint Rule is sometimes called the *Tangent Rule*. The area of the shaded trapezoid here equals the area of the shaded rectangle from the Midpoint Rule, in Fig. 13.4

$$\frac{1}{b-a}\int_a^b f(x)\,dx = \int_a^b f\big((1-t)a+tb\big)\,dt \le \int_0^1 (1-t)f(a)\,dt + \int_0^1 tf(b)\,dt$$

$$= f(a)\int_0^1 (1-t)\,dt + f(b)\int_0^1 t\,dt = \frac{f(a)+f(b)}{2}.$$

For the left-hand inequality, notice that since f is convex its graph lies on or above its tangent lines on $[a,b]$, by Lemma 8.7. So in particular, its graph lies above the tangent line at $(\frac{a+b}{2}, f(\frac{a+b}{2}))$. That is,

$$f(x) \ge f'(\tfrac{a+b}{2})\big(x - \tfrac{a+b}{2}\big) + f(\tfrac{a+b}{2}).$$

Integrating this with respect to x, we get

$$\int_a^b f(x)\,dx \ge f'(\tfrac{a+b}{2})\int_a^b \big(x - \tfrac{a+b}{2}\big)\,dx + f(\tfrac{a+b}{2})[b-a].$$

Now the reader may verify directly that $\int_a^b (x - \frac{a+b}{2})\,dx = 0$, or see Example 9.23 (or just draw a picture). Therefore

$$\int_a^b f(x)\,dx \ge f(\frac{a+b}{2})[b-a],$$

as desired. □

Remark 13.5. Denote by M the Midpoint Rule and by T the Trapezoid Rule, applied to some function f over $[a, b]$. For f strictly convex, the HH Inequalities (Lemma 13.4) say that for any $t \in (0, 1)$,

$$tM + (1 - t)T$$

is a better approximation to $\int_a^b f(x)\,dx$ than either of M or T. ○

Example 13.6. [7] Let $x, y > 0$ and denote by G, L, and A the Geometric, Logarithmic and Arithmetic Means, respectively, of x and y. Here we show another way of proving Lemma 6.20: $G \leq L \leq A$. Applying the HH Inequalities (Lemma 13.4) to the convex function $f(t) = \mathrm{e}^t$ gives

$$\mathrm{e}^{\frac{a+b}{2}} \leq \frac{1}{b-a} \int_a^b \mathrm{e}^t\,dt \leq \frac{\mathrm{e}^a + \mathrm{e}^b}{2}.$$

Now since $x, y > 0$, we may set $a = \ln(x)$ and $b = \ln(y)$ to get

$$\left(\mathrm{e}^{\ln(xy)}\right)^{1/2} \leq \frac{1}{\ln(y)-\ln(x)} \int_{\ln(x)}^{\ln(y)} \mathrm{e}^x\,dx \leq \frac{\mathrm{e}^{\ln(x)} + \mathrm{e}^{\ln(y)}}{2}.$$

That is,

$$\sqrt{xy} \leq \frac{y - x}{\ln(y) - \ln(x)} \leq \frac{x + y}{2},$$

just as we wanted to show. ◇

Remark 13.7. In Exercises 13.8 and 13.9 we see how the HH Inequalities (Lemma 13.4) can be applied to other functions to yield the same inequalities $G \leq L \leq A$. ○

Example 13.8. [26, 32] For the convex function $f(t) = 1/t$ on $[x, x + 1]$, the right-hand side of the HH Inequalities (Lemma 13.4) gives

$$\int_x^{x+1} \frac{1}{t}\,dt = \ln\left(\frac{x+1}{x}\right) \leq \frac{1}{2}\left(\frac{1}{x+1} + \frac{1}{x}\right) \quad \text{for } x > 0.$$

And the left-hand side of the HH Inequalities (Lemma 13.4) gives

$$\frac{2}{2x+1} \leq \int_x^{x+1} \frac{1}{t}\,dt = \ln\left(\frac{x+1}{x}\right) \quad \text{for } x > 0.$$

These two inequalities improve (13.1) considerably. Restricting to natural numbers, they imply

$$\frac{2n}{2n+1} \le \ln\left(1+\frac{1}{n}\right)^n \le \frac{2n+1}{2(n+1)} \qquad \text{for } n \in \mathbf{N}. \tag{13.2}$$

We shall see in the next section that these are pretty good inequalities. ◇

Example 13.9. The left-hand side of (13.2), but with $n+1$ instead of n, is

$$\frac{2n+2}{2n+3}.$$

Now the reader can easily verify that

$$\frac{2n+2}{2n+3} > \frac{2n+1}{2(n+1)},$$

which is the right hand side of (13.2). Therefore, as we have seen, $\left\{\left(1+\frac{1}{n}\right)^n\right\}$ is an increasing sequence. A very similar argument, which we leave for Exercise 13.20, shows that $\left\{\left(1+\frac{1}{n}\right)^{n+1}\right\}$ is a decreasing sequence. ◇

Example 13.10. [33] The right-hand inequality in (13.2) improves the right-hand inequality in (13.1), which we used in Sect. 6.7 to show that Euler's constant γ is ≥ 0. So with (13.2), we should be able to obtain a better lower bound for γ. Indeed, for $n = 1, 2, 3, \ldots,$

$$\ln(n) = \sum_{k=1}^{n-1} \ln\left(\frac{k+1}{k}\right) \le \frac{1}{2}\sum_{k=1}^{n-1}\left(\frac{1}{k}+\frac{1}{k+1}\right) = \frac{1}{2}\left(1+\sum_{k=2}^{n-1}\frac{2}{k}+\frac{1}{n}\right)$$

$$= \frac{1}{2}\left(2+\sum_{k=2}^{n-1}\frac{2}{k}+\frac{2}{n}-1-\frac{1}{n}\right) = \sum_{k=1}^{n}\frac{1}{k}-\frac{1}{2}\left(1+\frac{1}{n}\right).$$

Therefore $\gamma_n = \sum_{k=1}^{n} \frac{1}{k} - \ln(n) > \frac{1}{2}\left(1+\frac{1}{n}\right)$, and so in fact $\gamma \ge 1/2$. ◇

13.3 Stirling's Formula

For $n \in \mathbf{N}$, the **factorial function** is given by

$$n! = n(n-1)(n-2)\cdots(2)(1).$$

This function comes up in many areas of mathematics, like combinatorics, probability and algebra, as well as real and complex analysis. Factorials grows incredibly quickly as n gets large. (For a neat description of just how quickly $n!$ grows with n, see [29].) For example, the number of different ways of shuffling a standard deck of 52 cards is 52!, a number which has 68 digits.

Unfortunately, there is no shortcut formula for computing the factorial of a particular number—one really must do all the multiplying, i.e., get a calculator or computer to do it. But the huge numbers that result from these multiplications are difficult even for calculators and mathematical software to handle. So good estimates for $n!$ are very useful.

Using inequalities (13.2) we shall see below that for n large,

$$\sqrt{2\pi}\,\sqrt{n}\,\left(\frac{n}{\mathrm{e}}\right)^n \;\leq\; n! \;\leq\; \sqrt{2\pi}\,\sqrt{n}\,\left(\frac{n}{\mathrm{e}}\right)^n \mathrm{e}^{\frac{1}{4n}}. \tag{13.3}$$

These estimates are very good because $\mathrm{e}^{\frac{1}{4n}} \to 0$ as $n \to +\infty$. In particular, (13.3) implies the following result, named for Scottish mathematician James Stirling (1692–1770).

Theorem 13.11. (Stirling's formula) $\displaystyle\lim_{n\to+\infty} \frac{n!\mathrm{e}^n}{n^n\sqrt{n}} = \sqrt{2\pi}.$

Proof. We divide the proof into two parts, (i) and (ii). In part (i), as regards estimates (13.3), we show that there exists $L > 0$ such that

$$L\,\sqrt{n}\,\left(\frac{n}{\mathrm{e}}\right)^n \;\leq\; n! \;\leq\; L\,\sqrt{n}\,\left(\frac{n}{\mathrm{e}}\right)^n \mathrm{e}^{\frac{1}{4n}}, \tag{13.4}$$

which gives $\displaystyle\lim_{n\to+\infty} \frac{n!\mathrm{e}^n}{n^n\sqrt{n}} = L$. Then in part (ii), we show that $L = \sqrt{2\pi}$.

(i) First we rewrite the inequalities (13.2) in a form which will be more useful for our purposes. Multiplying (13.2) by $\frac{2n+1}{2n} = \frac{1}{n}(n+\frac{1}{2})$, we get

$$1 \leq \left(n+\frac{1}{2}\right)\ln\left(\frac{n+1}{n}\right) \leq \frac{(2n+1)^2}{4n(n+1)} = 1 + \frac{1}{4n} - \frac{1}{4(n+1)}.$$

Therefore

$$\mathrm{e} \leq \left(\frac{n+1}{n}\right)^{n+1/2} \leq \mathrm{e}^{1+\frac{1}{4n}-\frac{1}{4(n+1)}}. \tag{13.5}$$

Now, looking at (13.4), consider the sequence $\{a_n\}$ given by

$$a_n = \frac{\sqrt{n}}{n!}\left(\frac{n}{\mathrm{e}}\right)^n = \frac{n^{n+1/2}}{\mathrm{e}^n\,n!}.$$

Then $\{a_n\}$ is increasing $\Leftrightarrow \quad a_{n+1} \geq a_n \quad \Leftrightarrow$

$$\frac{(n+1)^{n+1+1/2}}{\mathrm{e}^{n+1}(n+1)!} \geq \frac{n^{n+1/2}}{\mathrm{e}^n\, n!} \quad \Leftrightarrow \quad \left(\frac{n+1}{n}\right)^{n+1/2} \geq \mathrm{e}\,,$$

which is the left-hand inequality in (13.5). Now consider the sequence $\{b_n\}$ given by

$$b_n = a_n\, \mathrm{e}^{\frac{1}{4n}} = \frac{n^{n+1/2}}{\mathrm{e}^n\, n!}\, \mathrm{e}^{\frac{1}{4n}}.$$

Here, $\{b_n\}$ is decreasing $\Leftrightarrow\ b_{n+1} \leq b_n \quad \Leftrightarrow$

$$\frac{(n+1)^{n+1+1/2}}{\mathrm{e}^{n+1}\,(n+1)!}\mathrm{e}^{\frac{1}{4(n+1)}} \leq \frac{n^{n+1/2}}{\mathrm{e}^n\, n!}\mathrm{e}^{\frac{1}{4n}} \quad \Leftrightarrow \quad \left(\frac{n+1}{n}\right)^{n+1/2} \mathrm{e}^{\frac{1}{4(n+1)}} \leq \mathrm{e}^{1+\frac{1}{4n}},$$

which holds by the right-hand side of (13.5).

Now $a_n < b_n$ for all n, and since $\{a_n\}$ is increasing and $\{b_n\}$ is decreasing, $\{[a_n, b_n]\}$ is a nested sequence of intervals, with

$$b_n - a_n = b_n\big(1 - \mathrm{e}^{\frac{1}{4n}}\big) < b_1\big(1 - \mathrm{e}^{\frac{1}{4n}}\big) \to 0.$$

So by the Nested Interval Property (Theorem 1.41) there is a point c which belongs to each of these intervals. And $c > 0$ because $a_1 = 1/\mathrm{e} > 0$. Therefore

$$a_n \leq c \leq a_n\, \mathrm{e}^{\frac{1}{4n}}.$$

That is,

$$\sqrt{n}\, \frac{n^n}{\mathrm{e}^n\, n!} \leq c \leq \sqrt{n}\frac{n^n}{\mathrm{e}^n\, n!}\, \mathrm{e}^{\frac{1}{4n}}.$$

which gives (13.4), with $L = 1/c$.

(ii) We show now that $L = \sqrt{2\pi}$. Replacing n with $2n$ in $\lim\limits_{n\to+\infty} \dfrac{n!\, \mathrm{e}^n}{n^n\sqrt{n}} = L$ then squaring both sides, we get

$$\frac{((2n)!)^2\, \mathrm{e}^{4n}}{(2n)^{4n}\, 2n} \to L^2 \quad \text{as } n \to +\infty.$$

And taking the 4th power of both sides of $\lim\limits_{n\to+\infty} \dfrac{n!\, \mathrm{e}^n}{n^n\sqrt{n}} = L$, we get

$$\frac{(n!)^4\, \mathrm{e}^{4n}}{n^{4n}\, n^2} \to L^4 \quad \text{as } n \to +\infty.$$

Therefore

$$\frac{((2n)!)^2\,\mathrm{e}^{4n}}{(2n)^{4n}\,2n} \;\div\; \frac{(n!)^4\,\mathrm{e}^{4n}}{n^{4n}\,n^2} \to L^2 \;\div\; L^4 \quad \text{as } n \to +\infty\,.$$

That is,

$$\frac{n\,((2n)!)^2}{(n!)^4\,2^{4n+1}} \to \frac{1}{L^2} \quad \text{as } n \to +\infty\,.$$

So to finish the proof, it remains to show that

$$\frac{n\,((2n)!)^2}{(n!)^4\,2^{4n+1}} \to \frac{1}{2\pi} \quad \text{as } n \to +\infty\,. \tag{13.6}$$

Now in our proof of Wallis's product (Lemma 12.1) we showed that

$$\begin{aligned} \frac{I_{2n}}{I_{2n+1}} &= \frac{\frac{2n-1}{2n}\,\frac{2n-3}{2n-2}\cdots\frac{3}{4}\,\frac{1}{2}}{\frac{2n}{2n+1}\,\frac{2n-2}{2n-1}\cdots\frac{4}{5}\,\frac{2}{3}} \\ &= (2n+1)\left(\frac{2n-1}{2n}\right)^2\left(\frac{2n-3}{2n-2}\right)^2\cdots\left(\frac{5}{6}\right)^2\left(\frac{3}{4}\right)^2\left(\frac{1}{2}\right)^2 \\ &\to \frac{2}{\pi} \quad \text{as } n \to +\infty\,. \end{aligned}$$

Looking at the numerators here, observe that

$$(2n-1)(2n-3)\cdots(5)(3)(1) = \frac{(2n)!}{(2n)(2n-2)(2n-4)\cdots(4)(2)} = \frac{(2n)!}{2^n n!}\,.$$

And looking at the denominators, observe that

$$(2n)(2n-2)\cdots(4)(2) = (2n)(2(n-1))(2(n-2))\cdots(2(2))(2(1)) = 2^n n!\,.$$

Therefore,

$$\frac{I_{2n}}{I_{2n+1}} = (2n+1)\frac{((2n)!)^2}{2^{4n}\,(n!)^4} \to \frac{2}{\pi} \quad \text{as } n \to +\infty\,.$$

This is equivalent to

$$\frac{(2n+1)}{2n}\,\frac{n\,((2n)!)^2}{2^{4n+1}\,(n!)^4} \to \frac{1}{2\pi} \quad \text{as } n \to +\infty\,,$$

from which (13.6) follows, and the proof is complete. □

The approach above was motivated in part by [10, 11, 28], but see also [8, 14, 17, 22]. We consider alternative ways to show that L exists and is positive, in Exercises 13.34 and 13.35. In Exercise 13.37 we see another way of computing L using Wallis's product (Lemma 12.1).

Remark 13.12. Less precise estimates for $n!$, like

$$\mathrm{e}\left(\frac{n}{\mathrm{e}}\right)^n \;\le\; n! \;\le\; \mathrm{e}\left(\frac{n+1}{\mathrm{e}}\right)^{n+1},$$

can be obtained more easily—see for example Exercise 13.2. There, the Rectangle Rules are used instead of the more accurate Midpoint and Trapezoid Rules. Depending on the context, such inequalities are often sufficient for estimating $n!$. ∘

13.4 Trapezoid Rule or Midpoint Rule: Which Is Better?

Neither. For some functions the Trapezoid Rule is better, and for others the Midpoint Rule is better. (The reader should agree with this statement, perhaps after making a few sketches; see Exercise 13.39.)

But we show below (following [41]) that for functions which are either convex or concave, the Midpoint Rule is always at least as good as the Trapezoid Rule [3,6,21,41,47]. See **Fig. 13.6**. (So, for example, the left-hand inequality in (13.2) is at least as sharp as the right-hand inequality in (13.2).)

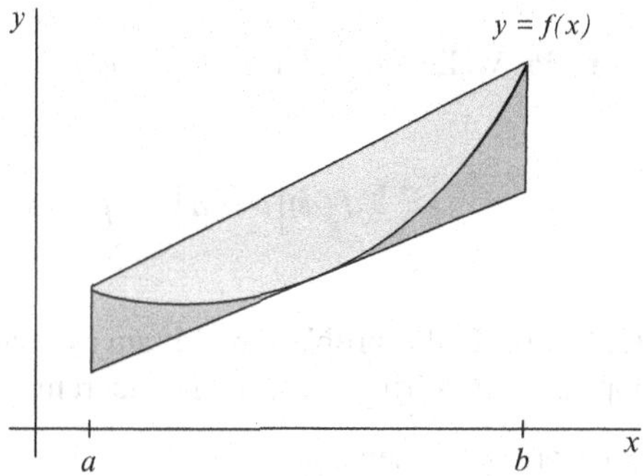

Fig. 13.6 Lemma 13.13, for a convex function. The lighter shaded area is the error for the Trapezoid Rule and the darker shaded area is the error for the Midpoint Rule

Lemma 13.13. *Let f be such that f'' exists on $[a,b]$. If f is convex then*

$$0 \;\le\; \int_a^b f(x)\,dx - f\left(\frac{a+b}{2}\right)[b-a] \;\le\; \frac{f(a)+f(b)}{2}[b-a] - \int_a^b f(x)\,dx\,,$$

and these inequalities are reversed for f concave.

Proof. We prove the statement for f being convex. (If f is concave then $-f$ is convex.) Let $c = (a+b)/2$. Applying the Trapezoid Rule on each of $[a,c]$ and $[c,b]$ we get, by the right-hand side of the HH Inequalities (Lemma 13.4),

$$\int_a^c f(x)\,dx + \int_c^b f(x)\,dx \le \frac{f(a)+f(c)}{2}[c-a] + \frac{f(c)+f(b)}{2}[b-c].$$

Then after some manipulations, using $c - a = b - c = (b-a)/2$, this reduces to

$$2\int_a^b f(x)\,dx \le f(c)[b-a] + \frac{f(a)+f(b)}{2}[b-a],$$

which is equivalent to the desired inequality. □

Again we denote by M the Midpoint Rule and by T the Trapezoid Rule, applied to a function f over $[a,b]$. In Remark 13.5 we saw that if f is strictly convex then for any $t \in (0,1)$, $tM + (1-t)T$ is a better approximation to $\int_a^b f(x)\,dx$ than either of M or T. Lemma 13.13 says that in the approximation $tM + (1-t)T$, we should take $t \ge 1/2$. We shall explore this matter further in Sect. 14.5.

Exercises

13.1. Prove Lemma 13.1: *If f is continuous and increasing on $[a,b]$, then*

$$f(a)[b-a] \le \int_a^b f(x)\,dx \le f(b)[b-a].$$

13.2. [46] In this problem we obtain estimates for $n!$ which are not as sharp as those appearing in Stirling's formula (Theorem 13.11), but are still of practical use.

(a) Verify that Lemma 13.1 gives

$$\ln(k-1) \le \int_{k-1}^{k} \ln(x)\,dx \le \ln(k) \qquad \text{for } k \in \mathbf{N}. \tag{13.7}$$

(b) Sum the right-hand side of (13.7) from $k = 2$ to n then integrate, to show that

$$\mathrm{e}\left(\frac{n}{\mathrm{e}}\right)^n \le n!$$

(c) Sum the left-hand side of (13.7) from $k = 3$ to n then integrate, to show that

$$n! \le e^2 \frac{n}{4} \left(\frac{n}{e}\right)^n.$$

(d) Sum the left-hand side of (13.7) from $k = 3$ to $n + 1$ to show that

$$n! \le e \left(\frac{n+1}{e}\right)^n.$$

(The rougher estimates $\left(\frac{n}{e}\right)^n \le n! \le \left(\frac{n+1}{2}\right)^n$ were obtained in Exercises 2.17, 2.27 and 8.55.)

13.3. [20] In Exercise 13.2 we saw that

$$e\left(\frac{n}{e}\right)^n \le n! \qquad \text{for } n = 3, 4, 5, \ldots.$$

Use this to show that for $n = 3, 4, 5, \ldots,$

$$\left[(n-1)!\right]^{n!} e \left(\frac{n}{e}\right)^{n!} < (n!)!.$$

13.4. [46] Show that for natural numbers $n \ge 3$,

$$n! \ln(n!) < n^n.$$

13.5. [12] Here's a way to show that $\left\{\left(1+\frac{1}{n}\right)^n\right\}$ is increasing, using Lemma 13.1.

(a) Apply Lemma 13.1 on $[n, n+1]$ to $f(x) = 1/x$ to show that

$$\frac{1}{n+2} < \int_1^{1+\frac{1}{n+1}} \frac{1}{x}\, dx.$$

(b) Verify that

$$\frac{1}{n+2} = \frac{n(n+1)}{n+2} \int_{1+\frac{1}{n+1}}^{1+\frac{1}{n}} 1\, dx > n \int_{1+\frac{1}{n+1}}^{1+\frac{1}{n}} \frac{1}{x}\, dx.$$

(c) Conclude that

$$n \ln\left(1 + \frac{1}{n}\right) = n \int_1^{1+\frac{1}{n}} \frac{1}{x}\, dx < (n+1) \ln\left(1 + \frac{1}{n+1}\right),$$

and so $\left\{\left(1 + \frac{1}{n}\right)^n\right\}$ is increasing.

(d) Modify the above analysis to show that $\{(1+\frac{1}{n})^{n+1}\}$ is decreasing.
Hint: $\frac{1}{n+1} = \frac{n+1}{1+\frac{1}{n}}\left[(1+\frac{1}{n}) - (1+\frac{1}{n+1})\right]$.

13.6. Denote by T the Trapezoid Rule and by M the Midpoint Rule, applied to a function f over some particular interval. Show that $(T+M)/2$ applied on $[a,b]$ is the same as T applied on $[a,(a+b)/2]$ and on $[(a+b)/2,b]$, then added together.

13.7. **(a)** Show that if the quadrature rule

$$\int_a^b f(x)\,dx \cong Af(a) + Bf(b)$$

is exact for $f(x) = 1$ and $f(x) = x$, then it's the Trapezoid Rule: $A = B = (b-a)/2$.

(b) Show that if the quadrature rule

$$\int_a^b f(x)\,dx \cong Af(\frac{a+b}{2})$$

is exact for $f(x) = 1$ and $f(x) = x$, then it's the Midpoint Rule: $A = (b-a)$.

13.8. [38] Let $x, y > 0$ and let $f(t) = x^t y^{(1-t)} = ye^{t\ln(x/y)}$.

(a) Verify that $f''(t) \geq 0$, so that f is convex.
(b) Apply the HH Inequalities (Lemma 13.4) on $[0,1]$ to give another proof of Lemma 6.20: $G \leq L \leq A$.

13.9. [2, 5, 42] Let $0 < x < y$. Apply the left-hand side of the HH Inequalities (Lemma 13.4) to $f(t) = 1/t$ on $[x,y]$ and the right-hand side of the HH Inequalities to the function $f(t) = 1/t$ on $[\sqrt{x},\sqrt{y}]$ to give another proof of Lemma 6.20: $G \leq L \leq A$.

13.10. [39] Denote by G, L, and A respectively, the Geometric, Logarithmic, and Arithmetic Means of $x, y > 0$. Show, as follows, that if $e^{3/2} \leq x < y$ then

$$A^L < G^A,$$

and that these inequalities are reversed for $0 < x < y \leq e^{3/2}$.

(a) Let $f(x) = \ln(x)/x$ for $x > 0$. Verify that f is convex on $(e^{3/2},\infty)$, and concave on $(0,e^{3/2})$.
(b) Apply the HH Inequalities (Lemma 13.4) for each case in (a).

13.11. The estimate $1/2 \leq \gamma$ (Euler's constant) from Example 13.10 was obtained by applying the right-hand side of the HH Inequalities (Lemma 13.4) to the convex function $1/x$ on $[a,b]$.

(a) Do this on each of $\left[a, \frac{a+b}{2}\right]$ and $\left[\frac{a+b}{2}, b\right]$ to show that $\gamma \geq 5/4 - \ln(2) \cong 0.556$.
(b) For only the most intrepid reader: Do this on each of the intervals $[a, (a+b)/4]$, $[(a+b)/4, (a+b)/2]$, $[(a+b)/2, 3(a+b)/4]$, and $[3(a+b)/4, b]$ to show that $\gamma \geq 47/24 - 2\ln(2) \cong 0.572$.

13.12. [27] For $0 < a < b$, the **Identric Mean** of a and b is

$$I(a,b) = \frac{1}{\mathrm{e}}\left(\frac{b^b}{a^a}\right)^{\frac{1}{b-a}}.$$

(a) Show that $I(a,b)$ is indeed a *mean*.
(b) How should we define $I(a,a)$?
(c) Show that

$$\left(\frac{\sqrt{a}+\sqrt{b}}{2}\right)^2 < I(a,b) < b^{\sqrt{b}}a^{\sqrt{a}}.$$

Hint: Apply the HH Inequalities (Lemma 13.4) to $f(t) = t\ln(t)$ on $\left[\sqrt{a}, \sqrt{b}\right]$.

13.13. [19] Let f be concave and differentiable on $[a,b]$. Find the point $c \in (a,b)$ such that the total area of the two inscribed trapezoids on $[a,c]$ and $[c,b]$ gives the best approximation to $\int_a^b f(x)\,dx$.

13.14. [44]

(a) Let f be differentiable on $[a,b]$. Use Integration by Parts to show that

$$\int_a^b \left(x - \tfrac{a+b}{2}\right) f'(x)\,dx = \tfrac{f(a)+f(b)}{2}(b-a) - \int_a^b f(x)\,dx.$$

(b) Show that if g is continuous and increasing on $[a,b]$, then

$$\int_a^b \left(x - \tfrac{a+b}{2}\right) g(x)\,dx \geq 0.$$

(c) Put these together to obtain the right-hand side of the HH Inequalities (Lemma 13.4).

13.15 ([18, 40]). Let f be a function with $f'' > 0$ on $[a,b]$. Show the best approximation to $\int_a^b f(t)\,dt$ by an *inscribed* trapezoid (the top of the trapezoid is tangent to $y = f(t)$ at $t = x$) occurs precisely when $x = \frac{a+b}{2}$. In this sense the left-hand side of the HH Inequalities (Lemma 13.4) is as good as it can be. See **Fig. 13.7.**

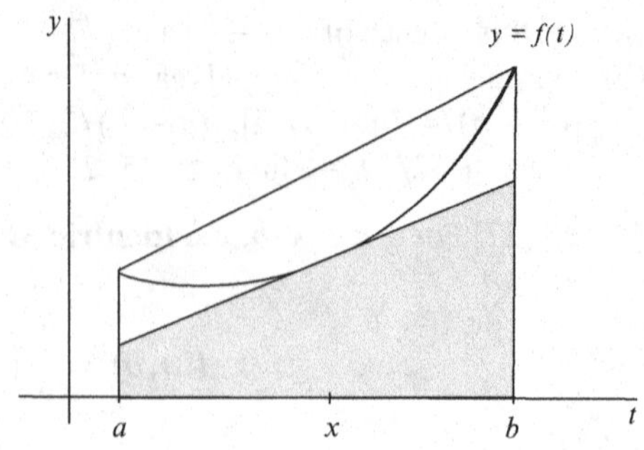

Fig. 13.7 For Exercise 13.15: For f convex, an inscribed trapezoid best approximates $\int_a^b f(x)\,dx$ when it is tangent to $y = f(x)$ at $x = (a+b)/2$

13.16. [15, 16] Let f be defined on $[a, b]$ with $f'' \geq 0$, so that f is convex.

(a) Fix $t \in [a, b]$ and integrate (with respect to x) the result of Lemma 8.7, i.e.,

$$f(x) \;\geq\; f'(t)(x-t) + f(t),$$

to obtain

$$\int_a^b f(x)\,dx \;\geq\; f'(t)(b-a)\big(\tfrac{a+b}{2} - t\big) + f(t)\big[b-a\big].$$

(b) What happens if $t = (a+b)/2$?

(c) Show that for any $t \in [a, b]$, the inequality in (a) is a refinement of the left-hand side of the HH Inequalities.

13.17. [15, 16] Let f be defined on $[a, b]$ with $f'' \geq 0$, so that f is convex.

(a) For $t \in [a, b]$, integrate (with respect to t) the result of Lemma 8.7, i.e.,

$$f(x) \;\geq\; f'(t)(x-t) + f(t),$$

to obtain

$$\int_a^b f(t)\,dt \;\leq\; \frac{(b-a)}{2} f(x) \;+\; \frac{bf(b) - af(a) - x\big(f(b) - f(a)\big)}{2}.$$

(b) What happens if $x = a$ or $x = b$?

(c) Show that for any $x \in [a, b]$, the inequality in (a) is a refinement of the right-hand side of the HH Inequalities.
Hint: Observe that

$$x = \frac{b-x}{b-a}a + \frac{x-a}{b-a}b,$$

so that

$$f(x) \le \frac{b-x}{b-a} f(a) + \frac{x-a}{b-a} f(b),$$

by convexity. Now write

$$\frac{1}{b-a}\left(\frac{(b-a)}{2} f(x) + \frac{bf(b)-af(a)-x(f(b)-f(a))}{2}\right) = \frac{f(x)}{2} + \frac{1}{2}\left[\frac{b-x}{b-a} f(b) + \frac{x-a}{b-a} f(a)\right].$$

13.18. [35] Let f and g be continuous on $[0, 1]$, with f decreasing and $0 \le g \le 1$. From Exercises 9.52 and/or 10.48 and/or 11.17 (also Exercise 11.36), **Steffensen's Inequalities** are:

$$\int_{1-\lambda}^{1} f(x)\,dx \le \int_{0}^{1} f(x)g(x)\,dx \le \int_{0}^{\lambda} f(x)\,dx, \quad \text{where } \lambda = \int_0^1 g(x)\,dx.$$

Apply Steffensen's Inequalities to f' and $g(t) = \begin{cases} t + 1/2 & \text{if } t \in [0, 1/2] \\ t - 1/2 & \text{if } t \in (1/2, 1] \end{cases}$ then to f' and $g(t) = t$ (in each case $\lambda = \int_0^1 g(x)\,dx = 1/2$), to obtain the HH Inequalities (Lemma 13.4). Can you do this on $[a, b]$?

13.19. **(a)** Show that (13.2) implies

$$\left(1 + \frac{1}{n}\right)^{n + \frac{n}{2n+1}} < \mathrm{e} < \left(1 + \frac{1}{n}\right)^{n+1/2} \qquad \text{for } n \in \mathbf{N}.$$

(b) Which is better, the left-hand side here or the left-hand side of (6.10), for $n \in \mathbf{N}$?

13.20. [32] Use (13.2) to show that $\left\{ \left(1 + \frac{1}{n}\right)^{n+1} \right\}$ is a decreasing sequence.

13.21. Recall that for $n = 1, 2, 3 \ldots$,

$$\gamma_n = \sum_{k=1}^{n} \frac{1}{k} - \ln(n).$$

Use (13.2) to show that $\{\gamma_n - \frac{1}{2n}\}$ is an increasing sequence. (Another approach to this can be found in [1].)

13.22. **(a)** Show that if $f'' \ge 0$ on $[a, b]$, then

$$f(a)[b-a] + f'(a)\frac{(b-a)^2}{2} \le \int_a^b f(x)\,dx \le f(b)[b-a] + f'(b)\frac{(b-a)^2}{2}.$$

(b) Draw a picture which shows what these estimates are saying geometrically.
(c) How do these estimates compare with the HH Inequalities (Lemma 13.4)?

13.23. [23] If f is increasing on $[0, 1]$, then clearly

$$\int_0^1 f(x)\,dx \;-\; \frac{1}{n}\sum_{k=1}^{n} f\left(\tfrac{k}{n}\right) \;\le\; 0.$$

Use the right-hand side of the HH Inequalities (Lemma 13.4) to sharpen this: *If f is continuous and convex on* $[0, 1]$, *then*

$$\int_0^1 f(x)\,dx \;-\; \frac{1}{n}\sum_{k=1}^{n} f\left(\tfrac{k}{n}\right) \;\le\; \frac{1}{2n}\big(f(0) - f(1)\big).$$

13.24. [4,34]

(a) Verify that for $x > 0$,

$$\int_1^x \frac{1-t}{t}\,dt = \ln(x) - x + 1.$$

(b) Use this to conclude that $\ln(x) \le x - 1$, with equality only for $x = 1$.
(c) Show that the integrand is convex.
(d) Apply the HH Inequalities (Lemma 13.4) to obtain the better estimates

$$\frac{(x-1)^2}{x+1} + \ln(x) \le x - 1 \le \frac{(x-1)^2}{2x} + \ln(x) \quad \text{for } 0 < x \le 1,$$

and

$$\frac{(x-1)^2}{2x} + \ln(x) \le x - 1 \le \frac{(x-1)^2}{x+1} + \ln(x) \quad \text{for } x > 1.$$

13.25. [45] Let f be continuous and differentiable on $[a, b]$.

(a) Apply Rolle's Theorem (Theorem 5.1) to

$$h(t) = \tfrac{f(t)+f(a)}{2}(t-a) - \int_a^t f(x)\,dx$$

to show that if the Trapezoid Rule is exact—that is, if

$$\tfrac{f(a)+f(b)}{2}(b-a) = \int_a^b f(x)\,dx,$$

then the conclusion of Flett's Mean Value Theorem (Theorem 7.4) holds.

(b) Draw a picture which shows what this is saying geometrically.

13.26. [9] Let f be such that f'' is continuous on $[a,b]$ and let $M = \max_{x\in[a,b]}\{f''(x)\}$. Fill in the details of the following proof of **Ostrowski's Inequality**:

$$\left| f(x) - \frac{1}{b-a}\int_a^b f(t)\,dt \right| \le M(b-a)\left(\tfrac{1}{4} + \left(\tfrac{x-\frac{a+b}{2}}{b-a}\right)^2\right).$$

(a) Use Integration by Parts to show that

$$\int_a^x (t-a)\,f'(t)\,dt = (x-a)\,f(x) - \int_a^x f(t)\,dt$$

and

$$\int_x^b (t-b)\,f'(t)\,dt = (b-x)\,f(x) - \int_x^b f(t)\,dt.$$

(b) Add these to get

$$(b-a)\,f(x) = \int_a^b f(t)\,dt + \int_a^b g_x(t)\,f'(t)\,dt,$$

where

$$g_x(t) = \begin{cases} t-a & \text{if } t\in[a,x] \\ t-b & \text{if } t\in(x,b]. \end{cases}$$

(c) Show that

$$\left|\int_a^b g_x(t)\,f'(t)\,dt\right| \le M\left(\tfrac{(x-a)^2+(b-x)^2}{2}\right) = M(b-a)\left(\tfrac{1}{4} + \left(\tfrac{x-\frac{a+b}{2}}{b-a}\right)^2\right).$$

13.27. [30] Let f and f' be continuous on $[a, b]$, with $|f'| < M$ there. Show that

$$\left| \int_a^b f(x)\,dx - \frac{f(a)+f(b)}{2}(b-a) \right| \le M \left(\frac{b-a}{2}\right)^2 .$$

Hint: Write $\int_a^b f(x)\,dx = \int_a^{(a+b)/2} f(x)\,dx + \int_{(a+b)/2}^b f(x)\,dx.$

13.28. We saw in Example 8.25 that for any $x \in \mathbf{R}$,

$$\lim_{n\to+\infty} \frac{x^n}{n!} = 0 .$$

Show this using Stirling's formula (Theorem 13.11).

13.29. [24] In our proof of Stirling's formula (Theorem 13.11), we saw that

$$n! > \left(\frac{n}{\mathrm{e}}\right)^n \sqrt{2\pi n} .$$

Use this to show that

$$\prod_{k=0}^{n-1} k! > \left(\frac{n}{\mathrm{e}}\right)^n .$$

13.30. [25] Stirling's formula (Theorem 13.11) is often written compactly as

$$n! \approx \sqrt{2\pi n} \left(\frac{n}{\mathrm{e}}\right)^n .$$

Use this to show that in fact

$$n! \approx \sqrt{2\pi} \left(\frac{n+1/2}{\mathrm{e}}\right)^{n+1/2} .$$

For n very large the error in this latter estimate is about half that of the former. It is really the latter estimate that Stirling obtained; the former was obtained by French mathematician Abraham DeMoivre (1667–1754) [31,43].

13.31. **(a)** Show that the fact that $\lim_{n\to\infty} \frac{n!\mathrm{e}^n}{n^n\sqrt{n}}$ exists, by Theorem 13.11, implies that

$$\lim_{n\to\infty} \frac{\sqrt[n]{n!}}{n} = \frac{1}{\mathrm{e}}.$$

We met this latter limit, in Exercises 6.12 and 10.47.

(b) If A_n is the Arithmetic Mean of the first n natural numbers, and G_n is their Geometric Mean, show that the result in (a) is the same as

$$\lim_{n\to\infty} \frac{G_n}{A_n} = \frac{2}{\mathrm{e}}.$$

13.32. [37] Apply the Trapezoid Rule and the HH Inequalities (Lemma 13.4) to $\int_k^{k+1} \ln(x)\,dx$ then sum from $k = 1$ to $n - 1$, to show that

$$n! \le \frac{n^n \sqrt{n}\,\mathrm{e}}{\mathrm{e}^n} \quad \text{for } n = 1, 2, 3, \ldots.$$

13.33. [37] Show, as follows, that

$$\frac{n^n \sqrt{n}\,\sqrt{\mathrm{e}}}{\mathrm{e}^n} \le n! \quad \text{for } n = 1, 2, 3, \ldots.$$

(a) Verify that $\int_k^{k+1} \ln(x)\,dx$ is bounded by the area of the trapezoid whose base is the interval $[k, k+1]$, and top is part of the tangent line to $f(x) = \ln(x)$ at $x = k$.

(b) Use the bound obtained in (a), sum from $k = 1$ to $n - 1$, then manipulate.

13.34. [33] Here's another way to show that the limit $L > 0$ in Stirling's formula (Theorem 13.11) exists. We showed in the proof that $\{a_n\}$ given by

$$a_n = \frac{n^n \sqrt{n}}{n!\,\mathrm{e}^n}$$

is an increasing sequence. So if we can show that $\{a_n\}$ is bounded above then we are finished, by the Increasing Bounded Sequence Property (Theorem 1.34).

(a) Use (13.2) to show that

$$\left(1 + \frac{1}{k}\right)^{k(1+\frac{1}{2k+1})} < \mathrm{e} < \left(1 + \frac{1}{k}\right)^{k+1/2}.$$

(b) Use the left-hand inequality from (a) to show that

$$\frac{(n+1)^n}{n!} \prod_{k=1}^{n} \left(\frac{k+1}{k}\right)^{k/(2k+1)} < \mathrm{e}^n.$$

(c) Use both inequalities from (a) to show that $\{a_n\}$ is bounded above.

13.35. cf. [37] Here's another way to show that the limit $L > 0$ in Stirling's formula (Theorem 13.11) exists. We showed in the proof that $\{q_n\} = \{1/a_n\}$ given by

$$q_n = \frac{n!\,e^n}{n^n\sqrt{n}}$$

is a decreasing sequence. So if we can show that $\{q_n\}$ is bounded below by a positive constant then we are finished.

(a) Show that the Midpoint Rule and the HH Inequalities (Lemma 13.4) give

$$\int_k^{k+1} \ln(x)\,dx \le \ln\left(\frac{2k+1}{2}\right).$$

(b) Show that

$$\ln\left(\frac{2k+1}{2}\right) = \ln(k)+\ln\left(\frac{2k+1}{2k}\right) = \ln(k)+\ln\left(1+\frac{1}{2k}\right) \le \ln(k)+\frac{1}{2k}.$$

(c) Conclude that

$$\int_k^{k+1} \ln(x)\,dx \le \ln(k)+\frac{1}{2k}.$$

(d) Sum from $k = 1$ to $n-1$, to get

$$\int_1^n \ln(x)\,dx \le \sum_{k=1}^{n-1} \ln(k) + \frac{1}{2}\big(\ln(n)+1\big).$$

(e) Evaluate the integral and do some rearranging to get

$$\frac{1}{2} \le \ln(n!) + n - \left(n+\frac{1}{2}\right)\ln(n).$$

(f) Conclude that $\{q_n\}$ is indeed bounded below, by $\sqrt{e}$.

13.36. (a) Apply the Midpoint Rule and the HH Inequalities (Lemma 13.4) to $\ln(x)$ on $[k, k+1]$ (for $k > 0$) then sum from $k = 1$ to n to show that

$$(n+1)^{n+1}\left(\frac{4}{e}\right)^n \le \frac{(2n+1)!}{n!}.$$

Note: The number

$$C_n = \frac{1}{n+1}\binom{2n}{n} = \frac{1}{n+1}\frac{(2n)(2n-1)\cdots(n+1)}{n!}$$

is called the nth *Catalan number*, named for Belgian/French mathematician Eugène Catalan (1814–1894). This number comes up often in combinatorics. (C_n is, for example, the number of ways one can *triangulate* an $n+2$ sided convex polygon i.e., cut the polygon into triangles by connecting its vertices with straight lines. See [13].) So the result shows that

$$C_n \geq \frac{(n+1)^n}{(2n+1)n!}\left(\frac{4}{e}\right)^n.$$

(b) Can you get a better estimate using Stirling's formula (Theorem 13.11)?

13.37. [28] Here's another way to show that $L = \sqrt{2\pi}$ in Stirling's formula (Theorem 13.11), using Wallis's product (Lemma 12.1)

$$\frac{\pi}{2} = \lim_{m\to+\infty} \frac{2}{1}\frac{2}{3}\frac{4}{3}\frac{4}{5}\frac{6}{5}\frac{6}{7}\cdots\frac{2m}{2m-1}\frac{2m}{2m+1}.$$

(a) Show that Wallis's product may be rewritten as

$$\frac{\pi}{2} = \lim_{m\to+\infty} \frac{2^2}{3^2}\frac{4^2}{5^2}\frac{6^2}{7^2}\cdots\frac{(2m-2)^2}{(2m-1)^2}2m.$$

(b) Conclude that

$$\sqrt{2\pi} = \lim_{m\to+\infty} \frac{2}{3}\frac{4}{5}\frac{6}{7}\cdots\frac{(2m-2)}{(2m-1)}2\sqrt{2}\sqrt{m}.$$

(c) Show that

$$\lim_{m\to+\infty} \frac{2}{3}\frac{4}{5}\frac{6}{7}\cdots\frac{(2m-2)}{(2m-1)}2\sqrt{2}\sqrt{m} = \lim_{m\to+\infty} \frac{2^{2m}(m!)^2}{(2m)!}\frac{\sqrt{2}}{\sqrt{m}}.$$

(d) Show that

$$L = \lim_{n\to+\infty} \frac{n!e^n}{n^{n+1/2}} = \lim_{m\to+\infty} \frac{(2m)!e^{2m}}{(2m)^{2m+1/2}}$$

$$= \lim_{m\to+\infty} \left(\frac{(2m)!\sqrt{m}}{2^{2m}(m!)^2\sqrt{2}}\right)\left(\frac{m!e^m}{m^{m+1/2}}\right)^2.$$

(e) Conclude that $L = \frac{1}{\sqrt{2\pi}}L^2$, and so $L = \sqrt{2\pi}$.

13.38. [47] Fill in the details, as follows, of another proof of Lemma 13.13, for f convex. (If f is concave then $-f$ is convex.)

(a) Set $g(x) = f(x) + f(a+b-x)$. Verify that g is convex, and symmetric with respect to $(a+b)/2$.
(b) Verify that for g nonnegative,

$$\int_a^b g(x)\,dx \;\le\; \frac{b-a}{2}\left[g(a) + g(\tfrac{a+b}{2})\right].$$

(c) Write down what (b) means, in terms of f.
(d) What if g is sometimes negative?

13.39. **(a)** Sketch the graph of a function on $[a,b]$ for which the Trapezoid Rule is better than the Midpoint Rule. **(b)** Sketch the graph of a function, which is neither convex nor concave on $[a,b]$, for which the Midpoint Rule is better than the Trapezoid Rule.

13.40. Show that $\left(1+\frac{1}{x}\right)^{x+1/2} - \mathrm{e} \;\le\; \mathrm{e} - \left(1+\frac{1}{x}\right)^{\sqrt{x(x+1)}}$ for $x > 0$.

References

1. Bracken, P., Plaza, Á.: Problem 1781. Math. Mag. **81**, 376–377 (2008)
2. Bruce, I.: The logarithmic mean. Math. Gaz. **81**, 89–92 (1997)
3. Bullen, P.S.: Error estimates for some elementary quadrature rules. Univ. Beograd. Publ. Elektrotehn. Fak. Ser. Mat. Fiz. **602–633**, 97–103 (1978)
4. Bullen, P.S.: Handbook of Means and Their Inequalities. Kluwer Academic, Dordrecht (2003)
5. Burk, F.: Mean inequalities. Coll. Math. J. **14**, 431–434 (1983)
6. Burk, F.: Behold! The midpoint rule is better than the trapezoid rule for convex functions (A proof without words). Coll. Math. J. **16**, 56 (1985)
7. Burk, F.: Geometric, logarithmic, and arithmetic mean inequality. Am. Math. Mon. **94**, 527–528 (1987)
8. Carlson, R.: A Concrete Approach to Real Analysis. Chapman & Hall/CRC, Boca Raton (2006)
9. Cerone, P., Dragomir, S.S.: Mathematical Inequalities: A Perspective. CRC, New York (2011)
10. Conrad, K.: Stirling's formula (2014). www.math.uconn.edu/~kconrad
11. Courant, R.: Differential and Integral Calculus, vol. 1. Wiley Classics Library, Hoboken (1988)
12. DeTemple, D.W.: An elementary proof of the monotonicity of $(1+1/n)^n$ and $(1+1/n)^{n+1}$. Coll. Math. J. **36**, 147–149 (2005)
13. Devadoss, S.L., O'Rourke, J.: Discrete and Computational Geometry. Princeton University Press, Princeton (2011)
14. Dobrescu, M.: A new look at the convergence of a famous sequence. Int. J. Math. Educ. Sci. Technol. **41**, 1079–1085 (2010)
15. Dragomir, S.S.: On Hadamard's inequalities for convex functions. Mat. Balk. **6**, 215–222 (1992)
16. Dragomir, S.S., Ionescu, N.M.: Some integral inequalities for differentiable convex functions. Coll. Pap. Fac. Sci. Kragujev. (Yugoslavia) **13**, 11–16 (1992)

17. Dutkay, D.E., Niculescu, C.P., Popovici, F.: Stirling's formula and its extension for the Gamma function. Am. Math. Mon. **120**, 737–740 (2013).
18. Eddy, R.H., Fritsch, R.: An optimization oddity. Coll. Math. J. **25**, 227–229 (1994)
19. Fischer, I.: Problem Q828. Math. Mag. **67**, 385–390 (1994)
20. Garfunkel, J., Pittie, H.: Problem E1816. Am. Math. Mon. **74**, 202 (1967)
21. Hammer, P.: The midpoint method of numerical integration. Math. Mag. **31**, 193–195 (1958)
22. Impens, C.: Stirling's series made easy. Am. Math. Mon. **110**, 730–735 (2003)
23. Jichang, K.: Some extensions and refinements of the Minc-Sathre inequality. Math. Gaz. **83**, 123–127 (1999)
24. Just, E., Waterhouse, W.C.: Problem E1652. Am. Math. Mon. **71**, 1043 (1964)
25. Keiper, J.B.: Stirling's formula improved. College Math. J. **10**, 38–39 (1979)
26. Khattri, S.K.: Three proofs of the inequality $e < \left(1+\frac{1}{x}\right)^{x+1/2}$. Am. Math. Mon. **117**, 273–277 (2010)
27. Kung, S., Perlman, M.D.: Problem 1365. Math. Mag. **65**, 61–63 (1992)
28. Lang, S.: A First Course in Calculus. Addison Wesley, Reading (1964)
29. Lipkin, L.J.: How large is $n!$? Coll. Math. J. **37**, 109 (2006)
30. Lupu, C., Lupu, T.: Problem 927. Coll. Math. J. **41**, 242 (2010); two solutions (different from that suggested in the hint) in Coll. Math. J. **42**, 236–237 (2011)
31. Maritz, P.: James Stirling: mathematician and mine manager. Math. Intell. **33**, 141–147 (2011)
32. Mercer, P.R.: On the monotonicity of $\{(1+1/n)^n\}$ and $\{(1+1/n)^{n+1}\}$. Coll. Math. J. **34**, 236–238 (2003)
33. Mercer, P.R.: On a precursor to Stirling's formula. Math. Gaz. **87**, 530–532 (2003)
34. Mercer, P.R.: Refined arithmetic, geometric, and harmonic mean inequalities. Rocky Mt. Math. J. **33**, 1459–1464 (2003)
35. Mercer, P.R.: Error Terms for Steffensen's, Young's, and Chebyshev's inequalities. J. Math. Inequal. **2**, 479–486 (2008)
36. Mitrinović, D.S., Lacković, I.B.: Hermite and convexity. Aequ. Math. **28**, 229–232 (1985)
37. Mountford, D.: Bounds for $n!$. Math. Gaz. **69**, 286–287 (1985)
38. Neuman, E.: The weighted logarithmic mean. J. Math. Anal. Appl. **18**, 885–900 (1994)
39. Neuman, E., Zhou, L.: Problem 10798. Am. Math. Mon. **108**, 178 (2001)
40. Pare, R.: A visual proof of Eddy and Fritsch's minimal area property. Coll. Math. J. **26**, 43–44 (1995)
41. Pecaric, J.E., Proschan, F., Tong, Y.L.: Convex Functions, Partial Orderings and Statistical Applications. Academic, New York (1992)
42. Pinker, A., Chow, T.Y., Merino, J.C.: Problem 209. Coll. Math. J. **14**, 353–356 (1983)
43. Roy, R.: Sources in the Development of Mathematics. Cambridge University Press, Cambridge/New York (2011)
44. Sandor, J.: Some simple integral inequalities. Oct. Math. Mag. **16**, 925–933 (2008)
45. Tong, J.: On Flett's mean value theorem. Int. J. Math. Educ. Sci. Technol. **35**, 936–941 (2004)
46. Velleman, D.J.: Exponential vs. factorial. Am. Math. Mon. **113**, 689–704 (2006)
47. Wąsowicz, S., Witkowski, A.: On some inequality of Hermite-Hadamard Type. Opusc. Math. **32**, 591–600 (2012)

Chapter 14
Error Terms

From error to error one discovers the entire truth.

—Sigmund Freud

Any inequality $A \leq B$ can be restated, at least in principle, as an equality $A = B - E$ where $E \geq 0$ is an *error term*. Of course, if $A \leq B$ is complicated, then the error term probably cannot be known exactly. But quite often in calculus, something can be said about the error term. When this is the case, interesting and useful things usually follow.

14.1 The Mean Value Theorem Again

Let f be a differentiable function defined on an open interval J and let $x_0 \in J$ be fixed. By the Mean Value Theorem (Theorem 5.2), for $x \in J$ (with $x \neq x_0$) there exists c between x and x_0 such that

$$f(x) - f(x_0) = f'(c)\big[x - x_0\big].$$

Or stated another way,

$$f(x) - f(x_0) = f'(c)\big[I(x) - I(x_0)\big],$$

where I is the identity function $I(x) = x$.

So the error $E = f(x) - f(x_0)$ that results from the approximation

$$f(x) \cong f(x_0)$$

is given by the *error term* $f'(c)[I(x) - I(x_0)]$.

Equivalently, the error in $f(x) \cong f(x_0)$ is given by that error which results from the approximation $I(x) \cong I(x_0)$, but scaled by a factor of $f'(c)$.

P.R. Mercer, *More Calculus of a Single Variable*, Undergraduate Texts in Mathematics, DOI 10.1007/978-1-4939-1926-0_14

If f is a constant function then the error $E = 0$, but if f is the identity function $I(x) = x$ then $E \neq 0$. This is why the error term involves the *first* derivative of f.

These simple observations show themselves in many results which follow from the Mean Value Theorem (Theorem 5.2). Here is one such example, which provides an error term for the Right Rectangle Rule.

Lemma 14.1. *Let f be defined on $[a, b]$ with f' continuous. Then there is $c \in [a, b]$ such that*

$$\int_a^b f(x)\,dx \; - \; f(b)\big[b - a\big] \; = \; -f'(c)\frac{(b-a)^2}{2}.$$

Proof. Let $x \in [a, b)$. By the Mean Value Theorem (Theorem 5.2), there is $\xi \in (a, b)$ such that

$$f(b) - f(x) = f'(\xi)(b - x).$$

Integrating from a to b we get

$$f(b)\big[b - a\big] - \int_a^b f(x)\,dx = \int_a^b f'(\xi)(b - x)\,dx.$$

Now $(b-x) \geq 0$ for $x \in [a, b]$ and f' is continuous, so by the Mean Value Theorem for Integrals (Theorem 9.14) there is $c \in (a, b)$ such that

$$f(b)\big[b - a\big] - \int_a^b f(x)\,dx = f'(c)\int_a^b (b - x)\,dx = f'(c)\frac{(b-a)^2}{2},$$

as desired. □

The reader may verify that for I being the identity function $I(x) = x$,

$$\int_a^b I(x)\,dx - I(b)\big[b - a\big] = \int_a^b x\,dx - b\big[b - a\big] = -\frac{(b-a)^2}{2},$$

so that the conclusion of Lemma 14.1 can be written as

$$\int_a^b f(x)\,dx - f(b)\big[b - a\big] = f'(c)\left[\int_a^b I(x)\,dx - I(b)\big[b - a\big]\right].$$

That is, the error that results from the approximation

$$\int_a^b f(x)\,dx \cong f(b)\big[b-a\big],$$

is precisely the error that results from the approximation $\int_a^b I(x)\,dx \cong I(b)\big[b-a\big]$, but scaled by a factor of $f'(c)$.

Also, the error $E = \int_a^b f(x)\,dx - f(b)\big[b-a\big]$ is zero if f is a constant function, but if f is the identity function $I(x) = x$ then $E \neq 0$. This is why the error term $f'(c)\left[\int_a^b I(x)\,dx - I(b)\big[b-a\big]\right]$ involves the *first* derivative of f.

We leave it for Exercise 14.1 so show that for the Left Rectangle Rule we get a similar error term (but with a different c):

$$\begin{aligned} f(a)\big[b-a\big] - \int_a^b f(x)\,dx &= -f'(c)\frac{(b-a)^2}{2} \\ &= f'(c)\left[I(a)\big[b-a\big] - \int_a^b I(x)\,dx\right]. \end{aligned}$$

So putting Lemma 14.1 and Exercise 14.1 together, we obtain the following.

Theorem 14.2. (Error Terms for the Rectangle Rules) *Let f be defined on $[a,b]$, with f' continuous. Then there exists $c_1, c_2 \in [a,b]$ such that*

$$f(a)\big[b-a\big] + f'(c_1)\frac{(b-a)^2}{2} = \int_a^b f(x)\,dx = f(b)\big[b-a\big] - f'(c_2)\frac{(b-a)^2}{2}.$$

Finally, recall that if $f' \geq 0$ then the Mean Value Theorem (Theorem 5.2) implies that $f(x) \geq f(x_0)$ for $x > x_0$. This is Lemma 5.6. Here, if $f' \geq 0$ then Theorem 14.2 gives the conclusion of Lemma 13.1:

$$f(a)\big[b-a\big] \leq \int_a^b f(t)\,dt \leq f(b)\big[b-a\big].$$

14.2 Jensen's Inequality Again

Let f be twice differentiable on an open interval J and let $x_0 \in J$ be fixed. By the Mean Value Theorem for the Second Derivative (Theorem 8.6), for each $x \in J$ (with $x \neq x_0$) there exists c between x and x_0 such that

$$f(x) - f(x_0) - f'(x_0)(x - x_0) = \frac{f''(c)}{2}(x - x_0)^2.$$

With $I_2(x) = x^2$, this reads (as the reader may verify):

$$f(x) - f(x_0) - f'(x_0)\big(x - x_0\big) \;=\; \frac{f''(c)}{2!}\big[I_2(x) - I_2(x_0) - I_2'(x_0)\big(x - x_0\big)\big].$$

This says that the error $E = f(x) - \big(f(x_0) + f'(x_0)(x - x_0)\big)$ that results from the approximation

$$f(x) \;\cong\; f(x_0) + f'\big(x_0\big)\big(x - x_0\big)$$

is the error that results from the approximation $I_2(x) \;\cong\; I_2(x_0) + I_2'(x_0)\big(x - x_0\big)$, but scaled by a factor of $f''(c)/2$.

If f is a constant function then the error $E = 0$, and $E = 0$ if f is the identity function $I(x) = x$. But $E \neq 0$ if f is the function $I_2(x) = x^2$. This is why the error term $(f''(c)/2!)[I_2(x) - I_2(x_0) - I_2'(x_0)(x - x_0)]$ involves the *second* derivative.

These observations show themselves in many results which follow from the Mean Value Theorem for the Second Derivative (Theorem 8.6). Here is an example which provides an error term for Jensen's Inequality (Theorem 8.17). Our proof follows [17]; see also [7,31].

Theorem 14.3. (Error Term for Jensen's Inequality) *Let f be defined on $[a, b]$, with f'' continuous. For $n \geq 2$, let $x_1, x_2, \ldots, x_n \in [a, b]$ and let $w_1, w_2, \ldots, w_n$ be positive, with $\sum_{j=1}^{n} w_j = 1$. Then there exists $c \in [a, b]$ such that*

$$f\left(\sum_{j=1}^{n} w_j x_j\right) - \sum_{j=1}^{n} w_j f(x_j) \;=\; -\frac{f''(c)}{2}\sum_{j=1}^{n} w_j (x_j - A)^2.$$

Proof. With $A = \sum_{j=1}^{n} w_j x_j$, by the Mean Value Theorem for the Second Derivative (Theorem 8.6) there is c_j between x_j and A such that

$$f(x_j) = f(A) + f'(A)(x_j - A) + \frac{1}{2} f''(c_j)(x_j - A)^2.$$

Multiplying by w_j and summing from 1 to n we get

$$\begin{aligned}\sum_{j=1}^{n} w_j f(x_j) &= \sum_{j=1}^{n} w_j f(A) + \sum_{j=1}^{n} w_j f'(A)(x_j - A) + \frac{1}{2}\sum_{j=1}^{n} w_j f''(c_j)(x_j - A)^2 \\ &= f(A)\sum_{j=1}^{n} w_j + f'(A)(A - A) + \frac{1}{2}\sum_{j=1}^{n} w_j f''(c_j)(x_j - A)^2 \\ &= f(A) + 0 + \frac{1}{2}\sum_{j=1}^{n} w_j f''(c_j)(x_j - A)^2.\end{aligned}$$

Now since f'' is continuous and $w_j(x_j - A)^2 \geq 0$ for each j, by the Mean Value Theorem for Sums (Theorem 3.22) there is $c \in (a, b)$ such that

$$\frac{1}{2}\sum_{j=1}^{n} w_j f''(c_j)(x_j - A)^2 = \frac{f''(c)}{2}\sum_{j=1}^{n} w_j (x_j - A)^2,$$

as desired. □

Observe that

$$\begin{aligned}\sum_{j=1}^{n} w_j (x_j - A)^2 &= \sum_{j=1}^{n} w_j x_j^2 - \sum_{j=1}^{n} 2w_j x_j A + \sum_{j=1}^{n} w_j A^2 \\ &= \sum_{j=1}^{n} w_j x_j^2 - 2A^2 + A^2 \\ &= \sum_{j=1}^{n} w_j x_j^2 - \left(\sum_{j=1}^{n} w_j x_j\right)^2 \\ &= \sum_{j=1}^{n} w_j I_2(x_j) - I_2\left(\sum_{j=1}^{n} w_j x_j\right),\end{aligned}$$

where $I_2(x) = x^2$. So the conclusion of Theorem 14.3 can be written as

$$f\left(\sum_{j=1}^{n} w_j x_j\right) - \sum_{j=1}^{n} w_j f(x_j) = \frac{f''(c)}{2}\left[I_2\left(\sum_{j=1}^{n} w_j x_j\right) - \sum_{j=1}^{n} w_j I_2(x_j)\right].$$

That is, the error that results from the approximation

$$f\left(\sum_{j=1}^{n} w_j x_j\right) \cong \sum_{j=1}^{n} w_j f(x_j)$$

is precisely the error that results from the approximation

$$I_2\left(\sum_{j=1}^{n} w_j x_j\right) \cong \sum_{j=1}^{n} w_j I_2(x_j),$$

but scaled by a factor of $f''(c)/2$.

If f is a constant function then the error $E = f\left(\sum_{j=1}^{n} w_j x_j\right) - \sum_{j=1}^{n} w_j f(x_j) = 0$, and $E = 0$ if f is the identity function $I(x) = x$. But $E \neq 0$ if f is the function $I_2(x) = x^2$. This is why the error term involves the *second* derivative of f.

These observations should not be surprising, in hindsight, because we used the Mean Value Theorem for the Second Derivative (Theorem 8.6) to obtain the error term. In the next sections we shall rely on this pattern to guess what error terms might look like in other contexts.

Finally, if we assume also that f is convex, i.e., $f'' \geq 0$, then Theorem 14.3 yields the conclusion of Jensen's Inequality (Theorem 8.17):

$$f\left(\sum_{j=1}^{n} w_j x_j\right) \leq \sum_{j=1}^{n} w_j f\left(x_j\right).$$

Remark 14.4. In Exercises 14.4 and 14.5 we look at other ways of obtaining the Error Term for Jensen's Inequality (Theorem 14.3). In Exercise 14.8 we see that Jensen's Integral Inequality (Theorem 9.29) has an error term completely analogous to that of Theorem 14.3. ∘

Example 14.5. [24] We saw in Example 8.18 that letting $f(x) = -\ln(x)$ in Jensen's Inequality (Theorem 8.17) then applying the exponential function to both sides, yields the weighted AGM Inequality (Theorem 6.15):

$$G = \left(\prod_{j=1}^{n} x_j^{w_j}\right) \leq \sum_{j=1}^{n} w_j x_j = A.$$

Here, the Error Term for Jensen's Inequality (Theorem 14.3) gives

$$G = Ae^{-\frac{1}{2}f''(c)\sum_{j=1}^{n} w_j (x_j - A)^2} = Ae^{-\frac{1}{2c^2}\sum_{j=1}^{n} w_j (x_j - A)^2}.$$

This gives a lower bound for, and a refinement of, the weighted AGM Inequality (Theorem 6.15) $G \le A$ as follows. If $0 < x_1 \le x_2 \le \cdots \le x_n$ then $c \in (x_1, x_n)$ and so

$$A\,e^{-\frac{1}{2x_1^2}\sum_{j=1}^{n} w_j (x_j - A)^2} \le G \le A e^{-\frac{1}{2x_n^2}\sum_{j=1}^{n} w_j (x_j - A)^2} \le A. \qquad \diamond$$

Let us now take $n = 2$ in the Error Term for Jensen's Inequality (Theorem 14.3), with $w_1 = 1 - t$ and $w_2 = t$ (for $t \in [0, 1]$). After some manipulations, which we leave for Exercise 14.6, we get the following error term for the convexity condition.

Corollary 14.6. *Let f be defined on $[a, b]$, with f'' continuous. Let $a \le x < y \le b$ and $0 \le t \le 1$. Then there exists $c \in [a, b]$ such that*

$$f\big((1-t)x + ty\big) - \big[(1-t)f(x) + tf(y)\big] = \frac{f''(c)}{2} t(t-1)(x-y)^2.$$

Proof. This is Exercise 14.6. □

Notice that in Corollary 14.6, since $0 \le t \le 1$, we have $t(t - 1) \le 0$. So for $f'' \ge 0$ we get the convexity condition

$$f\big((1-t)x + t\big) \le (1-t)f(x) + tf(y),$$

exactly as we should.

We have seen that the Mean Value Theorem (Theorem 5.2) is Taylor's Theorem (Theorem 8.20) with $n = 0$, and the Mean Value Theorem for the Second Derivative (Theorem 8.6) is Taylor's Theorem with $n = 1$. Error terms continue the pattern we have seen thus far, for larger n.

For example, for the $n = 2$ case, writing $I_3(x) = x^3$,

$$\begin{aligned} E &= \frac{f^{(3)}(c)}{3!}(x - x_0)^3 \\ &= \frac{f^{(3)}(c)}{3!}\Big[I_3(x) - I_3(x_0) - I_3'(x_0)(x - x_0) - \frac{I_3''(x_0)}{2!}(x - x_0)^2\Big]. \end{aligned}$$

This is the error that arises from the approximation

$$I_3(x) \cong I_3(x_0) + I_3'(x_0)(x - x_0) + \frac{I_3''(x_0)}{2!}(x - x_0)^2,$$

scaled by a factor of $f^{(3)}(c)/3!$. And for the $n = 3$ case, writing $I_4(x) = x^4$,

$$\begin{aligned} E &= \frac{f^{(4)}(c)}{4!}(x - x_0)^4 \\ &= \frac{f^{(4)}(c)}{4!}\Big[I_4(x) - I_4(x_0) - I_4'(x_0)(x - x_0) - \frac{I_4''(x_0)}{2!}(x - x_0)^2 - \frac{I_4^{(3)}(x_0)}{3!}(x - x_0)^3\Big], \end{aligned}$$

This is the error that arises from the approximation

$$I_4(x) \cong I_4(x_0) + I_4'(x_0)(x - x_0) + \frac{I_4''(x_0)}{2!}(x - x_0)^2 + \frac{I_4^{(3)}(x_0)}{3!}(x - x_0)^3,$$

scaled by a factor of $f^{(4)}(c)/4!$, etcetera. Verifying these facts is the content of Exercise 14.2. These general observations are manifest in many results which follow from Taylor's Theorem (Theorem 8.20).

14.3 The Trapezoid Rule Again

For f defined on $[a, b]$, the Trapezoid Rule is the approximation

$$\int_a^b f(x)\,dx \cong \frac{f(a) + f(b)}{2}[b - a].$$

Here, we get *equality* if f is a constant function or if $f(x) = I(x) = x$, but not if $f(x) = I_2(x) = x^2$. So we might expect that an error term for this inequality should involve $f''(c)/2$. And we might also expect that in such an error term, the multiplier of $f''(c)/2$ would be $\int_a^b f(x)\,dx - \frac{f(a)+f(b)}{2}(b - a)$, but with $f(x) = I_2(x)$.

Theorem 14.7. (Trapezoid Rule Error) *Let f be defined on $[a, b]$ with f'' continuous. Then there is $c \in [a, b]$ such that*

$$\int_a^b f(x)\,dx - \frac{f(a) + f(b)}{2}[b - a] = -f''(c)\frac{(b - a)^3}{12}$$

$$= \frac{f''(c)}{2}\left[\int_a^b I_2(x)\,dx - \frac{I_2(a) + I_2(b)}{2}[b - a]\right].$$

Proof. We begin with Corollary 14.6: For some $\xi \in [a, b]$,

$$f\big((1 - t)a + tb\big) - \big[(1 - t)f(a) + tf(b)\big] = \frac{f''(\xi)}{2}t(t - 1)(b - a)^2.$$

Integrating from $t = 0$ to $t = 1$ we get

$$\int_0^1 f\big((1 - t)a + tb\big)\,dt - \int_0^1 \big[(1 - t)f(a) + tf(b)\big]\,dt = \int_0^1 \frac{f''(\xi)}{2}t(t - 1)(b - a)^2\,dt$$

$$= \frac{(b - a)^2}{2}\int_0^1 f''(\xi)t(t - 1)\,dt.$$

For the first integral on the left-hand side above, we make the change of variables $x = (1-t)a + tb$ so that $dx = (b-a)\,dt$, and the second integral is easily evaluated:

$$\frac{1}{b-a}\int_a^b f(x)\,dx - \frac{f(a)+f(b)}{2} = \frac{(b-a)^2}{2}\int_0^1 f''(\xi)t(t-1)\,dt.$$

Now $t(t-1) \leq 0$ on $[0,1]$ and so by the Mean Value Theorem for Integrals (Theorem 9.14) there is $c \in [a,b]$ such that

$$\begin{aligned}\frac{1}{b-a}\int_a^b f(x)\,dx - \frac{f(a)+f(b)}{2} &= \frac{(b-a)^2}{2} f''(c)\int_0^1 t(t-1)\,dt\\ &= -\frac{(b-a)^2}{12} f''(c),\end{aligned}$$

which yields the first equality. For the second equality, the reader may verify that

$$\int_a^b I_2(x)\,dx - \frac{I_2(a)+I_2(b)}{2}[b-a] = \int_a^b x^2\,dx - \frac{a^2+b^2}{2}[b-a] = -\frac{(b-a)^3}{6}.$$

□

For $f'' \geq 0$, so that f is convex, the Trapezoid Rule Error (Theorem 14.7) implies the right-hand side of the HH Inequalities (Lemma 13.4):

$$\int_a^b f(x)\,dx \leq \frac{f(a)+f(b)}{2}[b-a].$$

Remark 14.8. In Exercises 14.11–14.15, 14.19, and 14.27, we look at other ways of obtaining the Trapezoid Rule Error (Theorem 14.7). ∘

Example 14.9. Applying the Trapezoid Rule to the convex function $f(t) = 1/t$ on $[x, x+1]$, the right-hand side of the HH Inequalities (Lemma 13.4) yields (see Example 13.8):

$$\ln\left(\frac{x+1}{x}\right) \leq \frac{1}{2}\left(\frac{1}{x+1} + \frac{1}{x}\right) = \frac{2x+1}{2x(x+1)} \quad \text{for } x > 0.$$

This is equivalent to

$$\left(1+\frac{1}{x}\right)^{x+\frac{x}{2x+1}} < \mathrm{e} \quad \text{for } x > 0. \tag{14.1}$$

Now the Trapezoid Rule Error (Theorem 14.7) gives, for some $c \in [x, x+1]$,

$$\ln\left(1+\frac{1}{x}\right) = \frac{2x+1}{2x(x+1)} - \frac{2}{c^3}\frac{1}{12} = \frac{2x+1}{2x(x+1)} - \frac{1}{6c^3}.$$

Therefore

$$\ln\left(1+\frac{1}{x}\right) \le \frac{2x+1}{2x(x+1)} - \frac{1}{6(x+1)^3}.$$

This is equivalent to

$$\left(1+\frac{1}{x}\right)^{x+\frac{3x^3+7x^2+3x}{6x^3+15x^2+11x+3}} < \mathrm{e} \qquad \text{for } x > 0, \tag{14.2}$$

which improves (14.1). ◇

Example 14.10. We saw in Sect. 6.7, that with $\gamma_n = \sum_{k=1}^{n} \frac{1}{k} - \ln(n)$,

$$\lim_{n\to\infty} \gamma_n = \gamma = \text{Euler's constant} \cong 0.577216\,.$$

As is suggested in [13], here we use the Trapezoid Rule Error (Theorem 14.7) to obtain bounds for approximating γ with γ_n. Specifically, we show that

$$\frac{1}{2n} - \frac{1}{12(n-1)^2} < \gamma_n - \gamma < \frac{1}{2n} \qquad \text{for } n = 2, 3, 4, \ldots.$$

Applying the Trapezoid Rule Error to $f(x) = 1/x$ on $[k-1, k]$ $(k > 1)$ we get, for some $c_k \in [k-1, k]$,

$$\int_{k-1}^{k} \frac{1}{x}\,dx - \frac{1}{2}\left(\frac{1}{k} + \frac{1}{k-1}\right) = -\frac{1}{6c_k^3}.$$

Summing from $k = 2$ to n, we get

$$\int_1^n \frac{1}{x}\,dx - \frac{1}{2}\sum_{k=2}^{n}\left(\frac{1}{k} + \frac{1}{k-1}\right) = \ln(n) - \sum_{k=1}^{n}\frac{1}{k} + \frac{1}{2}\left(1+\frac{1}{n}\right) = -\sum_{k=2}^{n}\frac{1}{6c_k^3}.$$

That is,

$$\gamma_n = \frac{1}{2}\left(1+\frac{1}{n}\right) + \sum_{k=2}^{n}\frac{1}{6c_k^3}.$$

Therefore (since we already know that γ exists),

$$\gamma = \frac{1}{2} + \sum_{k=2}^{\infty} \frac{1}{6c_k^3},$$

and

$$\gamma_n - \gamma = \frac{1}{2n} - \sum_{k=n+1}^{\infty} \frac{1}{6c_k^3} < \frac{1}{2n}.$$

On the other hand, since $k - 1 \leq c_k$ we also have

$$\sum_{k=n+1}^{\infty} \frac{1}{6c_k^3} \leq \sum_{k=n+1}^{\infty} \frac{1}{6(k-1)^3} = \frac{1}{6}\sum_{k=n}^{\infty} \frac{1}{k^3} < \frac{1}{6}\int_{n-1}^{\infty} \frac{1}{x^3}\,dx = \frac{1}{12(n-1)^2}.$$

Therefore

$$\gamma_n - \gamma > \frac{1}{2n} - \frac{1}{12(n-1)^2},$$

as we set out to show. The reader can verify that $\frac{1}{2n} - \frac{1}{12(n-1)^2} \geq \frac{1}{2(n+1)}$ for $n \geq 2$, and so we have the weaker but tidier estimates

$$\frac{1}{2(n+1)} < \gamma_n - \gamma < \frac{1}{2n}.$$

Another interesting approach to estimating $\gamma_n - \gamma$ can be found in [33]. ◇

14.4 The Midpoint Rule Again

For f defined on $[a, b]$, the Midpoint Rule is the approximation

$$\int_a^b f(x)\,dx \cong f(\frac{a+b}{2})[b-a].$$

Here again, we get *equality* if f is a constant function or if $f(x) = I(x) = x$, but not if $f(x) = I_2(x) = x^2$. So we might expect that an error term for this inequality should involve $f''(c)/2$. We obtain an error term below, again by way of the Mean Value Theorem for the Second Derivative (Theorem 8.6). And the error term is, as expected, $f''(c)/2$ multiplied by $f(\frac{a+b}{2})(b-a) - \int_a^b f(x)\,dx$, but with $f(x) = x^2$.

Theorem 14.11. (Midpoint Rule Error) *Let f be defined on $[a,b]$ with f'' continuous. Then there is $c \in [a,b]$ such that*

$$f(\tfrac{a+b}{2})[b-a] - \int_a^b f(x)\,dx = -f''(c)\frac{(b-a)^3}{24}$$

$$= \frac{f''(c)}{2}\left[I_2(\tfrac{a+b}{2})[b-a] - \int_a^b I_2(x)\,dx \right].$$

Proof. By the Mean Value Theorem for the Second Derivative (Theorem 8.6) we have, for some ξ between x and $(a+b)/2$,

$$f(x) = f(\tfrac{a+b}{2}) + f'(\tfrac{a+b}{2})(x - \tfrac{a+b}{2}) + \frac{f''(\xi)}{2}(x - \tfrac{a+b}{2})^2.$$

Integrating from a to b, we get

$$\int_a^b f(x)\,dx = f(\tfrac{a+b}{2})[b-a] + f'(\tfrac{a+b}{2})\int_a^b (x - \tfrac{a+b}{2})\,dx + \int_a^b \tfrac{f''(\xi)}{2}(x - \tfrac{a+b}{2})^2\,dx$$

$$= f(\tfrac{a+b}{2})[b-a] + 0 + \frac{1}{2}\int_a^b f''(\xi)(x - \tfrac{a+b}{2})^2\,dx.$$

Now $(x - \frac{a+b}{2})^2 \geq 0$ on $[a,b]$ and so by the Mean Value Theorem for Integrals (Theorem 9.14) there is $c \in [a,b]$ such that

$$\frac{1}{2}\int_a^b f''(c)(x - \tfrac{a+b}{2})^2\,dx = \frac{f''(c)}{2}\int_a^b (x - \tfrac{a+b}{2})^2\,dx = f''(c)\frac{(b-a)^3}{24}.$$

This gives the first equality. For the second equality, the reader may verify that

$$I_2(\tfrac{a+b}{2})[b-a] - \int_a^b I_2(x)\,dx = (\tfrac{a+b}{2})^2[b-a] - \int_a^b x^2\,dx = -\frac{(b-a)^3}{12}.$$

□

For $f'' \geq 0$, so that f is convex, the Midpoint Rule Error (Theorem 14.11) yields the left-hand side of the HH Inequalities (Lemma 13.4):

$$f(\tfrac{a+b}{2})[b-a] \leq \int_a^b f(x)\,dx.$$

Remark 14.12. In Exercises 14.17–14.19 and 14.27, we look at other ways of obtaining the Midpoint Rule Error (Theorem 14.11). ○

Example 14.13. Applying the Midpoint Rule to the convex function $f(t) = 1/t$ on $[x, x+1]$, the left-hand side of the HH Inequalities (Lemma 13.4) yields (see Example 13.8):

$$\frac{2}{2x+1} \le \ln\left(\frac{x+1}{x}\right) \quad \text{for } x > 0\,.$$

This is equivalent to the rather good inequality

$$\mathrm{e} < \left(1+\frac{1}{x}\right)^{x+1/2} \quad \text{for } x > 0. \tag{14.3}$$

Now the Midpoint Rule Error (Theorem 14.11) gives, for some $c \in [x, x+1]$,

$$\ln\left(\frac{x+1}{x}\right) = \frac{2}{2x+1} + \frac{2}{c^3}\frac{1}{24} = \frac{2}{2x+1} + \frac{1}{12c^3}.$$

Therefore

$$\ln\left(\frac{x+1}{x}\right) \ge \frac{2}{2x+1} + \frac{1}{12(x+1)^3} = \frac{24x^3+72x^2+74x+25}{12(2x+1)(x+1)^3}.$$

This is equivalent to the following improvement of (14.3):

$$\mathrm{e} < \left(1+\frac{1}{x}\right)^{x+\frac{12x^3+34x^2+35x+12}{24x^3+72x^2+74x+25}} \quad \text{for } x > 0.$$ ◇

14.5 Simpson's Rule

By Theorem 14.7, the error for the Trapezoid Rule is

$$\int_a^b f(x)\,dx - \frac{f(a)+f(b)}{2}\big[b-a\big] = -\frac{(b-a)^3}{12} f''(c),$$

and by Theorem 14.11, the error for the Midpoint Rule is

$$\int_a^b f(x)\,dx - f(\tfrac{a+b}{2})\big[b-a\big] = \frac{(b-a)^3}{24} f''(c).$$

These are of course different c's. But still, these error terms suggest that the Midpoint Rule may often be better than the Trapezoid Rule, by approximately a factor of two, at least if $b - a$ is very small. (We saw in Lemma 13.13 that for f either strictly convex or strictly concave, the Midpoint Rule is always better than the Trapezoid Rule.)

Looking again at these error terms, observe also that

$$\frac{1}{3}\left(-\frac{1}{12}\right) + \frac{2}{3}\left(\frac{1}{24}\right) = 0.$$

This suggests that the weighted mean $\frac{1}{3}$(Trapedoid Rule) + $\frac{2}{3}$(Midpoint Rule) might be a good approximation to $\int_a^b f(x)\,dx$. Indeed it is—this is **Simpson's Rule**:

$$\begin{aligned}\int_a^b f(x)\,dx \;&\cong\; \frac{1}{3}\left(\frac{f(a)+f(b)}{2}[b-a]\right) \;+\; \frac{2}{3}\left(f(\frac{a+b}{2})[b-a]\right)\\ &=\; \frac{1}{6}\left(f(a)+4f(\frac{a+b}{2})+f(b)\right)[b-a].\end{aligned}$$

Simpson's Rule turns out to be a *very* good quadrature rule, considering its relative simplicity. Its error term is given in the theorem below. We don't prove it here—instead we leave several proofs for the Exercises.

Theorem 14.14. (Simpson's Rule Error) *Let f be defined on $[a,b]$ with $f^{(4)}$ continuous. Then there is $c \in [a,b]$ such that*

$$\int_a^b f(x)\,dx - \frac{1}{6}\left(f(a)+4f(\frac{a+b}{2})+f(b)\right)[b-a] = -\frac{(b-a)^5}{2{,}880}f^{(4)}(c).$$

Proof. See any of Exercises 14.19, 14.24, 14.25, 14.26, or 14.27. □

Example 14.15. Applying Simpson's Rule to $f(t) = 1/t$ on $[x, x+1]$ we get

$$\ln\left(\frac{x+1}{x}\right) = \int_x^{x+1} \frac{1}{t}\,dt \cong \frac{12x^2+12x+1}{6x(x+1)(2x+1)}.$$

Here, $f^{(4)}(t) = 24/t^5 > 0$ and so the Simpson's Rule Error (Theorem 14.14), tells us that this is an overestimate. That is,

$$\ln\left(\frac{x+1}{x}\right) < \frac{12x^2+12x+1}{6x(x+1)(2x+1)}.$$

This in turn gives the following improvement of (14.2):

$$\left(1+\frac{1}{x}\right)^{x+\frac{6x^2+5x}{12x^2+12x+1}} < \mathrm{e} \quad \text{for } x > 0 . \qquad \diamond$$

Example 14.16. [26] Let $x, y > 0$. In Example 13.6 we applied the HH Inequalities (Lemma 13.4) to the convex function $f(t) = \mathrm{e}^t$ to get

$$\mathrm{e}^{\frac{\ln(x)+\ln(y)}{2}}\big(\ln(y)-\ln(x)\big) \le \int_{\ln(x)}^{\ln(y)} \mathrm{e}^t\,dt \le \frac{\mathrm{e}^{\ln(x)}+\mathrm{e}^{\ln(y)}}{2}\big(\ln(y)-\ln(x)\big).$$

This gives the conclusion of Lemma 6.20:

$$G \le L \le A,$$

where G, L, and A are the Geometric, Logarithmic and Arithmetic Means, respectively, of x and y. Here, $f^{(4)}(t) = \mathrm{e}^t > 0$ and so Simpson's Rule (Theorem 14.14) gives overestimate for L. Therefore

$$L \le \frac{2}{3}G + \frac{1}{3}A \le A.$$

We obtained this inequality differently in Exercise 6.43. See also Exercise 14.28. $\diamond$

The Midpoint Rule is obtained by interpolating f with the constant function $p_0(x)$ which passes through $((a+b)/2, f((a+b)/2))$, then integrating p_0 instead of f. (The reader may verify this fact.)

Likewise (as the reader may also verify), the Trapezoid Rule is obtained by interpolating f with the linear function $p_1(x)$ which passes through $(a, f(a))$ and $(b, f(b))$, then integrating p_1 instead of f.

Exercise 14.22 shows that Simpson's Rule is obtained by interpolating f with the quadratic function $p_2(x)$ which passes through the points $(a, f(a))$, $(\frac{a+b}{2}, f(\frac{a+b}{2}))$, and $(b, f(b))$, then integrating p_2 instead of f.

As such, Simpson's Rule must be exact when applied to a quadratic function. That is,

$$E = \int_a^b p_2(x)\,dx - \frac{1}{6}\left(p_2(a) + 4p_2(\frac{a+b}{2}) + p_2(b)\right)[b-a] = 0 .$$

But it is a fortunate fact that Simpson's Rule applied to a cubic polynomial also yields the exact answer. Here's why, but see also [9, 12, 32]:

Suppose that $p(x)$ is a cubic that agrees with the quadratic $p_2(x)$ at the points $x = a$, $x = (a+b)/2$, and $x = b$. Then $E(x) = p(x) - p_2(x)$ is a cubic with zeros at these three points. Therefore, for some constant C,

$$E(x) = C \cdot (x-a)(x-\tfrac{a+b}{2})(x-b).$$

So $E(x)$ is an *odd function with respect to* $\frac{a+b}{2}$. That is,

$$E(\tfrac{a+b}{2} + x) = -E(\tfrac{a+b}{2} - x) \quad \text{for } x \in [0, \tfrac{b-a}{2}].$$

Therefore (as the change of variables $t = x - (a + b)/2$) reveals),

$$\int_a^b E(x)\,dx = 0\,.$$

This is why, in Theorem 14.14, the error term for Simpson's Rule involves $f^{(4)}$ rather than the $f^{(3)}$ that one might expect. And looking at the error term there, wherein $2{,}880 = (120)4!$, the energetic reader may verify that (with $I_4(x) = x^4$):

$$\int_a^b I_4(x)\,dx - \frac{1}{6}\left(I_4(a) + 4I_4(\tfrac{a+b}{2}) + I_4(b)\right)\left[b - a\right] = -\frac{(b-a)^5}{120}.$$

Interpolating f with polynomials of degree n generally yields better quadrature rules as n gets larger. These are called *Newton-Cotes Quadrature Rules*. But the price is that the quadrature rules then become more complicated. And, as with Simpson's Rule, interpolating f with a polynomial of degree n when is n even always yields a quadrature rule which is exact for polynomials up to degree $n + 1$. See Exercise 14.23. As such, the error term involves $f^{(n+2)}$ (e.g., [14, 28, 32]).

Subdividing $[a, b]$ into smaller subintervals and applying a quadrature rule on each of the subintervals also generally leads to better approximations, but again, at the expense of simplicity. These are called *composite rules*. For a composite rule which comes from a variation of Simpson's Rule, see [27]. In [29] is an interesting quadrature rule which uses quadratics, but is different from Simpson's Rule.

14.6 Error Terms for Other Inequalities

The proof of many an inequality can be modified to obtain an equality which includes an error term. We finish here by looking at one more example; we explore some others in the exercises.

Young's Integral Inequality (Theorem 11.15) says that if f is continuous on $[0, A]$ and strictly increasing with $f(0) = 0$, then for any $a \in [0, A]$ and $b \in [0, f(A)]$,

$$ab \le \int_0^a f(x)\,dx + \int_0^b f^{-1}(x)\,dx.$$

This inequality is reversed for f strictly decreasing. And in either case, there is equality if and only if $b = f(a)$. Young's Integral Inequality has a geometric interpretation, as shown in **Fig. 14.1.**

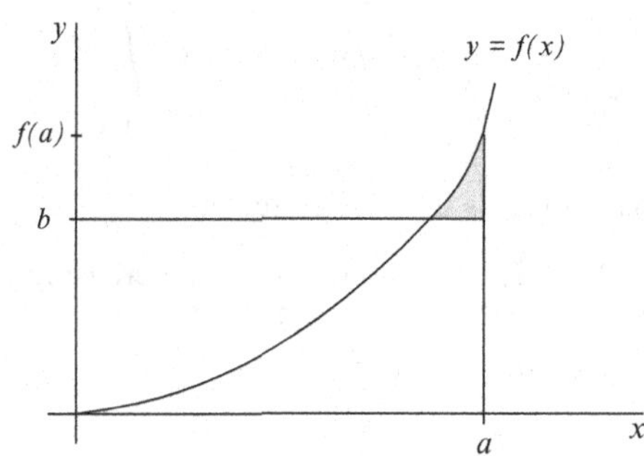

Fig. 14.1 For Young's Inequality (Theorem 11.15): the difference between the right-hand and left-hand sides is the area of the shaded region

The following result provides an error term, which indicates the size of the area of the shaded region in Fig. 14.1.

Theorem 14.17. (Error Term for Young's Integral Inequality) *Let f be continuous on $[0, A]$ with $f(0) = 0$. Suppose also that f' is continuous and that f is monotonic. Let $a \in [0, A]$ and let b be between 0 and $f(A)$. Then there is c between $f^{-1}(b)$ and a such that*

$$ab - \left(\int_0^a f(x)\,dx + \int_0^b f^{-1}(x)\,dx\right) = -\frac{f'(c)}{2}\left[a - f^{-1}(b)\right]^2.$$

Proof. Exactly as in the proof of Young's Integral Inequality (Theorem 11.15), we obtain

$$\int_0^a f^{-1}(x)\,dx + \int_0^b f^{-1}(x)\,dx - ab = \int_{f^{-1}(b)}^a \left[f(x) - b\right]dx.$$

We may assume that $f(a) \neq b$. Then we write

$$\int_{f^{-1}(b)}^a \left[f(x) - b\right]dx = \int_{f^{-1}(b)}^a \left[f(x) - f(f^{-1}(b))\right]dx.$$

By the Mean Value Theorem (Theorem 5.2) there is ξ_x is between x and $f^{-1}(b)$ such that

$$\int_{f^{-1}(b)}^a \left[f(x) - b\right]dx = \int_{f^{-1}(b)}^a f'(\xi_x)\left[x - f^{-1}(b)\right]dx.$$

Now in either of the cases $f^{-1}(b) < a$ or $a < f^{-1}(b)$, the term $[x - f^{-1}(b)]$ does not change sign over $[f^{-1}(b), a]$, and so by the Mean Value Theorem for Integrals (Theorem 9.14) there is c between $f^{-1}(b)$ and a such that

$$\int_{f^{-1}(b)}^{a} [f(x) - b]\,dx = f'(c) \int_{f^{-1}(b)}^{a} [x - f^{-1}(b)]\,dx = \frac{f'(c)}{2}[a - f^{-1}(b)]^2,$$

which yields the desired result. □

If f' is continuous and f is increasing (so that $f' \geq 0$), then the Error Term for Young's Integral Inequality (Theorem 14.17) implies Young's Integral Inequality (Theorem 11.15):

$$ab - \left(\int_0^a f(x)\,dx + \int_0^b f^{-1}(x)\,dx \right) = -\frac{f'(c)}{2}[a - f^{-1}(b)]^2 \leq 0.$$

Example 14.18. We saw in Sect. 11.3 that applying Young's Integral Inequality to $f(x) = x^2$ gives

$$ab - \frac{1}{2}a^2 - \frac{1}{2}b^2 \leq 0.$$

In this case, as the reader may verify, the error term provided by Theorem 14.17 is precisely $-\frac{1}{2}(b-a)^2$, so we get the identity (see **Fig. 14.2**)

$$ab - \frac{1}{2}a^2 - \frac{1}{2}b^2 = -\frac{1}{2}(b-a)^2.$$

This is equivalent to $(a+b)^2 - 4ab = (a-b)^2$, which is what we used to prove Lemma 2.7—the simplest version of the AGM Inequality. ◇

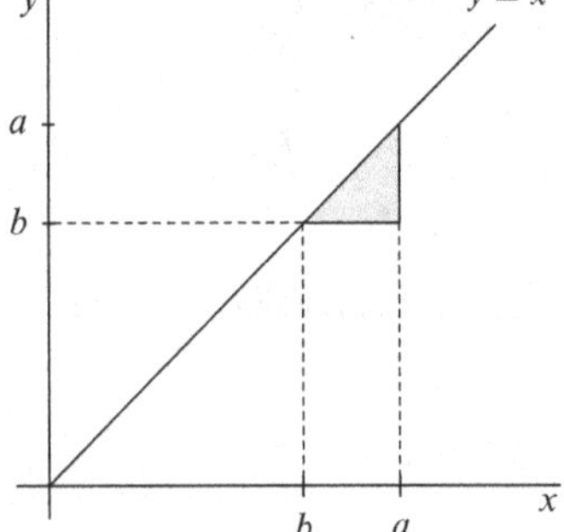

Fig. 14.2 For Example 14.18. The area of the shaded region is $\frac{1}{2}(b-a)^2$

Exercises

14.1. Theorem 14.2 (Error Terms for the Rectangle Rules) is: *Let f be defined on $[a,b]$, with f' continuous. Then there exists $c_1, c_2 \in [a,b]$ such that*

$$f(a)[b-a] + f'(c_1)\frac{(b-a)^2}{2} = \int_a^b f(x)\,dx = f(b)[b-a] - f'(c_2)\frac{(b-a)^2}{2}.$$

(a) We proved the right-hand side. Prove the left-hand side.
(b) Verify, for $I(x) = x$, that

$$-\frac{(b-a)^2}{2} = \left[I(a)[b-a] - \int_a^b I(x)\,dx \right].$$

14.2. **(a)** Let $I_3(x) = x^3$. Verify that the conclusion of Taylor's Theorem (Theorem 8.20) with $n = 2$ reads: There exists c between x and x_0 such that

$$f(x) - f(x_0) - f'(x_0)(x-x_0) - \frac{f''(x_0)}{2}(x-x_0)^2$$

$$= \frac{f^{(3)}(c)}{3!}\left[I_3(x) - I_3(x_0) - I_3'(x_0)(x-x_0) - \frac{I_3''(x_0)}{2}(x-x_0)^2 \right].$$

(b) Let $I_4(x) = x^4$. Verify that the conclusion of Taylor's Theorem (Theorem 8.20) with $n = 3$ reads: There exists c between x and x_0 such that

$$f(x) - f(x_0) - f'(x_0)(x-x_0) - \frac{f''(x_0)}{2}(x-x_0)^2 - \frac{f^{(3)}(x_0)}{3!}(x-x_0)^3$$

$$= \frac{f^{(4)}(c)}{4!}\left[I_4(x) - I_4(x_0) - I_4'(x_0)(x-x_0) - \frac{I_4''(x_0)}{2!}(x-x_0)^2 - \frac{I_4^{(3)}(x_0)}{3!}(x-x_0)^3 \right].$$

14.3. After the proof of the Error Term For Jensen's Inequality (Theorem 14.3), we saw that

$$\left(\sum_{j=1}^{n} w_j x_j \right)^2 - \sum_{j=1}^{n} w_j x_j^2 \le 0.$$

Use the Cauchy–Schwarz Inequality (Theorem 2.18) to show this directly.

14.4. [7,31] Fill in the details of another way to obtain the Error Term for Jensen's Inequality (Theorem 14.3).

(a) First, explain how we may suppose that there exist m and M such that

$$0 \leq m \leq f''(x) \leq M.$$

(b) Set $g(x) = \frac{1}{2}Mx^2 - f(x)$. Verify that g is convex, and apply Jensen's Inequality (Theorem 8.17) to g to get

$$\sum_{j=1}^{n} w_j f(x_j) - f(\sum_{j=1}^{n} w_j x_j) \leq -\frac{M}{2} \sum_{j=1}^{n} w_j (x_j - A)^2.$$

(c) Set $h(x) = f(x) - \frac{1}{2}mx^2$. Verify that h is convex, and apply Jensen's Inequality (Theorem 8.17) to h to get

$$\frac{m}{2} \sum_{j=1}^{n} w_j (x_j - A)^2 \leq \sum_{j=1}^{n} w_j f(x_j) - f(\sum_{j=1}^{n} w_j x_j).$$

(d) Now if m and M were chosen carefully, then an application of the Intermediate Value Theorem (Theorem 3.17) would finish the proof—explain.

14.5. [18] Fill in the details of another way to obtain the Error Term For Jensen's Inequality (Theorem 14.3). Let $A = \sum_{j=1}^{n} w_j x_j$, and for $0 \leq t \leq 1$, set

$$g(t) = \sum_{j=1}^{n} w_j f\big((1-t)x_j + tA\big).$$

(a) Apply the Mean Value Theorem for the Second Derivative (Theorem 8.6) to g on $[0, 1]$ to get, for some c between 0 and 1 :

$$g(0) = g(1) + g'(1)(0-1) + \frac{g''(c)}{2}(0-1)^2.$$

(b) Now write down what this says, in terms of f.

14.6. Prove Corollary 14.6.

14.7. Let $x_1, x_2, \ldots, x_n$ and $w_1, w_2, \ldots, w_n$ be positive numbers with $\sum_{j=1}^{n} w_j = 1$. For $r > 0$ their Power Mean is:

$$M_r = \left(\sum_{j=1}^{n} w_j x_j^r \right)^{1/r}.$$

(a) Verify that M_1 is the Arithmetic Mean, and M_2 is the Root Mean Square.
(b) Show that, by continuity, we should define M_0 to be the weighted Geometric Mean $G = x_1^{w_1} \cdot x_2^{w_2} \cdots x_n^{w_n}$.
(c) Apply the error term for Jensen's Inequality to $f(x) = x^{(s+1)/s}$ to obtain upper and lower bounds for $M_{s+1} - M_s$. See also Exercises 8.48, 9.51, and 10.18.

14.8. [25] Recall Jensen's Integral Inequality (Theorem 9.29) then obtain the following Error Term for Jensen's Integral Inequality. *Let f be continuous on $[a, b]$ with f'' continuous also. Let φ be such that $\varphi'' \geq 0$ on the range of f. Let $w \geq 0$ be continuous on $[0, 1]$, with $\int_0^1 w(x)\,dx = 1$. Then there is $c \in (a, b)$ such that*

$$\varphi\left(\int_0^1 w(x) f(x)\,dx\right) - \int_0^1 w(x)\varphi(f(x))\,dx$$

$$= \frac{f''(\xi)}{2}\left[\left(\int_0^1 w(x) f(x)\,dx\right)^2 - \int_0^1 w(x) f^2(x)\,dx\right].$$

14.9. Let f be a function defined on $[a, b]$ with f'' bounded (but not necessarily continuous), say $|f''| \leq M$. Show that

$$\left|\int_a^b f(x)\,dx - \frac{f(a) + f(b)}{2}[b - a]\right| \leq M\frac{(b-a)^3}{12}.$$

14.10. [13, 15, 22] Use the Trapezoid Rule Error (Theorem 14.7), as follows, to show that the limit in Stirling's formula (Theorem 13.11) $\lim_{n\to\infty} \frac{n!e^n}{n^n\sqrt{n}} = L$ exists and is positive. With

$$a_n = \frac{n^n\sqrt{n}}{n!e^n},$$

we showed in the first part of our proof of Theorem 13.11 that $\{a_n\}$ is increasing, so by the Increasing Bounded Sequence Property (Theorem 1.34), it is enough to show that $\{a_n\}$ is bounded above.

(a) Let $k > 1$. Show that for some $\xi_k \in [k-1, k]$,

$$\int_{k-1}^k \ln(x)\,dx - \frac{1}{2}\big[\ln(k-1) + \ln(k)\big] = \frac{1}{12\xi_k^2}.$$

(b) Sum from $k = 2$ to n and evaluate the resulting integral, to show that

$$n\ln(n) - n + 1 - \ln(n!) + \frac{1}{2}\ln(n) = \sum_{k=2}^{n} \frac{1}{12\xi_k^2}.$$

(c) Conclude that

$$\ln\left(\frac{n^n\sqrt{n}}{n!\mathrm{e}^n}\right) = \sum_{k=2}^{n} \frac{1}{12\xi_k^2} - 1.$$

(d) Show that

$$\sum_{k=2}^{n} \frac{1}{\xi_k^2} \le \sum_{k=1}^{n-1} \frac{1}{k^2},$$

and so $\{a_n\}$ is bounded above, as we wanted to show.

(e) Bonus: Show further, that

$$2.370 \cong \mathrm{e}^{1-\pi^2/72} \le L \le \mathrm{e}^{13/12-\pi^2/72} \cong 2.576.$$

14.11. [11] Here is another proof of the Trapezoid Rule Error (Theorem 14.7). Let $c = (a+b)/2$ and consider the function

$$F(t) = \int_{c-t}^{c+t} f(x)\,dx - t\big[f(c+t) + f(c-t)\big].$$

(a) Verify that

$$F\left(\tfrac{b-a}{2}\right) = \int_a^b f(x)\,dx - \frac{(b-a)}{2}\big[f(a) + f(b)\big].$$

(b) Compute $F'(t)$.

(c) Observe that $(c+t) - (c-t) = 2t$, then apply the Mean Value Theorem (Theorem 5.2) to f' to show that there is ξ_1 between $c-t$ and $c+t$ such that

$$\frac{F'(t)}{2t^2} = -f''(\xi_1).$$

(d) Apply Cauchy's Mean Value Theorem (Theorem 5.11) to $F(x)$ (numerator) and x^3 (denominator) on $[0, \frac{b-a}{2}]$ to show that there is $\xi_2 \in (0, \frac{b-a}{2})$ such that

$$\frac{F(\frac{b-a}{2})}{(\frac{b-a}{2})^3} = \frac{F'(\xi_2)}{3\xi_2^3}.$$

(e) Show that there is $\xi_3 \in (c - \frac{b-a}{2}, c + \frac{b-a}{2}) = (a, b)$ such that

$$\frac{F(\frac{b-a}{2})}{(\frac{b-a}{2})^3} = -\frac{2}{3} f''(\xi_3).$$

(d) Conclude that $F(\frac{b-a}{2}) = -\frac{2}{3}(\frac{b-a}{2})^3 f''(\xi_3) = -\frac{1}{12}(b-a)^3 f''(\xi_3)$.

14.12. [3] Here's another proof of the Trapezoid Rule Error (Theorem 14.7). Define the constant C by

$$\int_a^b f(x)\,dx - \frac{f(a) + f(b)}{2}(b-a) = C(b-a)^3.$$

(a) Verify that the function

$$F(t) = \int_a^t f(x)\,dx - \frac{f(a) + f(b)}{2}(t-a) - C(t-a)^3$$

satisfies $F(a) = F(b) = 0$. Conclude that F' vanishes somewhere in (a, b).

(b) Verify that $F'(a) = 0$. Conclude that F'' vanishes somewhere in (a, b).

(c) Write down what (b) means, and solve for C.

14.13. [8] Here's another proof of the Trapezoid Rule Error (Theorem 14.7). Denote by $L(x)$ the line through $(a, f(a))$ and $(b, f(b))$. Let $h = \frac{(b-a)}{2}$, fix $x \in (a, b)$, and set

$$F(t) = f(t) - L(t) - \frac{f(x) - L(x)}{x^2 - h^2}(t-h)^2.$$

(a) Verify that $F(a) = F(b) = F(x) = 0$.

(b) Apply Rolle's Theorem to show that F' vanishes at least twice and F'' vanishes at least once, say at $c = c_x$.

(c) Solve $F''(c) = 0$ for c to see that $f(x) - L(x) = \frac{(x-h)(x+h)}{2} f''(c_x)$.

(d) Now apply the Mean Value Theorem for Integrals (Theorem 9.14) to see that

$$\int_a^b f(x)\,dx - \frac{f(a) + f(b)}{2}(b-a) = -\frac{2}{3}h^3 f''(\xi) = -\frac{1}{12} f''(\xi)(b-a)^3$$

for some $\xi \in [a, b]$.

14.14. [2] Here's another proof of the Trapezoid Rule Error (Theorem 14.7).

(a) Apply Integration by Parts twice to show that

$$\int_a^b (x-a)(x-b)f''(x)\,dx = (b-a)\big[f(a)+f(b)\big] - 2\int_a^b f(x)\,dx.$$

(b) Apply the Mean Value Theorem for Integrals (Theorem 9.14) on the left side.

14.15. (L. Livshutz, private communication) Here's another proof of the Trapezoid Rule Error (Theorem 14.7).

(a) Apply Integration by Parts twice to show that

$$\int_a^b f(x)\,dx = f(x)\left(x-\frac{A}{2}\right)\Big|_a^b - f'(x)\left(\frac{x^2}{2}-\frac{A}{2}x+\frac{B}{2}\right)\Big|_a^b$$

$$+\frac{1}{2}\int_a^b f''(x)(x^2-Ax+B)\,dx,$$

where A and B are arbitrary constants.

(b) Set $A = a+b$, $B = ab$ and apply the Mean Value Theorem for Integrals (Theorem 9.14).

14.16. Let f be a function on $[a,b]$ with f'' bounded (but not necessarily continuous), say $|f''| \le M$. Show that

$$\left|\int_a^b f(x)\,dx - f(\tfrac{a+b}{2})[b-a]\right| \le M\frac{(b-a)^3}{24}.$$

14.17. [2] In this problem we use Corollary 14.6 to obtain the Midpoint Rule Error (Theorem 14.11).

(a) Verify that

$$f\left(\tfrac{a+b}{2}\right) = f\left(\tfrac{1}{2}\left((1-t)a+tb\right)+\tfrac{1}{2}(ta+(1-t)b)\right).$$

(b) Show that Corollary 14.6 gives

$$f\left(\tfrac{a+b}{2}\right) = \tfrac{1}{2}f\left((1-t)a+tb\right)+\tfrac{1}{2}f\left(ta+(1-t)b\right)$$

$$+\ \tfrac{f''(c)}{2}\tfrac{1}{2}(\tfrac{1}{2}-1)\left[\left((1-t)a+tb\right)^2-\left(ta+(1-t)b\right)^2\right].$$

(c) Simplify this to get

$$f(\tfrac{a+b}{2}) = \tfrac{1}{2}f\left((1-t)a+tb\right) + \tfrac{1}{2}f\left(ta+(1-t)b\right) - \tfrac{f''(c)}{8}(b-a)^2(2t-1)^2.$$

(d) Now integrate from $t = 0$ to $t = 1$, and use the Mean Value Theorem for Integrals (Theorem 9.14).

14.18. Here is another proof of the Midpoint Rule Error (Theorem 14.11), motivated by [11]. Let $c = (a+b)/2$, and consider the function

$$F(t) = \int_{c-t}^{c+t} f(x)\,dx - 2tf(c),$$

(a) Verify that

$$F\left(\tfrac{b-a}{2}\right) = \int_a^b f(x)\,dx - (b-a)f\left(\tfrac{a+b}{2}\right).$$

(b) Compute $F''(t)$.

(c) Observe that $(c+t)-(c-t) = 2t$, then apply the Mean Value Theorem (Theorem 5.2) to f' to show that there is ξ_1 between $c-t$ and $c+t$ such that

$$\frac{F''(t)}{2t} = f''(\xi_1).$$

(d) Apply Cauchy's Mean Value Theorem (Theorem 5.11) to $F(x)$ (numerator) and x^3 (denominator) two times on $[0, \frac{b-a}{2}]$ to show that there is $\xi_2 \in (0, \frac{b-a}{2})$ such that

$$\frac{F(\frac{b-a}{2})}{(\frac{b-a}{2})^3} = \frac{F''(\xi_2)}{6\xi_2}.$$

(e) Show that there is $\xi_3 \in (c - \frac{b-a}{2}, c + \frac{b-a}{2}) = (a,b)$ such that

$$\frac{F(\frac{b-a}{2})}{(\frac{b-a}{2})^3} = \frac{1}{3}f''(\xi_3).$$

(f) Conclude that $F(\frac{b-a}{2}) = \frac{1}{3}(\frac{b-a}{2})^3 f''(\xi_3) = \frac{1}{24}(b-a)^3 f''(\xi_3)$.

14.19. [4, 6] Here is a rather unified way to obtain the Trapezoid Rule Error (Theorem 14.7), the Midpoint Rule Error (Theorem 14.11), and Simpson's Rule Error (Theorem 14.14). Suppose that f'' and h are continuous and set $H(x) = \int_a^x h(t)\,dt$.

(a) Verify the Integration by Parts formula

$$\int_a^b f(x)h'(x)\,dx - h(x)f(x)\Big|_a^b = \int_a^b H(x)f''(x)\,dx - f'(x)H(x)\Big|_a^b.$$

(b) With $h(x) = x - (a+b)/2$, apply this formula then the Mean Value Theorem for Integrals (Theorem 9.14) to obtain the Trapezoid Rule Error.

(c) Set

$$h(x) = \begin{cases} x - a & \text{if } x \in [a, \frac{a+b}{2}] \\ x - b & \text{if } x \in (\frac{a+b}{2}, b]. \end{cases}$$

Apply the formula in (a) then the Mean Value Theorem for Integrals (Theorem 9.14) to obtain the Midpoint Rule Error.

(d) Simpson's Rule is $\frac{1}{3}$(Trapezoid Rule) $+\frac{2}{3}$(Midpoint Rule). So consider

$$\frac{1}{3}h_1(x) + \frac{2}{3}h_2(x),$$

where h_1 is the h from (b) and h_2 is the h from (c) to obtain the Simpson's Rule Error (Theorem 14.14). (Warning: This is very messy!)

14.20. [19] Suppose that f'' and h are continuous and set $H(x) = \int_a^x h(t)\,dt$.

(a) Verify the Integration by Parts formula

$$\int_a^b f(x)h'(x)\,dx - h(x)f(x)\Big|_a^b = \int_a^b H(x)f''(x)\,dx - f'(x)H(x)\Big|_a^b.$$

With $h(x) = (x-a)(b-x) - \frac{1}{6}(b-a)^2$, apply this formula then the Mean Value Theorem for Integrals (Theorem 9.14) to obtain the **Corrected Trapezoid Rule** error estimate

$$\int_a^b f(x)\,dx - \frac{f(a)+f(b)}{2}(b-a) - \frac{f'(b)-f'(a)}{12}(b-a)^2 = f^{(4)}(c)\frac{(b-a)^5}{720},$$

for some $c \in (a,b)$.

Note: The Corrected Trapezoid Rule is motivated by the fact that for $b-a$ small, the error in the Trapezoid Rule is

$$\frac{1}{12}f''(\xi)(b-a)^3 \cong \frac{1}{12}\left(f'(b) - f'(a)\right)(b-a)^2.$$

(b) We have seen that $L < A$, where L and A are respectively, the Logarithmic and Arithmetic Means of the positive numbers $x \neq y$. Apply the Corrected Trapezoid Rule to $f(t) = x^t y^{1-t}$ to obtain the refinement

$$L + \tfrac{1}{12} \ln\left(\tfrac{x}{y}\right)(x - y) < A .$$

14.21. In Exercise 14.10 we used the Trapezoid Rule Error (Theorem 14.7) to show that the $L > 0$ in Stirling's formula (Theorem 13.11) exists, and we obtained

$$2.370 \cong e^{1-\pi^2/72} \leq L \leq e^{13/12-\pi^2/72} \cong 2.576 .$$

Use the Corrected Trapezoid Rule Error (see Exercise 14.20)

$$\int_a^b f(x)\,dx - \tfrac{f(a)+f(b)}{2}\left[b-a\right] - \tfrac{f'(b)-f'(a)}{12}\left[b-a\right]^2 = f^{(4)}(\xi)\tfrac{(b-a)^5}{720},$$

for some $\xi \in (a, b)$, and the fact that $\sum_{k=1}^{\infty} 1/k^4 = \pi^4/90$, to obtain the better bounds

$$2.503 \cong e^{11/12+(6/720)(\pi^4/90-1)} \leq L \leq e^{11/12+(6/720)(\pi^4/90)} \cong 2.524 .$$

14.22. Show that Simpson's Rule is obtained by interpolating f with the quadratic function $p_2(x)$ which passes through the three points $(a, f(a))$, $(\frac{a+b}{2}, f(\frac{a+b}{2}))$, and $(b, f(b))$, then integrating p_2 instead of f.

14.23. [32] Suppose that $p(x)$ is a polynomial of degree $2k + 1$ with the $2k + 1$ zeros

$$x = -kh,\ x = -(k-1)h,\ \ldots,\ x = -h,\ x = 0,\ x = h,\ x = 2h, \ldots,\ x = kh.$$

(a) Show that $p(x)$ is an odd function: $p(x) = -p(-x)$ for all $x \in \mathbf{R}$.
(b) Conclude that

$$\int_{-h}^{h} p(x)\,dx = 0.$$

14.24. [8] Here is a proof of the Simpson's Rule Error (Theorem 14.14). Let $S(g)$ denote Simpson's Rule applied to the function g on $[-h, h]$.

(a) Let p be the unique cubic polynomial such that $p(\pm h) = f(\pm h)$, $p(0) = f(0)$, and $p'(0) = f'(0)$. Show that

$$\int_{-h}^{h} f(x)\,dx - S(f) = \int_{-h}^{h} f(x)\,dx - S(p) = \int_{-h}^{h} \big(f(x) - p(x)\big)\,dx.$$

(b) Fix $x \in (-h, h)$ and for $t \in (-h, h)$, consider

$$F(t) = f(t) - p(t) - \frac{f(x) - p(x)}{x^2(x^2 - h^2)} t^2(t^2 - h^2).$$

Verify that $F(\pm h) = F(0) = F(x) = 0$, and conclude that F' vanishes at least three times in $(-h, h)$.

(c) Verify that $F'(0) = 0$, and conclude that F' vanishes at least four times in $(-h, h)$.

(d) Show that $F^{(4)}$ vanishes at least once in $(-h, h)$, say at $t = c_1$, and conclude that

$$f(x) - p(x) = \frac{x^2(x^2 - h^2)}{4!} f^{(4)}(c_1).$$

Therefore,

$$\int_{-h}^{h} f(x)\,dx - S(f) = \int_{-h}^{h} \frac{x^2(x^2 - h^2)}{4!} f^{(4)}(c_1)\,dx.$$

(f) Apply the Mean Value Theorem for Integrals (Theorem 9.14) to obtain

$$\int_{-h}^{h} f(x)\,dx - S(f) = f^{(4)}(c_2) \int_{-h}^{h} \frac{x^2(x^2 - h^2)}{4!}\,dx = -\frac{1}{90} h^5 f^{(4)}(c_2).$$

14.25. [5] (see also [10]) Here's another proof of the Simpson's Rule Error (Theorem 14.14).

(a) Use Integration by Parts to verify that

$$72 \int_{-h}^{h} f(x)\,dx = \int_{-h}^{0} (x+h)^3(3x-h) f^{(4)}(x)\,dx + \int_{0}^{h} (x-h)^3(3x+h) f^{(4)}(x)\,dx.$$

(b) Use the Mean Value Theorem for Integrals (Theorem 9.14) then the Intermediate Value Theorem (Theorem 3.17) to show that there is $c \in (-h, h)$ such that

$$72\int_{-h}^{h} f(x)\,dx = 2f^{(4)}(c)\int_{0}^{h}(x-h)^3(3x+h)\,dx$$

$$= 2f^{(4)}(c)\frac{4}{5}h^5.$$

14.26. [11] Here's another proof of the Simpson's Rule Error (Theorem 14.14). Set $c = (a+b)/2$ and consider the function

$$F(t) = \int_{c-t}^{c+t} f(x)\,dx - \tfrac{t}{3}\big[f(c-t) + 4f(c) + f(c+t)\big].$$

(a) Verify that

$$F\left(\tfrac{b-a}{2}\right) = \int_{a}^{b} f(x)\,dx - \frac{(b-a)}{6}\big[f(a) + 4f(c) + f(b)\big].$$

(b) Verify that

$$F'(t) = \tfrac{2}{3}\big[f(c+t) - 2f(c) + f(c-t)\big] - \tfrac{t}{3}\big[-f'(c-t) + f'(c+t)\big],$$

$$F''(t) = \tfrac{1}{3}\big[f'(c+t) - f'(c-t)\big] - \tfrac{t}{3}\big[f''(c-t) + f''(c+t)\big], \quad \text{and}$$

$$F^{(3)}(t) = -\tfrac{t}{3}\big[f^{(3)}(c+t) - f^{(3)}(c-t)\big].$$

(c) Verify that $F(0) = F'(0) = F''(0) = F'''(0) = 0$ and conclude, by the Mean Value Theorem (Theorem 5.2), that there is $\xi \in (c-t, c+t)$ such that

$$F^{(3)}(t) = -\frac{2t^2}{3}f^{(4)}(\xi).$$

(d) Apply Cauchy's Mean Value Theorem (Theorem 5.11) three times, beginning with $F(t)$ and $g(t) = t^5$, to see that there is c between 0 and $\frac{b-a}{2}$ such that

$$\frac{F(\frac{b-a}{2}) - F(0)}{g(\frac{b-a}{2}) - g(0)} = \frac{F^{(3)}(c)}{g^{(3)}(c)} = \frac{-\frac{2c_3^2}{3}f^{(4)}(\xi)}{(5)(4)(3)c^2} = -\frac{1}{90}f^{(4)}(\xi).$$

(e) Show, finally, that

$$F(\tfrac{b-a}{2}) = g(\tfrac{b-a}{2})\Big(-\frac{1}{90}f^{(4)}(\xi)\Big) = -\frac{(b-a)^5}{2{,}880}f^{(4)}(\xi).$$

See [30] for an approach which unifies Exercises 14.11, 14.18 and 14.26.

14.27. (L. Livshutz, private communication) Here is a rather unified way to obtain the errors for the Trapezoid Rule (Theorem 14.7), the Midpoint Rule (Theorem 14.11) and Simpson's Rule (Theorem 14.14). For a continuous function f on $[-a, a]$, write $f(x) = f_E(x) + f_O(x)$, where

$$f_E(x) = \frac{f(x)+f(-x)}{2} \quad \text{and} \quad f_O(x) = \frac{f(x)-f(-x)}{2}.$$

For example, $e^x = \cosh(x) + \sinh(x)$.

(a) Verify that f_E is an even function and that f_O is an odd function.
(b) Show that

$$\int_{-a}^{a} f(x)\,dx = 2\int_{0}^{a} f_E(x)\,dx,\ f_E^{(2k-1)}(0) = 0,\ \text{and}\ f_E^{(2k)}(0) = f''(0)\ \text{ for } k = 1, 2, 3, \ldots.$$

(c) Integrate by parts twice to show that

$$\int_0^a f_E(x)\,dx = f_E(x)((x-a)+A)\Big|_0^a - \int_0^a f_E'(x)\left(\tfrac{(x-a)^2}{2} + A(x-a)\right)dx$$

$$= f_E(a)A - f_E(0)(A-a) - f_E'(x)\left(\tfrac{(x-a)^3}{6} + \tfrac{A(x-a)^2}{2}\right)\Big|_0^a$$

$$+ \int_0^a f_E''(x)\left(\tfrac{(x-a)^2}{2} + A(x-a)\right)dx$$

$$= f_E(a)A - f_E(0)(A-a) + \int_0^a f_E''(x)\frac{(x-a)}{2}(x-a+2A)\,dx.$$

(d) Set $A = 0$, apply the Mean Value Theorem for Integrals (Theorem 9.14), and write what you get in terms of $\int_{-a}^{a} f(x)\,dx$ to obtain the Midpoint Rule Error (Theorem 14.11).

(e) Set $A = a$, apply the Mean Value Theorem for Integrals (Theorem 9.14), and write what you get in terms of $\int_{-a}^{a} f(x)\,dx$ to obtain the Trapezoid Rule Error (Theorem 14.7).

(f) Integrate by parts as above, but four times, and take $A = a/3$, to show that

$$\int_0^a f_E(x)\,dx = f_E(x)\left(x - \frac{2}{3}a\right)\Big|_0^a + \frac{1}{72}\int_0^a f_E''(x)\,(x-a)^3\,(3x+a)\,dx.$$

(g) Apply the Mean Value Theorem for Integrals (Theorem 9.14), and write what you get in terms of $\int_{-a}^{a} f(x)\,dx$ to obtain the Simpson's Rule Error.

14.28. Let $x, y > 0$. Apply Simpson's Rule to $f(t) = x^t y^{(1-t)}$ on $[x, y]$ to obtain the result of Example 14.16 (also Exercise 6.43):

$$L \le \frac{2}{3}G + \frac{1}{3}A \le A,$$

where G, L, and A are respectively, the Geometric, Logarithmic and Arithmetic Means of x and y. The reader may want to look at Exercise 13.8 in which we applied the HH Inequalities (Lemma 13.4) to $f(t) = x^t y^{(1-t)}$ on $[x, y]$, to get

$$G \le L \le A.$$

14.29. [1, 16] **Simpson's 3/8 Rule** and its error term is given by

$$\int_a^b f(x)\,dx = \tfrac{1}{8}\left[f(a) + 3f\left(\tfrac{2a+b}{3}\right) + 3f\left(\tfrac{a+2b}{3}\right) + f(b)\right](b-a) - \tfrac{(b-a)^5}{6{,}480} f^{(4)}(\xi),$$

for some ξ between a and b. Apply this to $f(x) = \mathrm{e}^x$, to show that

$$L(x, y) = \tfrac{x-y}{\ln(x)-\ln(y)} \le \left(\tfrac{x^{1/3}+y^{1/3}}{2}\right)^3 = M_{1/3}(x, y).$$

$M_{1/3}(x, y)$ is called the **Lorentz Mean**; see also Exercise 6.46.

14.30. [20, 23] Fill in the details in obtaining the following error term for Chebyshev's Integral Inequality (see Exercises 9.44, 9.45, and 10.29): *Let f' and g' be continuous on $[a, b]$. Then there are $c_1, c_2 \in (a, b)$ such that*

$$\int_a^b f(x)g(x)\,dx - \frac{1}{b-a}\int_a^b f(x)\,dx \int_a^b g(x)\,dx = \frac{1}{12} f'(c_1)g'(c_2)(b-a)^3.$$

(a) Verify that

$$\int_a^b f(x)g(x)\,dx - \int_a^b f(x)\,dx \int_a^b g(x)\,dx = \frac{1}{2(b-a)} \int_a^b \int_a^b (f(x)-f(y))(g(x)-g(y))\,dxdy.$$

(b) Apply the Mean Value Theorem (Theorem 5.2) then the Mean Value Theorem for Integrals (Theorem 9.14) to show that for some $c_1, c_2 \in (a, b)$,

$$\int_a^b \int_a^b (f(x) - f(y))(g(x) - g(y))\,dxdy = f'(c_1)g'(c_2) \int_a^b \int_a^b (x - y)^2\,dxdy.$$

(c) Now evaluate the *double integral* in (b). (Work from the inside out: integrate with respect to x, then with respect to y.)

14.31. [21] In Exercise 8.44 we applied Jensen's Inequality (Theorem 8.17) to the function $f(1 - x) - f(x)$ to obtain **Levinson's Inequality** : *Let* $0 < x_1 < \cdots < x_n < 1/2$, *set* $y_j = 1 - x_j$, $A_1 = \sum_{j=1}^{n} w_j x_j$, *and* $A_2 = \sum_{j=1}^{n} w_j y_j = 1 - A_1$ *(where* $w_1, w_2, \ldots, w_n$ *satisfy* $w_j > 0$ *and* $\sum_{j=1}^{n} w_j = 1$*). Then for* $f^{(3)} \geq 0$ *on* $(0, 1)$,

$$f(A_2) - f(A_1) \leq \sum_{j=1}^{n} w_j f(y_j) - \sum_{j=1}^{n} w_j f(x_j).$$

Use the Error Term for Jensen's Inequality (Theorem 14.3) to obtain an error term for Levinson's Inequality.

14.32. [20] Fill in the details for obtaining the following error term for Steffensen's Inequality (see Exercises 9.52, 10.48, 11.17, and 11.36): *Let* f' *be continuous and let gbe continuous on* $[a, b]$, *with* $0 \leq g \leq 1$ *and* $\lambda = \int_a^b g(t)\,dt$. *Then there exists* $\xi \in (a, b)$ *such that*

$$\int_a^b f(t)g(t)\,dt - \int_a^{a+\lambda} f(t)\,dt = f'(\xi) \left[\int_a^b t g(t)\,dt - \lambda(a + \frac{\lambda}{2}) \right].$$

(a) Verify that

$$\int_a^b f(t)g(t)\,dt - \int_a^{a+\lambda} f(t)\,dt = \int_a^{a+\lambda} [f(a+\lambda) - f(t)][1-g(t)]\,dt + \int_{a+\lambda}^{b} [f(t) - f(a+\lambda)]g(t)\,dt.$$

(b) Apply the Mean Value Theorem (Theorem 5.2) then the Mean Value Theorem for Integrals (Theorem 9.14) to see that there are $p, q \in (a, b)$ such that

$$\int_a^b f(t)g(t)\,dt - \int_a^{a+\lambda} f(t)\,dt = f'(p)\int_a^{a+\lambda} [a+\lambda-t][1-g(t)]\,dt + f'(q)\int_{a+\lambda}^{b} [t-(a+\lambda)]g(t)\,dt.$$

(c) Apply the Intermediate Value Theorem to see that there is $\xi \in (a, b)$ such that this

$$= f'(\xi)\left[\int_a^{a+\lambda} [a+\lambda-t][1-g(t)]\,dt + \int_{a+\lambda}^{b} [t-(a+\lambda)]g(t)\,dt\right]$$

$$= f'(\xi)\left[\int_a^b tg(t)\,dt - \lambda(a+\frac{\lambda}{2})\right].$$

14.33. Use the error term for Steffensen's Inequalities from Exercise 14.32 to obtain the error term for Jensen's Inequality (Theorem 14.3). (See Exercise 10.48.)

14.34. Use the error term for Steffensen's Inequalities from Exercise 14.32 to obtain the error terms for the Trapezoid Rule (Theorem 14.7) and for the Midpoint Rule (Theorem 14.11). (See Exercise 13.18.)

14.35. In Exercise 1.12 we saw that the area A of a triangle T with vertices (x_1, y_1), (x_2, y_2), (x_3, y_3) is given by

$$A = \frac{1}{2}\Big|x_1(y_2 - y_3) + x_3(y_1 - y_2) + x_2(y_3 - y_1)\Big|.$$

Readers who know some Linear Algebra might recognize that $A = \frac{1}{2}|\det(M)|$, where M is the matrix $\begin{bmatrix} 1 & 1 & 1 \\ x_1 & x_2 & x_2 \\ y_1 & y_2 & y_3 \end{bmatrix}$. In Exercise 8.26 we saw that if $f'' \geq 0$ on $[a, b]$ then whenever $a \leq x_1 < x_2 < x_3 \leq b$,

$$\frac{1}{2}\big[x_1\big(f(x_2) - f(x_3)\big) + x_3\big(f(x_1) - f(x_2)\big) + x_2\big(f(x_3) - f(x_1)\big)\big] \geq 0.$$

Show that if f has continuous second derivative then there is $c \in [a, b]$ such that

$$\det\begin{bmatrix} 1 & 1 & 1 \\ x_1 & x_2 & x_3 \\ f(x_1) & f(x_2) & f(x_3) \end{bmatrix} = \det\begin{bmatrix} 1 & 1 & 1 \\ x_1 & x_2 & x_3 \\ x_1^2 & x_2^2 & x_3^2 \end{bmatrix} \frac{f''(c)}{2}.$$

Note: The conclusion of the Mean Value Theorem (Theorem 5.2) reads

$$\det\begin{bmatrix} 1 & 1 \\ f(x_1) & f(x_2) \end{bmatrix} = \det\begin{bmatrix} 1 & 1 \\ x_1 & x_2 \end{bmatrix} f'(c).$$

14.36. The content of Exercise 8.33 was: *Let f be convex on $[a, c]$ and let $a < b < c$. Then*

$$f(a - b + c) \leq f(a) - f(b) + f(c).$$

Suppose that f'' is continuous, then find and prove an error term for this inequality.

References

1. Burk, F.: Geometric, logarithmic, and arithmetic mean inequality. Am. Math. Mon. **94**, 527–528 (1987)
2. Cerone, P., Dragomir, S.S.: Mathematical Inequalities: A Perspective. CRC, New York (2011)
3. Courant, R., John, F.: Introduction to Calculus and Analysis, vol. I. Springer, New York (1989)
4. Cruz-Uribe, D., Neugebauer, C.J.: An elementary proof of error estimates for the trapezoidal rule. Math. Mag. **76**, 303–306 (2003)
5. Drazen, D.: Note on Simpson's rule. Am. Math. Mon. **76**, 929–930 (1969)
6. Fazekas, E., Jr., Mercer, P.R.: Elementary proofs of error estimates for the midpoint and Simpson's rules. Math. Mag. **82**, 365–370 (2009)
7. Fink, A.M.: Estimating the defect in Jensen's inequality. Publ. Math. Debr. **69**, 451–455 (2006)
8. Glaister, P.: Error analysis of quadrature rules. Int. J. Math. Educ. Sci. Techno. **35**, 424–432 (2004)
9. Greenwell, R.N.: Why Simpson's rule gives exact answers for cubics. Math. Gaz. **83**, 508 (1999)
10. Hai, D.D., Smith, R.C.: An elementary proof of the error estimates in Simpson's rule. Math. Mag. **81**, 295–300 (2008)

11. Hardy, G.H.: A Course of Pure Mathematics, 9th edn. Cambridge University Press, Cambridge (1948)
12. Jennings, W.: On the remainders of certain quadrature formulas. Am. Math. Mon. **72**, 530–531 (1965)
13. Johnsonbaugh, R.: The Trapezoid rule, Stirling's formula, and Euler's constant. Am. Math. Mon. **88**, 696–698 (1981)
14. Leach, E.B.: The remainder term in numerical integration formulas. Am. Math. Mon. **68**, 273–275 (1961)
15. Levrie, P.: Stirred, not shaken, by Stirling's formula. Math. Mag. **84**, 208–211 (2011)
16. Lin, T.P.: The power mean and the logarithmic mean. Am. Math. Mon. **81**, 879–883 (1974)
17. Mercer, A.McD.: An "error term" for the Ky Fan Inequality. J. Math. Anal. Appl. **220**, 774–777 (1998)
18. Mercer, A.McD.: Short proofs of Jensen's and Levinson's inequalities. Math. Gaz. **94**, 492–494 (2010)
19. Mercer, P.R.: Error estimates for numerical integration rules. Coll. Math. J. **36**, 27–34 (2005)
20. Mercer, P.R.: Error terms for Steffensen's, Young's, and Chebyshev's inequalities. J. Math. Ineq. **2**, 479–486 (2008)
21. Mercer, P.R., Sesay, A.A.: Error terms for Jensen's and Levinson's inequalities. Math. Gaz. **97**, 19–22 (2013)
22. Niizki, S., Araki, M.: Simple and clear proofs of Stirling's formula. Int. J. Math. Educ. Sci. Technol. **41**, 555–558 (2010)
23. Ostrowski, A.M.: On an integral inequality. Aequ. Math. **4**, 358–373 (1970)
24. Perisastry, M., Murty, V.N.: Bounds for the ratio of the arithmetic mean to the geometric mean. Coll. Math. J. **13**, 160–161 (1982)
25. Pecaric, J., Peric, I., Srivastava, H,M.: A family of the Cauchy type mean-value theorems. J. Math. Anal. Appl. **306**, 730–739 (2005)
26. Pinker, A., Shafer, R.E.: Problem 209. Coll. Math. J. **14**, 353–356 (1983)
27. Pinkham, R.: Simpson's rule with constant weights. Coll. Math. J. **32**, 91–93 (2001)
28. Ralston, A., Rabinowitz, P.: A First Course in Numerical Analysis. McGraw-Hill, New York (1978)
29. Richert, A.: A non-Simpsonian use of parabolas in numerical integration. Am. Math. Mon. **92**, 425–426 (1985)
30. Sandomierski, F.: Unified proofs of the error estimates for the midpoint, trapezoidal, and Simpson's rules. Math. Mag. **86**, 261–264 (2013)
31. Steele, J.M.: The Cauchy-Schwarz Master Class. Mathematical Association of America/Cambridge University Press, Washington DC - Cambridge/New York (2004)
32. Supowit, K.J.: Understanding the extra power of the Newton-Cotes formula for even degree. Math. Mag. **70**, 292–293 (1997)
33. Young, R.M.: Euler's constant. Math. Gaz. **75**, 187–190 (1991)

Appendix A
The Proof of Theorem 9.1

This morning, when I looked out of my window, I saw a problem standing outside the door of the house. When I went out, it was still standing there in exactly the same posture as before. In the afternoon I found it as I had left it. Only in the evening did it shift its weight from one foot to the other.

—*Modern Life*, by Slawomir Mrożek

Here we prove Theorem 9.1, which says that the average value of a continuous function on a closed interval exists. Then we prove some crucial properties of the definite integral, including Lemma 9.22. But we first do some preliminary work, on *subsequences* and *uniform continuity*. These notions are indispensable for any advanced study of calculus.

A.1 Subsequences

Given a sequence $\{a_n\}$ of real numbers, it is often useful to discriminate among various parts of the sequence. For example, the sequence

$$\{a_n\} = \left\{(-1)^n \left(\tfrac{n}{n+1}\right)\right\} = \left\{-\tfrac{1}{2},\, \tfrac{2}{3},\, -\tfrac{3}{4},\, \tfrac{4}{5},\, -\tfrac{5}{6},\, \tfrac{6}{7},\, \ldots\right\}$$

diverges. But it is worth pointing out that the *subsequences*

$$\{a_{2n-1}\} = \{a_1,\, a_3,\, a_5,\, \ldots\} = \left\{-\tfrac{1}{2},\, -\tfrac{3}{4},\, -\tfrac{5}{6},\, \ldots\right\}$$

and

$$\{a_{2n}\} = \{a_2,\, a_4,\, a_6,\, \ldots\} = \left\{\tfrac{2}{3},\, \tfrac{4}{5},\, \tfrac{6}{7},\, \ldots\right\}$$

have the properties that

$$a_{2n-1} \to -1 \quad \text{and} \quad a_{2n} \to 1.$$

Indeed it is *because* these two limits are not equal that $\{a_n\}$ diverges.

P.R. Mercer, *More Calculus of a Single Variable*, Undergraduate Texts in Mathematics, DOI 10.1007/978-1-4939-1926-0

To be precise, let $\{a_n\}$ be a sequence and let $n_1 < n_2 < n_3 < \cdots$ be natural numbers. Then $\{a_{n_1}, a_{n_2}, a_{n_3}, \ldots\}$ is called a **subsequence** of $\{a_n\}$. A subsequence of $\{a_n\}$ is denoted by $\{a_{n_j}\}$. And a subsequence of $\{a_{n_j}\}$ is denoted by $\{a_{n_{j_k}}\}$, etc.

Example A.1. $\left\{\frac{1}{2n}\right\}$, $\left\{\frac{1}{10^n}\right\}$, and $\left\{\frac{1}{n!}\right\}$ are each subsequences of

$$\left\{\frac{1}{n}\right\} = \left\{1, \frac{1}{2}, \frac{1}{3}, \frac{1}{4}, \frac{1}{5}, \frac{1}{6}, \frac{1}{7}, \ldots\right\}.$$

Neither

$$\left\{1, \frac{1}{3}, \frac{1}{2}, \frac{1}{4}, \frac{1}{5}, \frac{1}{7}, \frac{1}{6}, \frac{1}{8}, \ldots\right\}, \quad \text{nor} \quad \left\{1, 1, \frac{1}{2}, \frac{1}{3}, \frac{1}{4}, \frac{1}{5}, \ldots\right\}$$

is a subsequence of $\left\{\frac{1}{n}\right\}$. ◇

Remark A.2. Exercise A.1 contains the reasonable fact that a sequence $\{a_n\}$ converges to $A \in \mathbf{R}$ if and only if *every* subsequence of $\{a_n\}$ converges to A. ○

Remark A.3. The reader might look back at the proof of the Chain Rule, in Sect. 4.2. There we used a certain subsequence, we just didn't call it that at the time. ○

A **monotone** sequence is a sequence which is either increasing or decreasing. The following result is very useful. Our proof follows [9]. See also [6].

Lemma A.4. *Every sequence $\{a_n\}$ contains a monotone subsequence.*

Proof. If there is a number k such that the set $\{a_k, a_{k+1}, a_{k+2}, \ldots\}$ has no largest member, then $\{a_n\}$ clearly has an increasing subsequence. Otherwise, let A_1 be a largest member of $\{a_1, a_2, a_3, a_4, \ldots\}$, let A_2 be a largest member of $\{a_2, a_3, a_4, \ldots\}$, let A_3 be a largest member of $\{a_3, a_4, \ldots\}$, etc. Then $\{A_n\}$ is a decreasing subsequence of $\{a_n\}$ and the proof is complete. □

Example A.5. Here we prove the **Bolzano-Weierstrass Theorem**: *Every (infinite) bounded sequence $\{a_n\}$ in* $\mathbf{R}$ *contains a convergent subsequence.* By Lemma A.4, $\{a_n\}$ contains a monotone subsequence $\{a_{n_j}\}$. Since $\{a_n\}$ is bounded, $\{a_{n_j}\}$ must also be bounded. Therefore, by the Increasing Bounded Sequence Property of $\mathbf{R}$ (Theorem 1.34), $\{a_{n_j}\}$ converges. ◇

A.2 Uniform Continuity

Let f be a function defined on some interval $I \subset \mathbf{R}$. Recall now that *f is continuous at $x_0 \in I$* means that for any sequence $\{x_n\}$ in I for which $x_n \to x_0$ and for any $\varepsilon > 0$, there is a number N such that $|f(x_n) - f(x_0)| < \varepsilon$ for every $n > N$.

Careful consideration here reveals that N depends not only on ε, but also on x_0 in the following way. For a given $\varepsilon > 0$, there may be $y_0 \in I$ and a sequence $\{y_n\}$ with $|y_n - y_0| = |x_n - x_0| \to 0$, yet $|f(y_n) - f(y_0)| \geq \varepsilon$ for some value(s) of $n > N$. The stronger notion of uniform continuity removes the dependence on any particular point $x_0 \in I$.

We say that f is **uniformly continuous on** I to mean that for any pair of sequences $\{x_n\}, \{y_n\}$ in I for which $x_n - y_n \to 0$, it is the case that $f(x_n) - f(y_n) \to 0$. The reader should agree that if f is uniformly continuous on I then f is continuous at each $x_0 \in I$. Indeed, just take $y_n = x_0$ for every n.

Example A.6. The function $f(x) = 1/x$ is continuous on $(0, \infty)$ but not uniformly continuous on $(0, \infty)$: The sequences given by $x_n = \frac{1}{n}$ and $y_n = \frac{1}{n+1}$ each belong to $(0, \infty)$, and $x_n - y_n \to 0$, but $|f(x_n) - f(y_n)| = 1$ for every n. It is really the fact that $f(x) = 1/x$ is unbounded near 0 that spoils f being uniformly continuous on $(0, \infty)$. We look at this in Exercise A.7. ◇

The following important theorem is named for the German mathematician Heinrich Eduard Heine (1821–1881).

Theorem A.7. (Heine's Theorem) *Let f be continuous on the closed interval $[a, b]$. Then f is uniformly continuous on $[a, b]$.*

Proof. If f is not uniformly continuous on $[a, b]$, then there is $\varepsilon > 0$ and there are sequences $\{x_n\}, \{y_n\} \subset I$ with $x_n - y_n \to 0$, yet $\left|f(x_n) - f(y_n)\right| > \varepsilon$ for each n. By Lemma A.4, $\{x_n\}$ contains a monotone subsequence $\{x_{n_j}\}$. And again by Lemma A.4, $\{y_{n_j}\}$ contains a monotone subsequence $\{y_{n_{j_k}}\}$. So each of $\{x_{n_{j_k}}\}$ and $\{y_{n_{j_k}}\}$ is a monotone sequence and each is bounded, since they are in $[a, b]$. Then by the Increasing Bounded Sequence Property of **R** (Theorem 1.34), there are $c_1, c_2 \in [a, b]$ such that $x_{n_{j_k}} \to c_1$ and $y_{n_{j_k}} \to c_2$. But since $x_n - y_n \to 0$, we have $c_1 = c_2 = c$, say. Finally, f is continuous at c by hypothesis, so $f(x_{n_{j_k}}) \to f(c)$ and $f(y_{n_{j_k}}) \to f(c)$. This, as desired, contradicts $\left|f(x_n) - f(y_n)\right| > \varepsilon$ for each n. □

A.3 The Proof of Theorem 9.1

Let us recall the set-up for Theorem 9.1 Consider the closed interval $[a, b]$ and let $N \in \mathbf{N}$. We choose the points

$$a = x_0 < x_1 < x_2 < \cdots < x_{N-1} < x_N = b$$

according to

$$x_j = a + j\tfrac{(b-a)}{N} \quad \text{for} \quad j = 0, 1, 2, \ldots, N.$$

But we consider only N of the form $N = 2^n$, for $n = 0, 1, 2, 3, \ldots$.

This way, any partition (after $n = 0$) is a *refinement* of every previous partition: points x_j of $P_n = \{x_0, x_1, x_2, \ldots, x_{N-1}, x_N\}$ are also points of P_{n+1}. Finally, let x_j^* be any particular point of each subinterval $[x_{j-1}, x_j]$:

$$x_j^* \in [x_{j-1}, x_j] \quad \text{for} \quad j = 1, 2, \ldots, N.$$

With all of this notation in place, we recall and then prove Theorem 9.1.

Theorem 9.1. *Let f be continuous on $[a, b]$. With the notation as above (in particular $N = 2^n$),*

$$A_f\left([a,b]\right) = \lim_{N\to\infty}\left(\frac{1}{N}\sum_{j=1}^{N} f(x_j^*)\right) \quad \textit{exists.}$$

Proof. First, we set $\Delta x_N = \frac{(b-a)}{N}$ and observe that

$$\frac{1}{N}\sum_{j=1}^{N} f(x_j^*) = \frac{1}{(b-a)}\sum_{j=1}^{N} f(x_j^*)\frac{(b-a)}{N} = \frac{1}{(b-a)}\sum_{j=1}^{N} f(x_j^*)\Delta x_N.$$

By the Extreme Value Theorem (Theorem 3.23) there are $u_j, v_j \in [x_{j-1}, x_j]$ such that

$$f(u_j) \le f(x) \le f(v_j) \quad \text{for every } x \in [x_{j-1}, x_j].$$

(We should keep in mind that $u_j, v_j \in [x_{j-1}, x_j]$ depend on N as well as on j.) Notice also that if $J \subset I$, then

$$\min_{x\in I}\{f(x)\} \le \min_{x\in J}\{f(x)\} \qquad \text{and} \qquad \max_{x\in J}\{f(x)\} \le \max_{x\in I}\{f(x)\}.$$

Therefore, since each partition is a *refinement* of every previous partition, the sequence of intervals

$$\left\{\left[\frac{1}{(b-a)}\sum_{j=1}^{N} f(u_j)\Delta x_N\,,\ \frac{1}{(b-a)}\sum_{j=1}^{N} f(v_j)\Delta x_N\right]\right\}_{n=1}^{\infty}$$

is a *nested* sequence of intervals in $\mathbf{R}$.

We claim that

$$\frac{1}{(b-a)}\sum_{j=1}^{N} f(v_j)\Delta x_N - \frac{1}{(b-a)}\sum_{j=1}^{N} f(u_j)\Delta x_N \to 0 \quad \text{as } N\to\infty.$$

Then once the claim is verified, there must be a unique point ξ belonging to each of these intervals, by the Nested Interval Property of **R** (Theorem 1.41). And since

$$\frac{1}{(b-a)}\sum_{j=1}^{N} f(u_j)\Delta x_N \le \frac{1}{(b-a)}\sum_{j=1}^{N} f(x_j^*)\Delta x_N \le \frac{1}{(b-a)}\sum_{j=1}^{N} f(v_j)\Delta x_N,$$

we must have that $\xi = \lim_{N\to\infty}\left(\frac{1}{N}\sum_{j=1}^{N} f(x_j^*)\right) = A_f\big([a,b]\big)$ indeed exists.

To verify the claim, let $\varepsilon > 0$ be given. Since $\Delta x_N \to 0$ as $N \to \infty$, we have $v_j - u_j \to 0$ also. Being continuous on $[a,b]$, f is uniformly continuous there by Heine's Theorem (Theorem A.7), so for N large enough, $f(v_j) - f(u_j) < \varepsilon$ for each j.

Therefore

$$\begin{aligned}
\frac{1}{(b-a)}\sum_{j=1}^{N} f(v_j)\Delta x_N - \frac{1}{(b-a)}\sum_{j=1}^{N} f(u_j)\Delta x_N &= \frac{1}{(b-a)}\sum_{j=1}^{N}\big[f(v_j)-f(u_j)\big]\Delta x_N \\
&< \frac{\varepsilon}{(b-a)}\sum_{j=1}^{N}\Delta x_N \\
&= \frac{\varepsilon}{(b-a)}\sum_{j=1}^{N}\frac{(b-a)}{N} \\
&= \frac{\varepsilon}{(b-a)}(b-a) = \varepsilon,
\end{aligned}$$

and the proof is complete. □

In the proof of Theorem 9.1, notice that the factor $\frac{1}{b-a}$ appearing in the sequence of nested intervals doesn't really play an essential role. Indeed, the same sequence of partitions there gives rise to the sequence of nested intervals

$$\left\{\left[\sum_{j=1}^{N} f(u_j)\Delta x_N,\ \sum_{j=1}^{N} f(v_j)\Delta x_N\right]\right\}_{n=1}^{\infty}$$

which contains the single point

$$(b-a)A_f\big([a,b]\big) = \int_a^b f(x)\,dx\,.$$

To prove this, we would simply deal with $\varepsilon/(b-a)$ rather than with ε.

Notice also that in the proof of Theorem 9.1, the sequence of partitions under consideration is very specific. However, as long as the sequence of partitions is chosen in a reasonable way, the associated sequence of nested intervals always contains the *same* single point. This crucial fact is the content of the following.

Theorem A.8. *If f is continuous on $[a,b]$, then the number*

$$\int_a^b f(x)\,dx$$

is independent of the sequence of partitions used, as long as

(i) *Each partition P_n (after the first) is a refinement of the previous one, and*
(ii) *The length of the largest subinterval in each partition P_n tends to zero as $n \to \infty$.*

Proof. The argument is not overly difficult, but the necessary sea of symbols can make things confusing. We shall rely on the sequence of partitions from Theorem 9.1, but we change the notation very slightly, to accommodate what is to come. The sequence of partitions there is given by $P_n = \{x_0, x_1, x_2, \ldots, x_{p(n)-1}, x_{p(n)}\}$, where $p(n) = 2^n$, $a = x_0$ and $x_j = a + j\frac{(b-a)}{p(n)}$ for $j = 0, 1, 2, \ldots, p(n)$. That is,

$$a = x_0 < x_1 < x_2 < \cdots < x_{p(n)-1} < x_{p(n)} = b, \quad \text{and}$$

$$\Delta x_{p(n)} = x_j - x_{j-1} = \frac{(b-a)}{p(n)}, \quad \text{for } j = 1, 2, \ldots, p(n).$$

As we observed then,

$$\left\{\left[\sum_{j=1}^{p(n)} f(u_j)\Delta x_{p(n)}, \sum_{j=1}^{p(n)} f(v_j)\Delta x_{p(n)}\right]\right\}_{n=1}^{\infty}$$

is a sequence of nested intervals which contains the single point ξ_P. (The points $u_j, v_j \in [x_{j-1}, x_j]$ depend on $p(n)$ as well as on j.)

Now let $Q_n = \{x_0, x_1, x_2, \ldots, x_{q(n)-1}, x_{q(n)}\}$, where $q : \mathbf{N} \to \mathbf{N}$ is some increasing unbounded function, be another sequence of partitions of $[a,b]$ having the properties described in the statement of the theorem. Here,

$$a = x_0 < x_1 < x_2 < \cdots < x_{q(n)-1} < x_{q(n)} = b, \quad \text{and}$$

$$\Delta x_j = x_j - x_{j-1} \quad \text{for } j = 1, 2, \ldots, q(n).$$

We have written Δx_j here, instead of $\Delta x_{q(n)}$, because the subintervals within Q_n need not each have the same length. (That is, the partitions Q_n need not be *regular*.) In exactly the same way as for the sequence of partitions P_n in the proof of Theorem 9.1, we see that

$$\left\{ \left[\sum_{j=1}^{q(n)} f(\hat{u}_j)\Delta x_j, \sum_{j=1}^{q(n)} f(\hat{v}_j)\Delta x_j \right] \right\}_{n=1}^{\infty}$$

is sequence of nested intervals which contains the single point ξ_Q. (The points $\hat{u}_j, \hat{v}_j \in [x_{j-1}, x_j]$ depend on $q(n)$ as well as on j.) Now, taking the union of P_n and Q_n we get a *third* sequence of partitions R_n, and a *third* sequence of nested intervals, which contains the single point ξ_R. Again, if $J \subset I$, then

$$\min_{x \in I}\{f(x)\} \le \min_{x \in J}\{f(x)\} \qquad \text{and} \qquad \max_{x \in J}\{f(x)\} \le \max_{x \in I}\{f(x)\}.$$

So each interval in the third sequence of nested intervals is necessarily a subinterval of each corresponding interval arising from the first two partitions. Therefore we must have $\xi_R = \xi_P$ and $\xi_R = \xi_Q$, so that $\xi_P = \xi_Q$ as desired. □

We lean heavily on Theorem A.8 to prove the following important property of the definite integral, which was deferred in Sect. 9.4.

Lemma 9.22. *Let f be continuous on $[a, b]$ and let $c \in (a, b)$. Then*

$$\int_a^b f(x)\,dx = \int_a^c f(x)\,dx + \int_c^b f(x)\,dx\,.$$

Proof. Let P_n be any sequence of partitions of $[a, b]$, as in the proof of Theorem A.8: (i) each P_{n+1} is a refinement of P_n and (ii) the length of the largest subinterval in P_n tends to zero as $n \to \infty$. The reader should agree that if c happens to be a point of one of the partitions P_n (and hence all subsequent partitions), then the proof is immediate. Otherwise, we add c to each P_n to get another sequence of partitions with the same two essential properties and we denote it by P_n^c. By Theorem A.8, the associated sequences of nested intervals contain the *same* single point

$$\int_a^b f(x)\,dx\,,$$

whether we use P_n or P_n^c. But using P_n^c, together with each of the partitions $P_n^c \cap [a, c]$ and $P_n^c \cap [c, b]$,

$$\int_a^b f(x)\,dx = \int_a^c f(x)\,dx + \int_c^b f(x)\,dx.$$ □

The following result is often used in practice. It says that we can approximate the definite integral of a continuous functions as closely as we please with a Riemann sum, as long as the partition for the Riemann sum is *fine* enough. And the partition does not need to be regular.

Theorem A.9. *Let f be continuous on $[a, b]$, so that $\int_a^b f(x)\,dx$ exists, and let $\varepsilon > 0$. Then there is $\delta > 0$ such that for any partition $P = \{x_0, x_1, x_2, \ldots, x_{n-1}, x_n\}$ of $[a, b]$, and for any choice of points $x_j^* \in [x_{j-1}, x_j]$, it is the case that*

$$\left| \int_a^b f(x)\,dx - \sum_{j=1}^{n} f(x_j^*)\Delta x_j \right| < \varepsilon,$$

as long as $\max\limits_{1 \le j \le n} \{\Delta x_j\} < \delta$.

Proof. By the Extreme Value Theorem (Theorem 3.23) there are $u_j, v_j \in [x_{j-1}, x_j]$ such that

$$f(u_j) \le f(x) \le f(v_j) \quad \text{for all } x \in [x_{j-1}, x_j].$$

Then for *any* Riemann sum $\sum_{j=1}^{n} f(x_j^*)\Delta x_j$,

$$\sum_{j=1}^{n} f(u_j)\Delta x_j \le \sum_{j=1}^{n} f(x_j^*)\Delta x_j \le \sum_{j=1}^{n} f(v_j)\Delta x_j.$$

But we also have

$$\sum_{j=1}^{n} f(u_j)\Delta x_j \le \int_a^b f(x)\,dx \le \sum_{j=1}^{n} f(v_j)\Delta x_j.$$

Therefore

$$\left| \int_a^b f(x)\,dx - \sum_{j=1}^{n} f(x_j^*)\Delta x_j \right| \le \sum_{j=1}^{n} f(v_j)\Delta x_j - \sum_{j=1}^{n} f(u_j)\Delta x_j$$

$$= \sum_{j=1}^{n} [f(v_j) - f(u_j)]\Delta x_j.$$

Now since f is continuous on $[a, b]$ it is uniformly continuous there, by Heine's Theorem (Theorem A.7). So there is $\delta > 0$ such if each Δx_j is $< \delta$,

$$f(v_j) - f(u_j) < \frac{\varepsilon}{b-a}.$$

That is, if $\max\limits_{1 \le j \le n} \{\Delta x_j\} < \delta$, then

$$\left| \int_a^b f(x)\,dx - \sum_{j=1}^{n} f(x_j^*)\Delta x_j \right| = \sum_{j=1}^{n} [f(v_j) - f(u_j)]\Delta x_j$$

$$< \frac{\varepsilon}{b-a} \sum_{j=1}^{n} \Delta x_j = \varepsilon,$$

as desired. □

Exercises

A.1. Prove that $\{a_n\}$ converges to A if and only if every subsequence $\{a_{n_j}\}$ of $\{a_n\}$ converges to A.

A.2. The sequence $\{a_n\}$ is a **Cauchy sequence** if for any $\varepsilon > 0$, there is a number N such that $|a_n - a_m| < \varepsilon$ for $n, m > N$.

(a) Prove that if $a_n \to A$, then $\{a_n\}$ is a Cauchy sequence. (Roughly, this says that if the a_n's are getting close to A, then they must be getting close to each other.)
(b) Prove that if $\{a_n\}$ is a Cauchy sequence then $\{a_n\}$ is bounded.
(c) Prove that if $\{a_n\}$ is a Cauchy sequence then there exists $A \in \mathbf{R}$ such that $\{a_n\}$ converges to A.
(Roughly, this says that if the a_n's are getting close to each other, then they are getting close to some $A \in \mathbf{R}$.) Hint: Use Exercise A.1 and Lemma A.1.

A.3. The sequence $\{x_n\}$ is a **Cauchy sequence** if for any $\varepsilon > 0$, there is a number N such that $|x_n - x_m| < \varepsilon$ for $n, m > N$. Show that if $\{x_n\}$ is a Cauchy sequence in I and f is uniformly continuous on I, then $\{f(x_n)\}$ is a Cauchy sequence.

A.4. Prove that f defined on some interval I is uniformly continuous if and only if for any $\varepsilon > 0$ there is a $\delta > 0$ such that whenever $x, y \in I$ with $|x - y| < \delta$, we have $|f(x) - f(y)| < \varepsilon$.

A.5. Show that if f is uniformly continuous on $\mathbf{R}$, then so is $|f|$.

A.6. Show that $f(x) = x^2$ is continuous at each point of $[0, \infty)$ but is not uniformly continuous on $[0, \infty)$.

A.7. Show that if f is uniformly continuous on $(0, 1)$, then f is bounded. Does an Extreme Value Theorem (Theorem 3.23) hold in this case?

A.8. [3]

(a) Show that if $f(x)$ is continuous and has a bounded derivative on (a, b), then f is uniformly continuous on (a, b).
(b) For $x \in [-1, 1]$, consider the function

$$f(x) = \begin{cases} x^{4/3}\sin(\frac{1}{x}) & \text{if } x \neq 0 \\ 0 & \text{if } x = 0. \end{cases}$$

Show that f is uniformly continuous on $[-1, 1]$, yet has unbounded derivative there.

A.9. Verify that $f(x) = x$ and $g(x) = \sin(x)$ are uniformly continuous on $(-\infty, +\infty)$, but their product is not.

A.10. [5]

(a) Consider the closed interval $[a, b]$ and suppose that $\mathscr{C}$ is a collection of open intervals such that $[a, b] \subset \bigcup_{I \in \mathscr{C}} I$. Prove **Borel's Theorem**, named for French mathematician Émile Borel (1871–1956): *There is a finite subcollection $I_1, I_2, \ldots I_n$ of intervals from $\mathscr{C}$ such that* $[a, b] \subset \bigcup_{j=1}^{n} I_j$. Hint: Assume the conclusion does not hold, then employ a bisection algorithm, keeping at each step the half-interval which cannot be covered by a finite subcollection.
(b) What happens if the interval is $[a, b)$?

A.11. Prove Heine's Theorem (Theorem A.7) using Borel's Theorem from Exercise A.10 above.

A.12. Use a bisection algorithm to prove the Bolzano-Weierstrass Theorem from Example A.5.

A.13. The sequence $\{a_n\}$ is a **Cauchy sequence** if for any $\varepsilon > 0$, there is a number N such that $|a_n - a_m| < \varepsilon$ for $n, m > N$.

(a) Prove that if $a_n \to A$, then $\{a_n\}$ is a Cauchy sequence. (Roughly, this says that if the a_n's are getting close to A, then they must be getting close to each other.)
(b) Prove that if $\{a_n\}$ is a Cauchy sequence then $\{a_n\}$ is bounded.
(c) Use the Bolzano-Weierstrass Theorem from Example A.5 to prove that if $\{a_n\}$ is a Cauchy sequence then there is $A \in \mathbf{R}$ such that $a_n \to A$. (Roughly, this says that if the a_n's are getting close to each other, then they are getting close to some $A \in \mathbf{R}$.)

A.14. The sequence $\{a_n\}$ is a **Cauchy sequence** if for any $\varepsilon > 0$, there is a number N such that $|a_n - a_m| < \varepsilon$ for $n, m > N$. Prove that if $\{f(a_n)\}$ is a Cauchy sequence whenever $\{a_n\}$ is a Cauchy sequence in I, then f is uniformly continuous on I. Suggestion: Use the Bolzano-Weierstrass Theorem from Example A.5.

A.15. Let f be continuous on $[a, b]$ and let $c \in (a, b)$. Show that the average value of f on $[a, b]$ is a weighted average of the average value of f on $[a, c]$ and the average value of f on $[c, b]$.

A.16. [1] Let $a > 0$. Show, as follows, that

$$\lim_{n\to\infty} n\big(1 - a^{-1/n}\big) \;=\; \lim_{n\to\infty} n\big(a^{1/n} - 1\big) = \ln(a).$$

The case $a = 1$ is trivial. We may assume that $a > 1$. (Otherwise, consider $1/a$).

(a) Consider the *irregular* partition $P_n = \{a_0, a_1, a_2, \ldots, a_{n-1}, a_n\}$ of $[1, a]$ given by

$$a_j = a^{j/n} \quad \text{for} \quad j = 0, 1, 2, \ldots, n.$$

(b) Use the Right and Left Rectangle Rules to show that

$$n\big(1 - a^{-1/n}\big) \;<\; \int_1^a \frac{1}{x}\, dx \;<\; n\big(a^{1/n} - 1\big).$$

(c) Verify that $0 < a^{1/n} - 1 < \frac{a-1}{n}$, then use this and (b) to show that

$$0 \;<\; \ln(a) - n\big(1 - a^{-1/n}\big) \;<\; n\big(a^{1/n} - 1\big) - n\big(1 - a^{-1/n}\big) \;< n \left(\frac{a-1}{n}\right)^2$$

and

$$0 \;<\; n\big(a^{1/n} - 1\big) - \ln(a) \;< n \left(\frac{a-1}{n}\right)^2.$$

(d) Finally, let $n \to \infty$.

(e) Bonus: Show that if we double the number of points in any given partition *as above*, then the new points which result are those given by the Mean Value Theorem (Theorem 5.2) applied to $f(x) = 1/x$ on each subinterval of the original partition.

A.17. [4]

(a) Consider the partition $P_n = \{x_0, x_1, x_2, \ldots, x_{n-1}, x_n\}$ of $[0, x]$ given by

$$x_j = \frac{j^2 x}{n^2} \quad \text{for} \quad j = 0, 1, 2, \ldots, n.$$

Use a limit of Riemann sums, with

$$x_j^* = x_j \quad \text{for} \quad j = 1, 2, \ldots, n$$

to show that

$$\int_0^x \sqrt{t}\,dt = \frac{2x^{3/2}}{3}.$$

(b) Extend the idea in (a) to show that $\int_0^x t^{4/3}\,dt = \frac{3x^{7/3}}{7}$.

(c) How about $\int_0^x t^{p/q}\,dt$?

A.18. Let $a = x_0$ and $x_j = a + j\frac{(b-a)}{n}$ for $j = 0, 1, 2, \ldots, n$. Let $x_j^*, y_j^* \in [x_{j-1}, x_j]$. Show that if f and g are continuous on $[a, b]$, then

$$\lim_{n\to\infty}\left[\frac{1}{n}\sum_{j=1}^{N} f(x_j^*)g(y_j^*)\right] \quad \text{exists.}$$

Hint: Write $f(x_j^*)g(y_j^*) = f(x_j^*)g(x_j^*) + f(x_j^*)[g(y_j^*) - g(x_j^*)]$.

A.19. A set S which is bounded above has a **least upper bound** $lub(S)$ if (i) $lub(S)$ is an upper bound, and (ii) any upper bound U for S satisfies $lub(S) \le U$.

(a) Show that if S has a least upper bound, then it is unique.
(b) Show that if we assume that every nonempty set in **R** which is bounded above has a *least* upper bound (this is called the **Least Upper Bound Property**), then **R** has the Increasing Bounded Sequence Property (Theorem 1.34).
(c) Show that **R** having the Increasing Bounded Sequence Property implies that **R** has the Least Upper Bound Property. Hint: Begin with a set S that has upper bound b and let $a \in A$. Employ a bisection algorithm on $[a, b]$, keeping at each step the half-interval whose left-hand endpoint is not an upper bound for S and whose right-hand endpoint is an upper bound for S. (Make sure that you can justify this too.)

A.20. The **Completeness Axiom** for **R** says that every Cauchy sequence (see Exercises A.2 or A.13) in **R** converges to some $A \in$ **R**. The short phrase for this is: **R** *is complete*. Show that in **R**, the Completeness Axiom implies the Increasing Bounded Sequence Property (Theorem 1.34).

A.4 An Epilogue

Exercises A.2 (or A.13), A.19, and A.20 together show that in **R**, the following are equivalent:

(*A*) Increasing Bounded Sequence Property,
(*B*) Least Upper Bound Property,
(*C*) Completeness Axiom.

That (C) follows from either of (A) or (B) is because in **R**, either of (A) or (B) implies the **Archimedean Property**: For any $a \in \mathbf{R}$ with $a > 0$, there is $n \in \mathbf{N}$ such that $1/n < a$.

Roughly speaking, a **field F** is a set in which we can sensibly add, subtract, multiply, and divide. Each of **Q** and **R** is a field; **Z** is not a field. A field is **ordered** if there is also a sensible notion of "<" defined on it. For example, **Q** and **R** are ordered fields but the set of complex numbers **C** is a field which is not ordered.

For any *ordered* field **F**, the following are equivalent (e.g., [2,7,8]):

(A') Increasing Bounded Sequence Property,
(B') Least Upper Bound Property,
(C') Completeness Axiom + Archimedean Property.

Still, either of (A') or (B') implies the Archimedean Property, but there are complete ordered fields which are not Archimedean. The Completeness Axiom is sometimes preferable to either of (A') or (B'), because these require the field to be ordered. For example, **C** is a complete field which is not ordered.

Finally, if any of (A') or (B') or (C') holds for an ordered field **F**, then **F** is really just **R**. That is, there is an order preserving (i.e. $\leq$ is preserved) one-to-one and onto mapping between **F** and **R**. So **R** is a pretty good place to do mathematics.

References

1. Burk, F.: The logarithmic function and Riemann sums. Coll. Math. J. **32**, 369–370 (2001)
2. Dobbs, D.E.: On characterizations of the ordered field of real numbers. Int. J. Math. Educ. Sci. Technol. **32**, 299–306 (2001)
3. Kaptanoglu, H.T.: In praise of $x^{\alpha}\sin(1/x)$. Am. Math. Mon. **108**, 144–150 (2001)
4. Matthews, J.H.: The integral of $x^{1/2}$. Coll. Math. J. **25**, 142–144 (1994)
5. Natanson, I.P.: Theory of Functions of a Real Variable, vol. 1. Frederick Ungar, New York (1961)
6. Newman, D.J., Parsons, T.D.: On monotone subsequences. Am. Math. Mon. **95**, 44–45 (1988)
7. Propp, J.: Real analysis in reverse. Am. Math. Mon. **120**, 392–408 (2013)
8. Teismann, H.: Toward a more complete list of completeness axioms. Am. Math. Mon. **120**, 99–114 (2013)
9. Thurston, H.: Math bite: a simple proof that every sequence has a monotone subsequence. Math. Mag. **67**, 344 (1997)

Index

P.R. Mercer, *More Calculus of a Single Variable*, Undergraduate Texts in Mathematics, DOI 10.1007/978-1-4939-1926-0

Printed in the United States
By Bookmasters